BARBAROSSA

Other books by the same author:

The Donkeys: A Study of the BEF in 1915
The Fall of Crete

When BARBAROSSA begins, the world will hold its breath.

Hitler, OKW Conference Minutes, 3rd February, 1941

BARBAROSSA | The Russian-German Conflict, 1941-45 | by Alan Clark |

QUILL | NEW YORK

For my father

Inquiries should be addressed to Permissions Department, William Morrow and Company, Inc., 105 Madison Ave., New York, N.Y. 10016.

Library of Congress Cataloging in Publication Data

Clark, Alan, 1928–
Barbarossa : the Russian-German conflict, 1941–1945.

Reprint. Originally published: New York : Morrow, 1965. With new introduction.
Bibliography: p.
Includes index.
1. World War, 1939–1945—Campaigns—Eastern.
I. Title.
D764.C55 1985 940.54'21 85-502
ISBN 0-688-04268-6 (pbk.)

Printed in the United States of America

3 4 5 6 7 8 9 10

CONTENTS

LIST OF MAPS AND CHARTS

Maps in black and white:

Charts:

AUTHOR'S NOTE FOR 1985 EDITION

It is twenty years since *Barbarossa* was finished. I no longer write history, although often I pine to do so. As Under-Secretary of State, I see its raw material—sludge, with occasional streaks of ore, cross my desk.

I contemplated making revisions to the book, for many millions of words have since been written on and round the subject by survivors, participants and commentators. But I rejected it. Writers cannot return to their subjects, as artists do. We are allowed only one canvas. And so, on rereading *Barbarossa,* I have decided to alter nothing. From Preface to Epilogue I stand by the judgments, the interpretations and the conclusions in their entirety.

I will note, though, two personal reflections that occur strongly to me today. I feel regret that the Soviet Union, whose heroism I have chronicled in this book, should still be relegated—now, it seems in perpetuity—to the role of our ritual adversary; that the Russian peoples' qualities, their patience, their inventiveness, their romanticism, should be excluded from the family of nations. And I feel also a certain remorse at my portrayal of the German character. Researching, at a young age and with the papers fresh in my mind, I allowed, I suppose, my indignation at what I saw to show too often. Now I can see the rejection of the Judaeo-Christian ethic as part of a yearning for the primitivism of the Teutonic knights and the *Ordensburgen.* It is a recurrent theme in the history of that great nation whose denial does its people no credit and diminishes from their culture.

In these last twenty years human beings have been killing one another almost continuously. The wars have been bitter and inconclusive—as in Vietnam, Lebanon, the Gulf. Or short and spectacular—as over Yom Kippur and in the Falkland Islands. Weapons technology has made no difference to the verities of battle. Incompetence, corruption, brutality and waste are inseparable and take their toll. And yet, as a student of war, I am left with one conclusion, the secondary theme of this book, pagan, inescapable but uplifting:

> Two things have altered not
> Since first the world began
> The beauty of the wild green earth
> And the bravery of Man.

—ALAN CLARK

Saltwood
November 11, 1984

PREFACE

This book is devoted to the greatest and longest land battle which mankind has ever fought. Its outcome recast the world balance of power and completed the destruction of the old Europe, which World War I had begun. Its victor emerged as the only power capable of challenging—perhaps even of defeating—the United States in those very fields of technology and material power in which the New World had become accustomed to pre-eminence.

The subject, taken as a whole, has been neglected by historians. The Soviet authorities have lately begun to release their own official histories, but these, while lavish with minor detail, remain tantalisingly silent at certain points of crisis, and there is no official material comparable to that which the British and United States authorities allow to students of their campaigns. The scale of such other works as exist is often very small, or else it consists of personal memoirs, and is subject to the limitations of viewpoint and objectivity that are inseparable from this form.

Neither side has produced anything truly impartial. The Germans, who were defeated, have evolved a variety of excuses. With the passage of time these have become formalised under two distinct heads—inferiority of numbers and material, and the frustrations arising from Hitler's continual interference with his generals. Yet this study will show, I hope, that there were occasions when neither of these excuses had validity. The Russians, although their official accounts are, in the main, clear and factual, have their reservations. In common with other authoritarian regimes, the Soviets have reputations that must not be disturbed and mythology to cherish.

This book has its heroes, although they fall in with the classic tradition more easily than with the clear-cut "good" and "bad" categories of modern Western society.

Foremost must come the ordinary Russian soldier; abominably led, inadequately trained, poorly equipped, he changed the course of history by his courage and tenacity in the first year of fighting. There are individuals, too, who deserve an honourable mention. General Guderian, whom I have criticised for his impulsiveness and disobedience in the opening battles, emerges in the last years of the war as the one man who might have saved the Eastern front and who applied himself almost singlehanded to that end. And there is poor General Vlasov, one of the ablest commanders in the Red Army, betrayed by his superiors, swimming against the tide of history with his plans for an army of "Russians against Stalin." And General Chuikov, directing the hopeless energies of the Stalingrad garrison by candlelight in the Tsaritsa bunker, and three years later destined personally to accept the surrender of Berlin.

Finally, if it is not premature to do so, I have tried to suggest a reassessment of Hitler's military ability. His capacity for mastering detail, his sense of history, his retentive memory, his strategic vision—all these had flaws, but considered in the cold light of objective military history, they were brilliant nonetheless. The Eastern campaign, above all, was his affair, and his violent and magnetic personality dominated its course, even in defeat. Since the war Hitler has been a convenient repository for all the mistakes and miscalculations of German military policy. But a study of events in the East will show that occasions when Hitler was right and the General Staff wrong are far more numerous than the apologists of the German Army allow.

Seduction by personalities is at the same time the peril and the delight of the military historian, who should by right confine himself to the field of battle, the outline of armament, logistics, and deployment. But in the assessment of the campaign in the East, which was in truth a war between two absolute monarchies, the interplay of personal rivalries is often of critical importance. Human frailties—greed and ambition, fear and cruelty —can be seen acting directly on the conduct of operations.

Conversely, unless the book were to be intolerably long, many battles of only secondary importance to a strategic evaluation of the campaign have had to be omitted. I have tried to isolate four points of crisis—Moscow in the winter of 1941, Stalingrad, the Kursk offensive of 1943, and the last struggles on the Oder at the beginning of 1945—and hung the narrative around them. This has meant that some sectors of the war, such as the Crimea, the later stages of the siege of Leningrad, and the Caucasian campaign of 1942 are not described in detail. Nor does the development of the book unfold at the same pace as the passage of time. For nearly one third of its length is devoted to the summer and autumn of 1941, when every day was critical, less than two chapters to the wearisome German retreat across European Russia in 1944.

From this study is one left with any general conclusions? I believe the answer is yes, but they are not of a kind from which we in the West can derive much comfort. It does seem that the Russians could have won the war on their own, or at least fought the Germans to a standstill, without any help from the West. Such relief as they derived from our participation —the distraction of a few enemy units, the supply of a large quantity of material—was marginal, not critical. That is to say, it affected the duration but not the outcome of the struggle. It is true that once the Allies had landed in Normandy the drawing-off of reserves assumed critical proportions. But the threat, much less the reality, of a "second front" became a factor only after the real crisis in the East had passed.

It is often asked, could not the Germans have won the war if they had not made certain mistakes?

The general answer, I believe, is that the Russians, too, made mistakes. Which is the more absurd—to allow, with the wisdom of hindsight, an immaculate German campaign against a Russian resistance still plagued by those blunders and follies that arose in the heat and urgency of battle, or to correct both and to reset the board in an atmosphere of complete fantasy, with each side making the correct move like a chess text, when "white must win"?

I have discussed the question of sources at the head of the Bibliography, but there are some acknowledgments that I should like to make here. Although I have said that *taken as a whole,* the period has been neglected by historians, there are major works dealing with certain of its aspects, and from these I have drawn freely, both as to material and inspiration.

Sir John Wheeler-Bennett's classic on the Germany Army in politics could never be out of my mind; and Mr. Alexander Dallin's penetrating study, *German Rule in Russia,* is an essential backdrop to any serious work on the subject; No one who is concerned with the dark complexities of the Allgemeine and the Waffen SS can afford to do without Mr. Gerald Reitlinger's authoritative study, nor can any book that touches on the last days in Berlin avoid standing in the debt of Professor Trevor-Roper's masterly description of that dramatic period.

I should like to pay tribute to Colonel Leyderrey of the Swiss Army, who was the first to tackle the complexities of the Eastern front records, and to express my thanks to Captain B. H. Liddell Hart for help both from his files and his memory. Colonel Diem of the German Army and Colonel Vinnikov, the Soviet Military Attaché in London, have been of particular help in providing documents and material, and Virginia Kyro of William Morrow and Company has served beyond the call of duty as the tireless editor of this book. The Historical Section of the University of Pennsylvania was kind enough to supply me with microfilm of the transcripts of the Führer conferences. I should also like to express my gratitude to the librarian and staff of the Imperial War Museum in London.

Alan Clark

July 1964

BOOK I | The "Eastern Marshals"

. . . That's what we have army corps commanding generals for. What is lacking at the top level [i.e., Hitler] is that confidence in the executive commands which is one of the most essential features of a command organisation, and that is because it [i.e., he] fails to grasp the co-ordinating force that comes from the common schooling and education of an officer corps.

Halder, 3rd July, 1941

The majority of them are out to make their careers, in the lowest sense.

Hassell

Appoint a Commander in Chief . . . What would be the use? Even I cannot get the field marshals to obey me!

Hitler to Manstein, January 1944

For the convenience of the reader, the Appendices beginning on page 467 contain *A Chronology of Developments in the Eastern Campagin 1941-45;* important *Facts about the Russian and German Leaders* arranged under their alphabetically listed names; a *Waffen SS Rank Conversion Table,* a *Glossary of Abbreviations,* and the *Text of Führer Directive #34.*

one | THE STATE OF THE WEHRMACHT

On the afternoon of Sunday, 5th November, 1939, it was raining in Berlin. Through the empty streets a single black Mercedes, without escort, brought the Commander in Chief of the German Army from Zossen to the Chancellery, where he was to receive, at his own request, an audience with Hitler.

General (as he then was) Walther von Brauchitsch was suffering from a painful attack of "nerves"—an unexpected complaint for a commander whose armies had lately completed a rapid, victorious, and almost bloodless campaign. The source of his apprehensive condition was to be found in a bulky memorandum which lay in his briefcase and which, as he had promised to his colleagues on the *Generalstab,* he would personally read out to the Führer. This document, though it bore the signature of Brauchitsch, had been prepared by many hands and rambled over diverse subjects in the military field. Its purported motif was to "recommend" against launching an attack in the West that autumn, but in essence it was a historical throwback, an attempt to formulate an ultimatum whose substance was as much political as military and whose purpose was to assert the primacy of the Army over all the other organs of government in the Reich.

This was a particularly embarrassing task for Brauchitsch. One, indeed, which he had been urged by his colleagues to undertake on several occasions in the past, and which he had always managed to sidestep. Brauchitsch, who owed his appointment to Hitler, and who saw more of the Führer than any other soldier outside the immediate Nazi entourage, can have had few illusions about the value of any protest he might be allowed to utter or, indeed, concerning the violence of the reaction which it would provoke. Why, then, having evaded it so often in the past, did Brauchitsch now consent to take on the Führer face to face?

The development which had succeeded in uniting those elements in the Army which were opposed to the Nazi regime and the more strictly professional soldiers who concerned themselves exclusively with military efficiency arose out of the Führer's interference in the planning and conduct of military operations. Hitler had insisted on being shown every order, down to regimental level, for the first three days of the Polish campaign in September. Many he criticised, some he altered, one—the operation to seize the bridgehead at Dirschau—he completely recast in a more audacious pattern, against the advice of every officer along the chain of command which finally led up to Colonel General Halder,[1] Chief of Staff of the Army and, effectively, No. 2 under Brauchitsch.

The generals, who had already suffered the rebuttal of their traditional claim to be heard in matters of state that impinged on military policy, now sensed a direct threat to their most jealously guarded precinct—the details of tactical combat planning—and this on the very first occasion that the Army had taken up arms since 1918. And their distaste cannot have been lessened by the fact that in every case Hitler's revisions had been justified in battle. Brauchitsch, therefore, had found himself (and not for the last time) in a most delicate position: suspended between the unanimous protestations of his colleagues and the certain wrath of his Führer.

Hitler, who may have suspected that something was afoot, received his Commander in Chief in the main conference room of the Chancellery, under the bust of Bismarck, instead (as would have been more usual) of one of the smaller antechambers. After a certain amount of verbal shadowboxing, in an atmosphere that must have been anything but comfortable, Brauchitsch declared that "OKH[2] would be grateful for an understanding that it, and it alone, would be responsible for the conduct of any future campaign."

This suggestion was received "in icy silence." Brauchitsch then went on, with one of those curious and mendacious impulses which sometimes

[1] Colonel General Franz Halder, Chief of the Army General Staff; appointed 1938, dismissed September 1942.

[2] *Oberkommando des Heeres,* the High Command of the German Army.

seized him (and of which other examples will be found in this book), to say that ". . . the aggressive spirit of the German infantry was sadly below the standard of the First World War" and that there had been "certain symptoms of insubordination similar to those of 1917-18."

By this time the interview had already lost all semblance of an exchange between equals—much less the *deus-ex-machina* quality which was the traditional attribute of an encounter between the head of state and the Commander in Chief of the Army. Brauchitsch never really got started on his main purpose. As his peevish complaints died away, Hitler started to work up a tremendous rage. He accused the General Staff, and Brauchitsch personally, of disloyalty, sabotage, cowardice, and defeatism. For ten, fifteen, twenty minutes the Führer poured forth a torrent of abuse upon the head of his timorous and bewildered army commander, creating a scene which Halder, with truly English understatement, has recorded as being "most ugly and disagreeable."

It was the first of the occasions on which Hitler abused his generals. They were to occur more frequently, last longer, and be more "disagreeable" in the years to come. This was also the first occasion on which Brauchitsch remonstrated with his Führer, and the last. The Commander in Chief drove shakily back to his headquarters, where ". . . he arrived in such poor shape that at first he could only give a somewhat incoherent account of the proceedings."

Brauchitsch's fundamental error—or rather that of the conservative army generals whose emissary he was—was the error latent in all measures that are based on a historical throwback. It arose from a blindness to the pattern of evolution and, in particular, to the manner in which the power structure within the Reich had developed. For this structure was no longer a duumvirate, shared between the façade of civil administration and the authority of the military, but a lumpish hexagonal pyramid with Hitler at its summit. Obedient to the Führer but in deadly rivalry with one another, were four major private empires within the Reich administration and a host of secondary ones, revolving around personalities, crackbrained schemes, forgotten sectors of the economy or administration, whose numbers were to proliferate as the war lengthened.

One of the most rational and intelligent of these personalities, Albert Speer,[3] has said, "Relations between the various high leaders can be un-

[3] Hitler's architect. Promoted to be Minister of Armament and War Production 1942.

derstood only if their aspirations are interpreted as a struggle for the succession to Adolf Hitler." While (as was the case in the early 1940's) the prospect of succession was a remote one, the Nazi Diadochi[4] competed with one another to win Hitler's favour and to enlarge their own dominion. The result was that in addition to the Army there were many other foci of power, none of which was indispensable, yet each of which was manipulated by the Führer to preserve the internal balance.

First, there was the Nazi Party machine itself, controlled by Martin Bormann[5] and enjoying through him and Hess the privilege of daily access to the Führer. The Party had its own press, controlled education, regional government, and a variety of paramilitary organisations such as the *Hitlerjugend*.[6] Then there was the SS hierarchy,[7] presided over by Himmler[8] and including the Gestapo, the RSHA (Reich Central Security Office), the assassination squads of the SD, and the notorious "asphalt soldiers" of the Waffen SS. A third enclave was the personal creation of Goering and included the entire Luftwaffe, all the productive capability that supplied it, and the administrative organisation of the Four-Year Plan, of which Goering was the director.

Beside these three the conservative officers and gentlemen of the *Heeresleitung*, the German Army Command, carried no exceptional weight or authority. If it came to a showdown, Hitler had at his disposal a highly armed police, an air force and ground organisation, and a regional administrative machine. And as the stresses of the war multiplied, so did the fragmentation of the German body politic, so that there came to be nearly a dozen primary foci of power whose departmental rivalry was aggravated by personal animosities (Goebbels[9] hated Bormann, Goe-

[4] The Diadochi were the surviving generals of Alexander the Great, who quarrelled over the division of his empire. Several authorities (e.g., Trevor-Roper and Alexander Dallin) have adopted the term to describe the senior Nazi leaders.

[5] Bormann was Chief of the Party Office after Hess's flight to Britain in May 1941; Hitler's personal secretary from April 1943. Disappeared April 1945 after Hitler's suicide.

[6] The Hitler Youth organisation, paramilitary in character, was for boys and girls of submilitary age.

[7] For the various subdivisions of the SS readers are referred to the chart on p. 000.

[8] Heinrich Himmler was Reichsführer SS 1929; Police President, Bavaria, 1933; Chief of the Reich Political Police 1936; Minister of the Interior 1943; C. in C. of the Home Army, July 1944; C. in C. of the Rhine and Vistula armies December 1944-March 1945. Committed suicide at British Interrogation Centre, Lüneburg, 23rd May, 1945.

[9] Paul Joseph Goebbels. Reich Propaganda Minister, Gauleiter for Berlin. Plenipotentiary for Total War after 20th July, 1944. Committed suicide in Berlin 1st May, 1945.

ring[10] despised Ribbentrop[11] and mistrusted Himmler, Rosenberg[12] was not on speaking terms with Himmler and Koch,[13] and so on)—and which were co-ordinated only through their direct allegiance to the Führer.

Yet when all this has been said the fact remains that the German failure in the East was essentially a military failure. The Army proved unequal to the task which the state demanded of it, and so the state, living by the sword, could not survive when the sword was shattered. A fundamental cause of this failure was the continuous tension between the senior officers and staff of the Army (OKH) and the organisation of the Supreme Command (OKW) headed by Hitler. While military operations were uniformly successful, this tension was dormant. But once the Wehrmacht came under strain, relations between the two started to go sour. Hitler despised the generals for their caution, he resented their class-consciousness, and he believed (with some reason) that they were the only potential source of political opposition left in Germany. The generals, for their part, distrusted the Nazi Party because of its proletarian origins and its evident irresponsibility in matters of state. As individuals, it is true, several of them were converted by Hitler's "ideals" during the heyday of Nazi success, but under the stress of failure Party and military alike were to undergo a disastrous polarisation.

Thus, in analysing the causes of this failure and the tensions which aggravated it, we must first look outside the balance sheet of purely military affairs, of battalions and equipment, of brilliant tactics, bravery in combat, misguided strategy, and take up the clues from the history of the Army in the period between the wars.

All was not well with the German Army. A curious malaise had crept over that magnificent body, having its origins in the progressive erosion of its powers of decision. In the 1920's, under the brilliant and calculating Seeckt,[14] the German Army had enjoyed undisputed sovereignty as the

[10] Hermann Goering, Prime Minister of Prussia from 1930. C. in C. of the Luftwaffe; chairman of the Reich Defence Council; held the rank of Reichsmarschall, senior officer of the armed services. Committed suicide in Nuremberg 15th October, 1946.

[11] Joachim von Ribbentrop, Ambassador in London 1936-38. Foreign Minister 1938-45. Hanged at Nuremberg, October 1946.

[12] Alfred Rosenberg, Chief of foreign political section in the Party office. Minister of Eastern Territories from April 1941. Hanged Nuremberg, October 1945.

[13] Erich Koch, Gauleiter of East Prussia 1930-45. Reichskommissar of the Ukraine 1941-44. (See in particular p. 64.) Extradited to Poland 1950 and disappeared.

[14] Colonel General Hans von Seeckt, Commander in Chief of the German Army 1919-30.

arbiter of governments and policies. But in the 1930's extraneous factors had begun to make themselves felt. Partly these were technological—the advent of new weapons and new services threatened the primacy of the well-drilled soldier; partly they were political—in the shape of Adolf Hitler, the Nazi Party, and their private brigades of well-armed rowdies, the SA.

During this period Hitler had substantial popular support, but not a majority. Already Chancellor, he was determined to succeed Hindenburg[15] as President, and to achieve this he needed the support of the Army. The Army itself was anxious to reassert its power in domestic politics, and believed that in Hitler it might have found an acceptable protégé—provided that he fulfilled certain conditions. What followed, the Deutschland Compact, was a classic example of an agreement, not uncommon in history, in which each side believes that it has gained the advantage because of its simultaneous (but undeclared) resolve to double-cross the other within the framework of the agreement. In the mistaken belief that because the support of the Army would make Hitler, withdrawal of that support could at any time unmake him, the War Minister, Blomberg, had agreed to back Hitler's claim to succeed the ailing Hindenburg in the presidency. In return he extracted a promise that Hitler would curb the SA, and "assure the hegemony of the Reichswehr on all questions relative to military matters."

The two men had made their agreement in the secrecy of a wardroom on the *Deutschland* steaming between Kiel and Königsberg at the start of the spring manoeuvres of 1934. When these passed and spring turned to early summer with no move by Hitler to fulfil his part of the bargain, it was felt by many in the Army that the impertinent little Chancellor (he had been in office less than a year) was "unreliable." In June a political crisis, deeper in shadow than substance, blew up, and "the unity of the Reich" seemed to the military—or they professed it to seem—in jeopardy.

Hitler's experience on this occasion can have done nothing to temper his private resolve to subordinate the Army as soon and as ruthlessly as was feasible. The nominal head of the executive, he was sent for by Blomberg, who met him on the steps of the castle at Neudeck. The War Minister was in full uniform, and immediately (while remaining stand-

[15] Field Marshal Paul von Hindenburg, 1847-1934. Chief of the General Staff 1916-19; President of Germany 1925-34.

ing at a superior level to the Chancellor) delivered a cold and formal speech ". . . If the Government of the Reich could not of itself bring about a relaxation of the present state of tension, the President would declare martial law and hand over the control of affairs to the Army." Hitler was allowed exactly four minutes with Hindenburg, who woodenly recited a summary of Blomberg's caution while Blomberg remained standing at his side. Hitler was then dismissed.

This was the last occasion on which the Army exercised real power in the politics of the Third Reich. Within ten days the Nazis had shown that they were its equal in merciless application of the rules, and moreover, that they changed these rules to suit themselves as their grip on the national policy tightened. After warning the High Command that "civil action" was going to be taken against "certain disruptive elements" and arranging that the Reichswehr be placed in a state of general alert and confined to barracks, Hitler struck out—placing his own catholic interpretation upon the term "disruptive." To do the killing Hitler used his personal bodyguard, the black-uniformed SS. In 1934 there were only a few thousand of them, but surprise and the passivity of the Army more than made up for their lack of numbers. It was not from their own comrades in arms that the SA were expecting trouble.

By the time the Army came to its senses "order" was restored and the blood was being swabbed out of the execution cellars. The SA had gone, but so had nearly every figure of distinction, be he right, liberal, or even as Schleicher[16] and Bredow[17] of the *Generalstab,* who had opposed the rise of the Nazi Party. From that day on it was plain that whosoever opposed Hitler risked not simply his career but his life; and the instrument of execution, the SS, had emerged, by the very terms of its confinement to "police" duties, as the true arbiter of internal security.

For a few weeks, as the scope of the purge and the threat to its own position became apparent, the Army hesitated over what action it should take. Its discontent, diverted by the death of the aged Marshal Hindenburg, rumbled on into the following year, and then Hitler had opened the toy cupboard.

[16] Colonel General Kurt von Schleicher, a "political" soldier who held office as Minister of Defence in the twenties and as Chancellor in 1932. Advised Hindenburg against dealing with Hitler.

[17] Major General Kurt von Bredow; succeeded Schleicher as War Minister 1932; of similar political views.

The declaration of general rearmament and military conscription gave every professional soldier so much work and such glittering prospects as to effectively smother any desire he may have felt to dabble in politics. In any case, to what purpose would such dabbling be directed? The Army seemed to have achieved its every goal. Its "hegemony in military affairs" had been bloodily asserted, and all limits on its own development had been torn down. Blomberg spoke for all in his speech at the German Heroes Remembrance Day celebrations on 17th March, 1935:

"It was the Army, removed from political conflict, which laid the foundations on which a God-sent architect could build. Then this man came, the man who with his strength of will and spiritual power prepared for our discussions the end that they deserved, and made all good where a whole generation had failed."

But if Blomberg had forgotten the interview at Neudeck, Hitler had not. Nor had the Führer (as he now was) accepted the supercilious posturing of the then Commander in Chief, Fritsch, and his obstructive attitude to the SS; or the flagrant manner in which Fritsch harboured political suspects within the ranks of the Army. Both these men were marked for removal, and while their files at Gestapo headquarters accumulated detail and Himmler's web was spun, Hitler employed a number of psychological—indeed, totemic—devices to bind the Army to him more closely. It is in the history of this second period of Hitler's subjugation of the Army that the seeds of those unseemly and at times catastrophic disputes which were to plague the conduct of the Eastern campaign were sown.

One of the "concessions" Hitler had extracted from Blomberg at the time of the Deutschland Compact was the introduction of the Nazi emblem into the make-up of every soldier's uniform. From that time on the traditional German eagle held within its claws a tiny swastika, and soon the sign began to appear in larger scale—on regimental colours, in flags, over the entry arches to barracks, stencilled on the turrets of armoured vehicles. Regardless of the political detachment of the senior officers, this measure served to identify the ordinary soldier with the Nazi Party in the minds of the people and in their own consciences. This identity was reinforced by the terms of the fealty oath—sworn by every member of the armed forces in August 1934, which superseded the old form of oath sworn to the constitution under the Republic.

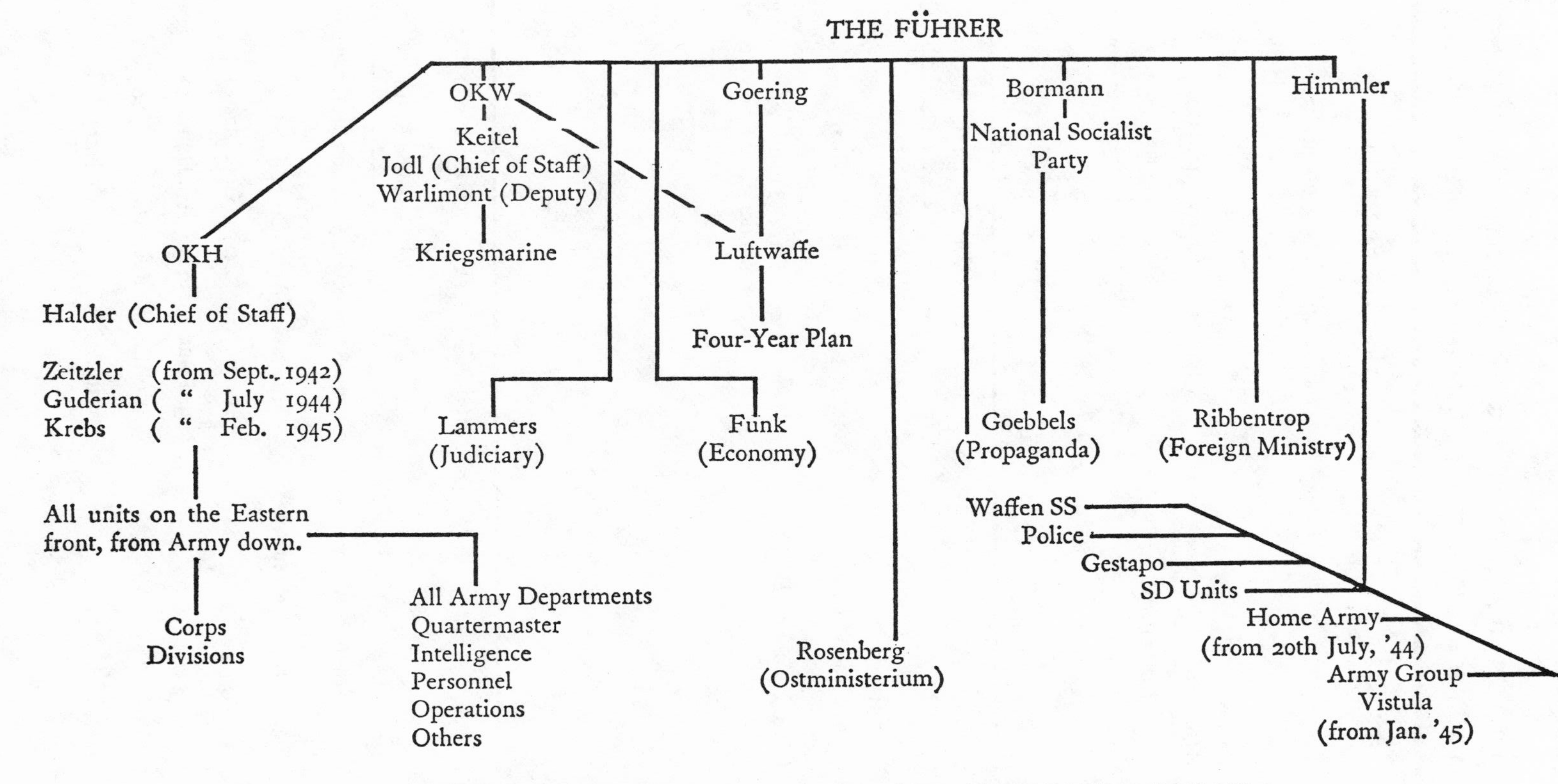

POWER AND PERSONALITIES IN THE THIRD REICH

ORDER OF BATTLE OF THE GERMAN ARMIES AT THE START OF *Barbarossa,* 22ND JUNE, 1941

Army Groups (AG)	Armies (A) & Panzer Armies (Pz A)	
	18th A	(Gen. von Küchler)
	16th A	(Col. Gen. Busch)
Army Group North (F.M. *Ritter von Leeb*)	4th Pz A	(Col. Gen. Hoepner)
	3rd Pz A	(Col. Gen. Hoth)
	9th A	(Col. Gen. Strauss)
Army Group Centre (F.M. *Fedor von Bock*)	4th A	(F.M. von Kluge)
		2nd Pz A (subordinate to Kluge), commander Col. Gen. Guderian
	6th A	(F.M. von Reichenau)
	1st Pz A	(Col. Gen. von Kleist)
Army Group South (F.M. *Gerd von Rundstedt*)	17th A	(Col. Gen. von Stülpnagel)
	Hungarian Army Corps	
	Italian Army Corps	(Gen. Masse)
	3rd Rumanian A	(Gen. Dumitrescu)
	4th Rumanian A	(Gen. Ciuperca)
	11th A	(Gen. von Schobert)

Infantry Strength	Panzer Corps (Pz K), Divisions, & SS	
11 Divisions	41st Pz K	(Gen. Reinhardt) 1st & 6th Panzer, 36th Motorised, 269th Infantry.
10 Divisions	56th Pz K	(Gen. von Manstein) 8th Panzer, 3rd Motorised, 290th Infantry.
	Pz A Reserve	SS *Totenkopf*

15 Divisions	39th Pz K	(Gen. Schmidt) 7th & 20th Panzer, 20th & 14th Motorised.
17 Divisions	57th Pz K	(Gen. Kuntzen) 12th & 19th Panzer, 18th Motorised.

Panzergruppe Guderian (from 27th July, 1941)

24th Pz K (Gen. Freiherr von Geyr)	46th Pz K (Gen. von Vietinghoff)	47th Pz K (Gen. Lemelsen)
3rd & 4th Panzer, 10th Motorised, 1st Cavalry.	10th Panzer, SS *Das Reich,* Inf. Reg. *Gross Deutschland.*	17th & 18th Panzer, 29th Motorised.

18 Infantry Divisions	3rd Pz K	(Gen. von Manteuffel) 14th Panzer, 44th & 298th Infantry.
12 Divisions		
2 Divisions	14th Pz. K	(Gen. von Wietersheim) 13th Panzer.
3 Divisions	48th Pz K	(Gen. Kempf) 11th Panzer, 54th & 75th Infantry.
22 Divisions	Pz A Reserve	16th & 19th Panzer, 16th & 25th Motorised, SS *Adolf Hitler,* SS *Viking.*
6 Divisions		

> I swear before God to give my unconditional obedience to Adolf Hitler, Führer of the Reich and of the German people, Supreme Commander of the Wehrmacht, and I pledge my word as a brave soldier to observe this oath always, even at peril of my life.

In 1937, Fortune gave Hitler the opportunity to rid himself of Blomberg, at the very moment Himmler's "frame" around Fritsch was complete. In one headlong rush of brilliant exploitation Hitler brought the Army, stunned and breathless, to heel.

The War Minister had proposed the luxury of taking to himself *en deuxième noce* a notorious prostitute. This indiscretion, though committed in all innocence, could not be tolerated by the doctrinaire standards of the officer corps. Hitler thus found himself in the impregnable position of being able to dismiss the Army's nominee while claiming that he was prompted solely by a consideration of its interest. Into this atmosphere of sexual scandal the Gestapo hastily flung its file on the Commander in Chief, accusing him of unnatural vice with a notorious Bavarian convict.

Poor Fritsch! He had no idea how to combat these charges, of which he was completely innocent, save the conventional resort of his caste: he challenged Himmler to a duel. In the subterranean jungle of Nazi politics such a gesture had as little effect as a peacock spreading his tail feathers at a python. Hitler pressed his advantage ruthlessly. Sixteen senior generals were dismissed (among them Rundstedt,[18] who had been injudicious enough as to suggest Fritsch as Blomberg's successor during the brief interval between the resignation of one and the charge against the other) and another forty-four were transferred from their commands.

But harassing and humiliating as these moves were, they were slight beside the formal administrative changes which were promulgated at the same time. By decree of 4th February, 1938, the three service ministries —of which that of the Army was naturally the senior—were unified and subordinated to a single commander, Hitler himself.

> From henceforth I exercise personally the immediate command over the whole armed forces. The former Wehrmacht office in the War Ministry becomes the High Command of the Armed Forces [OKW] and comes immediately under my command as my military staff. At the head of the

[18] Field Marshal Karl Rudolf Gerd von Rundstedt, "the Field Marshal who never lost a battle [until Normandy 1944]." Commanded Army Group South in Russia 1941. Dismissed after ordering the evacuation of Rostov.

> staff of the High Command stands the former chief of the Wehrmacht office [Keitel]. He is accorded the rank equivalent of Reich Minister. The High Command of the Armed Forces also takes over the functions of the War Ministry, and the Chief of the High Command exercises, as my deputy, the powers hitherto held by the Reich War Minister.

The creation of OKW and the consequent subordination of the Army to a small executive that came, as has been seen, increasingly under the Führer's technical control as well as subject to his personal influence was a political device, and as is so often the case with measures that are expedient from the aspect of domestic politics, it ran counter to the strict requirements of military efficiency.

It was the final blow in the struggle between the civil power (if the Nazi Party may be so described) and the Army. It meant that the *Generalstab,* which had already lost the broader power of judgment over the "best interests of the Reich" and of intervention in its domestic politics, was now deprived of its historic and fundamental prerogative—the decision as to when, and how, to make war. OKH was reduced in status to a department, specialising in army affairs and subordinate to a staff composed of men who were themselves the nominees of, and directly responsible to, the Führer. The result was that the orthodox procedure whereby strategic doctrine was evolved no longer functioned. In the place of study and consultation between experts there were the Führer conferences—little better than audiences at which Hitler, after listening with more or less good grace to "reports," hectored the assembled company with his mind already made up—and the Führer directives,[19] documen-

[19] The clearest description of the way the directives originated has been given by Professor Trevor-Roper in his introduction to their English translation (London 1964); although this, of course, refers to procedure in wartime:

"Every day, at noon, Hitler held his *Lagevortrag* or 'situation conference,' at which Jodl [the Chief of Staff] submitted a report which had been prepared for him by Warlimont [Deputy Chief of Staff at OKW]. Hitler would listen, discuss the situation, and then, after it had been fully debated, issue his orders. These orders, together with a full account of the discussion, were then passed by Jodl to Warlimont to be converted into formal documents and issued to the appropriate authorities."

It is true that the Commander in Chief, Brauchitsch, did have access to Hitler. He and, on occasion, individual army commanders, were summoned to Führer headquarters. But as Professor Trevor-Roper points out, ". . . their visits were not regular, and they could not compete with the constant presence of the regular courtiers. Besides, Hitler preferred to deal with them through Keitel and Jodl. He disliked new faces. He liked Keitel and Jodl, who gradually sank into the position of mere orderlies . . . and Keitel and Jodl liked the monopoly of power which their industry and subservience ensured to them. Consequently both Keitel and Jodl, while they

tary orders concerning which no dispute, query, or emendation was permitted. In this way the immense fund of technical expertise of which the *Generalstab* was the repository was canalised into tactical and substrategic planning. The broad outlines of war policy, the co-ordination of theatres, even the evolution of new weapons and the assessment of priorities in supply were settled without reference to its opinion. There was no permanent consultative body of experts preparing appreciations and alternatives, no equivalent to the Chiefs of Staff Committee or the Joint Chiefs of Staff in the West.

And, indeed, once the war began, policy in the military sense can hardly be said to have existed outside Goering's comment, *"Wenn wir diesen Krieg verlieren, dann möge uns der Himmel gnädig sein"* (If we lose this war, then God help us).[20] War aims, together with the detail and timing of their achievement, were decided by Hitler. Such discussion as took place was usually confined to the Führer's immediate entourage of Party cronies, Himmler and Bormann, Hess and Goering; men who could keep the same nocturnal hours and talk the same language of racialism and "destiny." Of these it was Goering to whom Hitler listened most often. But even Goering attained no more than a negative influence—preferential treatment for the Luftwaffe, and in later stages of the war his influence declined and he saw Hitler less and less often.

There is no evidence that Hitler ever changed his mind on questions of strategy either at the persuasion of his intimates in the Party or the senior officers of the Army. He carried on his own back the responsibility for every decision of importance and formulated in his own mind the development of his strategic ambition in its entirety.

This facility, which those who have reasons for belittling have compendiously labelled Hitler's "intuition" (pronounced with a sneer), was truly prodigious and, for many years, infallible. The Devil's hand guided Hitler, just as later on it was to protect his life. But with the outbreak of war, as the pressure intensified and responsibilities widened, the absence of a permanent consultative body began to make itself felt.

---

became increasingly indispensable to Hitler, became increasingly odious to the generals in the OKH and in the field."

After the dismissal of Halder, in September 1942, the proceedings of these conferences were recorded verbatim by a body of stenographers, and the surviving fragments of their records are exceedingly valuable source material. (See, in particular, Chs. 15 and 18.)

[20] When the English ultimatum of 3rd September, 1939, was delivered (Schmidt, Paul, *Statist auf diplomatischer Buehne, 1923-1945.* Bonn 1945).

The most serious, as also one of the earliest, examples of lacunae in strategic planning had followed immediately on the collapse of France. Not only was there no plan in existence for the invasion of the British Isles, but over a month passed before the *Sea Lion* directive—the order to prepare such a plan—was issued.

And the disadvantages of Hitler's practice of bypassing orthodox channels applied as much in matters of detail as in grand strategy. For example, after the campaign in France, Hitler had ordered that the 37-mm. gun in the Pz III tank be replaced with a 50-mm. L 60. However, for reasons which shall never be clear (but which owed much to there being no permanent body which could see a directive of this kind through to fulfilment and supervise the responsible officers at the Ordnance Office), the specification was altered to 50-mm. L 42. The result was that the most successful tank of the war was equipped with a gun of markedly lower range and muzzle velocity than Hitler had ordered, which if fitted would have preserved its technical ascendancy for another year at least.

After the French surrender Hitler approved an OKH suggestion to demobilise a number of divisions,[21] which is scarcely consistent with his own plan to attack what was believed to be the largest army in the world within the coming year. The only explanation is that in the absence of a proper supervisory body and procedure the order somehow leaked past. Yet at almost the same moment Hitler was directing that the number of Panzer divisions in the Army was to be doubled and tank production raised to a level of eight hundred to a thousand units per month. Once again the Ordnance Office intervened, with a report that an expansion of this kind would cost over two billion marks, and would require an additional one hundred thousand skilled workers and specialists. Hitler agreed to its postponement "for the time being," but the reorganisation of the Panzer divisions had gone ahead, so that the net effect was that the tank strength of each division was halved. In the result there was some compensation in their increased fire power and the gradual substitution of the heavier PzKw III for the PzKw II, but the Panzer divisions were never to recover the numerical strength and mobility with which they had begun the battle of France. Hitler had also directed

[21] The number of men scheduled for demobilisation has been put by some estimates as high as four hundred thousand, but it is unlikely that this number was actually discharged.

that the number of motorised divisions be doubled, but without making any provision for an increase in the production of the vehicle industry. The result was that many of the new formations had to equip themselves with captured or requisitioned trucks, which were to prove unreliable and difficult to service under severe conditions.

Examples of this kind could be multiplied, and it is true that the deficiencies in vigour, authority, and scope of the so-called OKW Chiefs of Staff were to make themselves increasingly felt as the war proceeded. But it would be less than just to claim for the generals of OKH (as they themselves are not slow to do in their own works on the subject) a particular but thwarted prescience in matters of grand strategy.

Hitler's sense of history was limited, but highly coloured, and he drew upon it to justify his assumption of a single and exclusive responsibility. In the Great War (he would argue) the German General Staff, directing its country's strategy for four years without hindrance, had made one error of judgment after another: it pressed the introduction of unrestricted submarine warfare, thereby hastening the entry of the United States into the war; it cast away any hope of a separate peace with Tsarist Russia after Galicia-Tarnow by its insistence on the establishment of a kingdom of Poland; then achieved the same result in 1917, when its annexationist attitude to France and Belgium ruined the chance of the Papal peace proposals being carried further. Finally there was its responsibility for the most catastrophic single action of the century—the despatch of Lenin and his colleagues from Switzerland to Russia in the famous "sealed train." Even in the exclusively military sphere it had made grave errors, mishandling the only two serious attempts to defeat the Western powers in the field. Falkenhayn had allowed the course of the attrition battle at Verdun to escape his control and thereby missed the chance of knocking France out of the war in 1916. Ludendorff's diminuendo sequence in April 1918 drew so heavily on his armies' blood and morale that they were incapable of offering prolonged resistance to the Allied counteroffensives which followed.

When Hitler became Chancellor he found that OKH was still free with advice, and that its attitude was sadly repetitive in two particulars—in the unanimity of the views of its members and the mistakenness (as it invariably emerged) of their appreciations.

The first expansionist move undertaken by the Reich, the reoccupation of the Rhineland, had called forth a whole sequence of protests from the General Staff. First Beck[22] proposed that the entry of German troops be accompanied by a declaration that the area would not be fortified. Hitler rejected the notion out of hand. Then Blomberg was persuaded by the General Staff to put forward a suggestion that the troops sent across the Rhine be withdrawn on condition that the French agree to withdraw four to five times as many men from their own borders. He was "bluntly and brutally snubbed" for his pains. Finally after a lethargic concentration of thirteen French divisions had been observed in the Maginot Line, Beck and Fritsch together had made Blomberg urge the withdrawal of the three battalions which had entered the demilitarised zone. Again Hitler refused, and again he was proved right.

The generals were nonplussed. They laid no claim to an understanding of the subtleties of international politics. But they had before them the figures of relative strengths. Did common sense and the simple calculations of a military balance sheet count for nothing? Answer, no. What counted was the will, and that, with its full *appareil de mystique,* was held in monopoly by Hitler. "It is my unalterable will to smash Czechoslovakia by military action in the near future," he told them, and throughout the summer of 1938 the preparations for this had gone ahead, without regard to the protesting bleats from almost every senior officer in OKH.

The original intention of the dissident generals had been that the Commander in Chief, Brauchitsch, be forced by their unanimous recommendation to go to Hitler and pronounce to him the magic words of Hindenburg and Seeckt: that he "no longer enjoyed the confidence of the Army." Fritsch might have done this; but Brauchitsch, never. In despair, the Chief of Staff of the Army, General Beck, resigned. None of Beck's colleagues followed this example, but many did allow themselves to become privy to a plot for kidnapping Hitler and declaring a military government. This coup was planned for the very last moment of peace, when it was established that Hitler had fixed a zero hour for the attack on the Czechs. It was thwarted (and the whole course of history perverted) by the Franco-British betrayal at Munich, but the commanders who had planned it—Witzleben, Helldorf, Schulenburg, Hoepner—remained in office.

[22] Colonel General Ludwig Beck, Chief of the General Staff 1935-38. Later nominated as head of the new German state by the 20th July plotters; committed suicide when the plot failed.

Thus can be detected two separate but complementary elements in the decline of the *Generalität.* It had been outmanoeuvred politically, forsaking one foothold after another in a downhill retreat from the pinnacle of influence it had occupied for the preceding half century. And the swift and bewildering march of events on the international stage had shown it up (it seemed) as a timid clique at fault in assessing its own strength and hesitant over using it.

Many factors perpetuated this unhappy condition. None of them were vital when considered in isolation, but they formed a sum of perplexity and disillusion; of confused loyalties, considered self-interest, and escapist devotion to the narrow technicalities of its appointment.

It is not easy to feel sympathy for the members of the *Generalität* because the ultimate source of their discontent was in their own lack of moral fibre. What affronted them about Hitler's conduct was not its immorality but its irresponsibility. Hence their tendency to hang back, to procrastinate on whatever excuse, and watch to see if the risk "came off." Furthermore, Hitler's success in curtailing their independent power had been achieved without alienating the bulk of the officer corps or disturbing the foundations of professional efficiency which had been laid by Seeckt. This meant that those who wished to alter the course of events must dabble in politics—a field which they entered no longer as arbiters but as participants, hampered by scruple, plagued by disunity, and burdened with a lingering contempt for civilians which was for long to frustrate all efforts to co-ordinate the two separate elements of the opposition.

Out of their depth in this unfamiliar element, the generals groped and fumbled. Some intrigued actively against the regime. Others, nearly all, listened with sympathy to those who were intriguing, yearned for the days of decision, and watched for a change in fortune. Others, and they included the majority in both these categories, sublimated their frustration in work. The result was a quality of staff work and a tactical brilliance unequalled by any other army.

Hitler had effectively shut out the Army from politics, and the price he paid seemed at first to be even less than the pittance he had promised Blomberg on board the *Deutschland.* But in one important respect the Army held out for its rights. It steadfastly and persistently refused all efforts by the Nazi Party to penetrate into the conduct and administration of its internal affairs. The generals clung to their privilege (more formal than real) of being the "sole bearers of arms within the Reich," and they twice resisted with success major efforts at infiltration by Himmler (once through a campaign by the SS to deprive army chaplains of their military status, on another occasion when it was proposed to institute "voluntary" classes of Nazi indoctrination in place of religious services). The Army became a haven for all those discontented with the regime, a loose fraternal body—politically inert, it was true, but where the writ and dossier of the SS never ran.

The result was, quite literally, fantastic. The whole of the Abwehr (the military intelligence branch) was riddled with dissent. Admiral Canaris, its head, and his lieutenants, Oster and Lahousen, not only allowed the organisation to be freely used as a medium of communication and movement by the various malcontents, but perpetrated the most incredible acts of treachery—Oster warning the Danish Military Attaché ten days before the impending invasion of that country and of Norway in April 1940; and doing the same to the Netherlands before the attack on the Low Countries.

Another department headed by a general steadfastly hostile to the regime was the *Wi Rü Amt,* the economic and armaments branch of OKW under Georg Thomas. Neither Thomas nor Canaris allowed his sympathies to affect the day-to-day running of his department, any more than their brother officers allowed their own feelings to intrude on the ruthless efficiency with which they planned and fought. But the effect, a certain inner weakness, was lasting. The "conspirators" (by which is meant those who were actively plotting for a change of regime), although hardly worthy of such a title at this stage, suffered no restrictions in such an atmosphere. Passes, movement orders, transfers, all these could be arranged at an instant's notice. They would receive early warning, too, of plans and proscriptions that might affect them.

Was this a form of reinsurance by the generals? Or was it simply the code of the officer and gentleman that allowed them to continue in the dangerous practice of tolerating seditious conversations in their presence, of not reporting the continuous and sometimes farcical indiscretions with which the conspirators bored them? Once the war had started, the practice of sedition was confined to the medium levels of the Army. The senior commanders regarded such activity with no more than a tolerant interest. For too long they had offered opinions, to one another and to Hitler, and had seen their validity compromised by the perverse tricks of circumstance. Like ultraconservative bankers during an inflationary boom, they could no longer bring themselves to utter the conventional warnings which had so often led to disappointing investment policies.

A time was approaching when orthodoxy and sober calculation were to assume their rightful importance, as it was when the fussing of the conspirators was to become an altogether more troublesome phenomenon; but dazzled by the brilliance of the Führer's achievement, the generals could no longer see that far ahead. To a man they would have echoed Brauchitsch when he told Otto John after the war, "I could have had Hitler arrested easily. I had enough officers loyal to me to carry out his arrest. But that was not the problem. Why should I have taken such action? It would have been an action against the German people. I was well informed, through my son and others. The German people were all for Hitler. And they had good reason to be . . ." [23]

These, then, were the infirmities that afflicted the German Army. But in that period of victorious euphoria, when the first strands of the *Barbarossa* plan were woven, they lay dormant. The generals were bathed in glory, and generously rewarded by their Führer. Decorations, pensions, gratuities, building permits, estates in East Prussia, were heaped on them. In disgust Hassell wrote, ". . . the majority are out to make careers in the lowest sense. Gifts and field-marshals' batons are more important to them than the great historical issues and moral values at stake."

[23] This conversation is taken from the John Memorandum, quoted in Wheeler-Bennett, *The Nemesis of Power,* 492. The background illustrates the surprising freedom with which sedition was discussed in the highest circles of the Army. Popitz (Johannes Popitz [1884-1944] Prussian Minister of State and Finance, a close friend of General von Schleicher, who had been killed by the SS in 1934, and one of the earliest members of the "Resistance circle") had visited Brauchitsch—the Commander in Chief, let it be remembered—in the autumn of 1939 and "besought him to take action for the honour of the Army in rescuing Germany from the talons of the Black *Landsknechte* [the SS]." Brauchitsch had remained "virtually silent" throughout the interview, but at the end he had asked if there was still a chance of securing a decent peace for Germany. Later General Thomas came to him with some details of "terms" on which the Pope was prepared to act as intermediary for an understanding with Britain. The Commander in Chief's reaction was surprisingly mild. Although complaining that "the whole thing was plain high treason," he did nothing further than tell Thomas that "if he persisted in seeing him in this connection he would place him under arrest."

At this stage, the winter of 1940, it is probably true that the Army would have followed Hitler wherever he led it, in spite of its deep-rooted fear of a direct confrontation with Russia. Only one senior member of OKW, Admiral Raeder, went on record at the time as being against it, and "All the men of the OKW and the OKH with whom I spoke," wrote Guderian,[24] ". . . evinced an unshakeable optimism and were quite impervious to criticism or objections."

These convictions were largely the result of personal inspiration from Hitler, whose strategic argument seemed unanswerable:

> . . . Britain's hope lies in Russia and the United States. If Russia drops out of the picture, America, too, is lost for Britain, because the elimination of Russia would greatly increase Japan's power in the Far East. Decision: Russia's destruction must be made a part of this struggle—the sooner Russia is crushed the better.

And in fact during the autumn of 1940 this strategic bias had received detailed support from a number of political developments in the Balkans. The differences between the two powers accumulated so rapidly that by November it had been necessary for Soviet Foreign Minister Molotov to visit Berlin. The ensuing conference, at which the last exchanges took place between the two tyrannies, had not been a happy affair. The purported occasion for its gathering was "the apportionment of the British Empire as a gigantic estate in bankruptcy," but in fact this subject was hardly mentioned (except by Ribbentrop, who spoke of nothing else).

When it was suggested to him that their latent differences be papered over by Russia joining the tripartite alliance, Molotov had replied, ". . . paper agreements do not suffice for the Soviet Union; rather, she must insist on effective guarantee for her security." The Russian Foreign Minister then went on to press a number of delicate points: What were German troops doing in Russia? And in Finland? What if the Soviets were to guarantee Bulgaria in the same terms as the German guarantee to Rumania? His intransigence had been emphasised by a "personal" letter from Stalin after the conference broke up in which the Russian dictator "insisted" on an immediate withdrawal of German troops from Finland, a long-term lease of a base for Soviet land and naval forces within range of the Bosphorus, and certain concessions from the Japa-

[24] Colonel General Heinz Guderian, the leading exponent of the Panzer arm in the German Army. Commander of the 2nd *Panzergruppe* 1941; Inspector General of Panzer forces 1943; Chief of the Army General Staff 1944.

nese in North Sakhalin. Stalin also warned of an imminent pact of mutual assistance between the Soviet Union and Bulgaria.

The tone of the November conference made a profound impression on the German Army when the details were made known to it—and this Hitler lost no time in doing. Many who had believed that diplomacy should keep the Russians at arm's length for as long as possible now swung around to the view that a preventive war could not be avoided. But it is wrong to claim, as many German writers do, that the November conference accelerated, or even initiated, the planning of the campaign in the East. This was already fixed for the spring of 1941—the earliest date at which it would be physically possible to move and deploy the whole army. Stalin's letter may have strengthened Hitler's resolution, and it gave him a convenient justification; but his mind had been made up during the battle of France, when he had seen what the Panzers did to the French Army.

The date most conveniently ascribed to the start of German planning for war with Soviet Russia is 29th July, 1940. On this day a conference was held at Bad Reichenhall, under conditions of the utmost secrecy, at which Jodl[25] addressed a few hand-picked planners drawn from the staff and the economic administration of the Reich, on the Führer's "expressed wishes." Some weeks earlier, while the battle of France was still being fought, Hitler had told Jodl, "I will take action against this menace of the Soviet Union the moment our military position makes it at all possible," and this resolve had been expanded in a series of private meetings at the Berghof between Hitler, Keitel, Jodl, and Goering in the days following the armistice. The first directive, Operation *Aufbau Ost,* was issued in August, with its intentions camouflaged under a plethora of code names and generalities, and from that time the widening circles of planning spread rapidly across the pool of Nazi administration, so that when the new quartermaster of OKH took up his appointment on 8th September he found in his files "a still incomplete operational plan dealing with an attack on the Soviet Union."

A further directive (No. 18), issued in November, was more explicit. In it Hitler wrote:

> Political discussions have been initiated with the aim of clarifying Russia's attitude for the time being [Molotov was actually visiting Berlin at the time]. *Irrespective of the results of these discussions all preparations*

[25] Colonel General Alfred Jodl, Chief of Staff at OKW 1938-45.

> *for the East which have been verbally ordered will be continued.* Instructions on this will follow as soon as the general outlines of the Army's operational plans have been submitted to me and received my approval.

Less than a month later Halder had submitted the OKH plan, and on 18th December the Führer, in his famous Directive No. 21, set out the strategic objectives and gave to the unborn child conceived that summer a name, Operation *Barbarossa.*

But although the summer of 1940 saw the start of the planning, the intention can be traced even earlier than this, to Hitler's celebrated Berghof conference of 22nd August, 1939. Of all the speeches and all the occasions in the history of the Nazis it is this "private" conference which illustrates most vividly their devilish character. Hitler had exulted that day, "There will probably never again be a man with such authority or who has the confidence of the whole German people as I have. . . . Our enemies are men below average, not men of action, not masters. They are little worms." In any case, he told his listeners, the Western powers would not move to defend Poland for that morning Ribbentrop had flown to Moscow to sign the nonaggression pact with the Soviets. "I have struck this instrument from their hands. Now we can strike at the heart of Poland—I have ordered to the East my Death's Head units [of the SS] with the order to kill without pity or mercy all men, women, and children of Polish race or language."

At this point, we are told, Goering jumped on the table, and after offering "bloodthirsty thanks and bloody promises . . . danced around like a savage." [26] "My only fear," Hitler said to his audience, "is that at the last moment some *Schweinhund* will make a proposal for mediation." As to the future, "There is no time to lose. War must come in my lifetime. My pact was meant only to stall for time, and, gentlemen, to Russia will happen just what I have practised with Poland—we will crush the Soviet Union."

With this last pronouncement the euphoria generated by Hitler's drum-

[26] These antics must have been rendered the more impressive by Goering's attire. "He was dressed in a soft-collared white shirt, worn under a green jerkin adorned with big buttons of yellow leather. In addition he wore grey shorts and long grey silk stockings that displayed his impressive calves to considerable effect. This dainty hosiery was offset by a pair of massive laced boots. To cap it all, his paunch was girded by a sword-belt of red, richly inlaid with gold, at which dangled an ornamental dagger in an ample sheath of the same material.

"Up till now I had assumed we were here for a serious purpose," was the acid comment of Manstein [at that time a Colonel on the planning staff], "but Goering appeared to have taken it for a masked ball."

beating was sensibly diminished, and at the close of the address, "A few doubtful ones [among the audience] remained silent." For here, let out quite casually, it seemed, was the one unpardonable military heresy that all had agreed must be eschewed forever—the "war on two fronts."

German military opinion was about evenly divided on the desirability of fighting Russia—the "Prussian school," which favoured an Eastern alliance, still balanced those whose ideological convictions were compounded by an imagined strategic necessity, the need for the raw materials and *Lebensraum;* but the most enthusiastic of these had never considered attacking Russia while a Western front was in being. Even in *Mein Kampf* this was held up as the cardinal error, the one fatal move which would annul every step in the ascent of the Reich to world domination. The General Staff had for long been uneasy about the weight and quality of Russian equipment,[27] concerning which their intelligence reports were so alarming that they usually adopted the practice of dismissing them as "plants" by the MVD. Every senior officer in the German Army had, at some time or other, warned Hitler about the danger of attacking Russia while still engaged in the West, and both Brauchitsch and Rundstedt claimed that he had given them an understanding never to do this.

But when, almost exactly a year later, the idea began to acquire the bones and flesh of operational planning, Hitler could with some reason contend that the Western front existed no longer. The French had collapsed and made peace, and the British were confined to their own territory, where they licked their wounds in impotence. The battle of Britain, that miraculous victory so light in blood and so limitless in consequence, could hardly have been foreseen—much less the Italian defeats in Africa and all the strategic complications and distractions that were to flow from them. In the warm afterglow of the battle of France,

[27] Guderian has related an incident which exemplifies this (*Panzer Leader* 143): "In the spring of 1940 Hitler had specifically ordered that a Russian military mission be shown over our tank schools and factories; in this order he had insisted that nothing be concealed from them. The Russian officers firmly refused to believe that the Panzer IV was in fact our heaviest tank. They said repeatedly that we must be hiding our newest models from them, and complained that we were not carrying out Hitler's orders to show them everything. They were so insistent on this point that eventually our manufacturers and the Ordnance Office officials concluded, 'It seems that the Russians must already possess better and heavier tanks than we do.' It was at the end of July 1941 that the T 34 tank appeared at the front and the riddle of the new Russian model was solved."

with absolute dominion over the whole of the European mainland, there was some substance to Hitler's argument that an invasion of Russia would be not a second but a first, and last, front.

As so often happens in global affairs of state, the planning, once set in motion, matured inexorably, while around it the circumstances in which it had originated altered in character and emphasis. The Luftwaffe, hitherto supreme, met its match. Certain regions of the European sky were closed to it. Operational control and many items of its equipment were shown to be deficient. The Navy had been seriously unbalanced by the losses sustained during the Norwegian campaign. The U-boat programme was retarded and poorly planned—in the summer of 1940 there were only fourteen submarines with the endurance to sail west of the Killarney Bluff.

These things made it difficult to strike at Britain and, if she remained obstinate in her choice of war, impossible to subdue her without a long period of revised priorities and careful preparation. But time was short, or so Hitler believed: ". . . I can be eliminated at any moment by a criminal or a lunatic." The Army was ready and undefeated. Alone of the three services it had risen to every demand which the German people had made of it. How preposterous to suggest that this magnificent machine be allowed to run down; that the armed forces be recast in an amphibious pattern to tackle a maritime power in her own element! The ascendancy Hitler had established over his generals in politics was now absolute, and he had no fear that their exploits in the field, however magnificent, could threaten this. Indeed, the Führer seems to have felt that his personal authority over the Army would be confirmed in such a campaign, with its powerful ideological overtone, and justified by the close attention he intended to devote to its conduct.

In 1930, Hitler had written, "Armies for the preparation of peace do not exist. They exist for triumphant exertion in war." And in the spring of 1941 the Wehrmacht stood victorious, hardly blooded; trained and equipped to perfection; a beautifully balanced and co-ordinated fighting machine now at a pinnacle of martial achievement. Where was it to go from there? Sheer gravitational pull must, it would seem, direct it against its one remaining opponent in the European land mass; draw it like Napoleon's armies, which also had stood in frustration on the Channel, eastward, to the dark unconquered steppe of Russia.

two | MOTHER RUSSIA

In the summer of 1941 the Red Army presented an enigma as much to the Western intelligence services as to those in Germany. Every facet by which military quality is assessed seemed to have an opposite. Its equipment, by all accounts, was lavish (in fact, it disposed of more tanks and as many aircraft as the rest of the world put together), but how much of this machinery was up to date, and how capable were the Soviet commanders of handling it? Its reserves of manpower seemed inexhaustible, but sheer mass was valueless without proper leadership, and Communist timeservers chosen for their political reliability would be as ineffective on the field of battle as the court favourites who had enjoyed the patronage of the Tsar. Even the innate courage and resilience of the Russian soldier, to which successive European wars bore testimony, was thought by some to have been jeopardised by political indoctrination. The "ordinary Russian," it was claimed, would show himself only too anxious to escape, by laying down his arms, from the menacing supervision of the commissars.

These problems faced foreign observers in 1941, and even today, with all the advantages of hindsight, it is not easy to resolve so many apparent contradictions. There are three distinct elements which must be considered: first, the paper strength of the Red Army, the state of its training, and its tactical doctrine; second, the impact of Party control on its leadership and its strategic posture; third, the reality of Soviet strength, as demonstrated by operational experience in the period immediately preceding the German invasion.

The modern Red Army was essentially the creation of two architects, Trotsky[1] and Tukhachevski[2] (both of whom were to pay with their lives for achieving such prominence). Trotsky had imposed form and discipline upon an amorphous proletarian rabble. Tukhachevski had evolved tactical and strategic doctrines which, although not so revolutionary as those of some British tank experts,[3] were nonetheless far in advance of current thinking in other European armies. However, in the late 1930's domestic politics and the shifting orientation of the Soviet Union in the European power complex led to corresponding (and damaging) changes in its military attitude.

The problem of defending Russia was dominated by the physical characteristics of her western frontier and the fact that Soviet economic and administrative centres were concentrated in a relatively small part of the country—within five hundred miles of this same western frontier. Furthermore, the eastern zone was effectively divided into two halves by the Pripet Marshes—a sprawling region of reed and forest, nearly two hundred miles across, which covers the area where the great rivers of European Russia take their source.

Besides their value as an obstacle to the invader the Marshes pose problems to the defence. For they effectively break the western zone into two halves, each of which must operate independently, being served by different rail complexes and protecting separate objectives. On a front of such length it is impossible to maintain strength everywhere, and the problem which had always confronted the Russian staff, and which was aggravated by the growing concentration of industrial power in the eastern Ukraine, was the according of priorities between the defence of the north, the twin capitals of Leningrad and Moscow, and the south, whence the country drew the bulk of its food, its machinery, and its armament.

In the early 1930's Marshal Tukhachevski had drawn up a master plan for the conduct of this defence, and this scheme, curiously, survived the

[1] Leon Trotsky, Commissar for War in March 1918. Ousted 1925; exiled 1928; assassinated 1940.

[2] Marshal M. N. Tukhachevski, Chief of the Red Army Staff 1926-28. Other appointments, and promoted to the Military Soviet (*q.v.*) 1934; demoted May 1937; executed June 1937.

[3] Notably Captain B. H. Liddell Hart and Major General J. F. C. Fuller, whose writings had greatly influenced the formation of the first Panzer division in the German Army (although they made little impression on the German General Staff until much later).

execution of its author on a charge of German-inspired espionage. He had suggested a relatively light concentration in the north, with the bulk of the mobile forces to be placed on the Dnieper, where they could menace the right flank of an invader and, if all went well, undertake a rapid occupation of the Balkans.

By this reckoning it was estimated that the sheer physical difficulties of distance and supply would protect the capital; the enemy would be drawn into a wide and desolate corridor between the Pripet Marshes and the fortress area of Leningrad, and the defence would be given time to regroup and to select its point for counterattack. This notion was originally conceived in the context of a threat from Poland or, at worst, an alliance between Poland and the rump of the German Army that remained after Versailles. But by 1935 three new factors had altered the scope of the appreciation. The pace of German rearmament under Hitler was rapidly accelerating, the emphasis in German training was on mobility and the use of armour, and the political attitudes of the other Western powers seemed clearly to indicate their hope for and encouragement of a move by Germany against the Soviet Union at some point in the future.

It was accordingly decided that a fortification system be extended southward from the Baltic to the northern fringe of the Pripet Marshes, and this work was started in 1936. At this time the doctrine of the all-powerful defence was firmly rooted in the armies of the West. The theory of the deep armoured thrust, although it originated in England, had taken rôot only among a few of the more enlightened of the officers in the German Army. The whole of military science was applied to the problems of devising and perfecting permanent defence systems against which the opponent would batter himself to exhaustion—systems which found their exemplar, if not their most perfect consummation, in the Maginot Line. Many details of the Maginot system were disclosed to the Russians, who had enjoyed intermittent good relations with France, at both military and diplomatic levels, for periods during the thirties, and it was not difficult for their intelligence to collect additional material from elements among the French military and the administration that were sympathetic to the Soviet ideology.

The result was that the Russians were able, by starting several years later, and with a considerable mass of data and experience at their finger tips, and with unlimited space and depth of ground to use, to construct a system—it was known as the Stalin Line—that was in places even more formidable than its French prototype. An appreciation by OKH intelligence made after the line had been overrun described it as:

> A dangerous combination of concrete, field works and natural obstacles, tank traps, mines, marshy belts around forts, artificial lakes enclosing defiles, cornfields cut according to the trajectory of machine-gun fire. Its whole extent right up to the positions of the defenders was camouflaged with a consummate art. . . . Along a front of 120 kilometres, no less than a dozen barriers, carefully camouflaged and proofed against light bombs and shells of 75 and 100 mm. had been constructed and sited in skilfully chosen fire positions. Thousands of pine trunks masked ditches which the attacker could not discover until it was too late. About three kilometres behind, over stretches of ten or twelve kilometres, three ranges of pines had been driven more than a metre into the ground. Behind this obstacle stretched out abatis made of trees sawn to within a metre of the ground, and whose tops, turned towards the enemy, had been entangled with barbed wire. Concrete pyramids strengthened this barrage.

But although stretches of the Stalin Line were extremely formidable, it was in no sense a continuous belt of fortification. Certain areas—notably around Lake Peipus, and between the Pripet Marshes and the upper Dniester; and the approaches to a number of key cities near the frontier zone—Pskov, Minsk, Korosten, Odessa—were heavily protected. The fortified districts were not linked, however, by any connecting strip of fieldworks, and the term "line," although it may have denoted an ultimate goal, was, in 1941, no more than a geographical illusion founded on the presence of a sequence of fortified districts all in roughly the same longitude.

Then, following the Nazi-Soviet pact of August 1939 and the agreement to partition Poland, the Red Army had deserted its fixed defences in White Russia and pushed westward, up to and beyond the line of the river Bug. And in July of the following year the Russians annexed Bessarabia and Bukovina. These measures, together with the "absorption" of the Baltic states in the north, advanced the western frontiers of the Soviet Union by hundreds of miles, and keeping step with the new geography, the Army went forward also, leaving empty its old training areas, its supply dumps, and the permanent emplacements of the Stalin Line.

Stalin believed that space was more important than fixed defences, but he ignored the fact that the Army was not trained in the sort of fluid defensive battle that alone makes the use of space profitable. And if any of the Red Army generals disagreed with him they had the sense, by 1939, to keep their thoughts to themselves. For still more important than his obsession with space (but equally disastrous) was the Russian dictator's conviction that the primary requirement in an army, and particularly in its senior officers, was that of political reliability. Communism teaches that the internal enemy is the most dangerous, and in a society as repressive as was prewar Russia the presence of three million men permanently under arms could become a source of anxiety to the regime unless they and their officers were ruthlessly disciplined into toeing the Party line.

In theory, the chain of command ran downward from the Committee of the Defence of the State (GOKO), which was presided over by Stalin and included Molotov,[4] Voroshilov,[5] Malenkov,[6] and Beria.[7] Subordinate to this was the *Stavka,* a kind of GHQ. Nominally a "committee of equals," the *Stavka* comprised eight army officers, with four commissars (among whom was Bulganin) to keep an eye on them; in fact, administrative control of the *Stavka* was in the hands of the Chief of Staff, Marshal Shaposhnikov, and his deputy, General Zhukov, both of whom consulted directly with Stalin. Neither GOKO nor the *Stavka* impeded the direct and autocratic sovereignty of Stalin himself, nor could they diminish the power of Beria and the NKVD, whose dossiers and firing squads, reaching along the web of commissars and "political education officers," were what kept the Army in line. They had been introduced in wartime in an attempt to revive the Red Army from the lethargic and apprehensive torpor into which it had fallen in the period following the great purges of 1937-38.

The high point of the Red Army's prestige and influence can be fixed at

[4] Vyacheslav M. Molotov, Soviet Foreign Minister 1939-52. A convinced Stalinist disgraced by Khrushchev after the 20th Party Congress.

[5] Marshal K. E. Voroshilov, an early member of the Communist Party (1903), fought with distinction in the civil war. Commissar for Defence 1934-40. Commanded Leningrad armies 1941 (see Ch. 6). Removed from operational command 1942 and held various ceremonial posts until disgraced by Khrushchev 1959.

[6] G. M. Malenkov, successful career in security and political sides of the military organisation, 1920-41. An intimate of Stalin's; succeeded him briefly as Premier after Stalin's death. Ousted by Khrushchev and Bulganin, and branded as "anti-Party" at 20th Congress.

[7] L. P. Beria, Chief of the NKVD 1935-52. Circumstances of death mysterious, but believed to have been shot on Khrushchev's orders, the only member of the anti-Party group to be "liquidated."

22nd September, 1935, when a decree introduced formal distinctions and marks of rank to its officers. Majors and above were granted immunity from civil arrest, and the political commissars were obliged from that time on to pass the exams of the normal military school. And at the pinnacle, to flaunt this new professionalism, were created five "Marshals of the Soviet Union." These were Blücher, the "Emperor of the Far East"; Yegorov and Tukhachevski; and those two sly and durable toadies of Stalin's, Budënny and Voroshilov.

Among the marshals Tukhachevski had stood paramount. And in the year following the September Decree he was allowed to travel extensively in Western Europe. On his tour Tukhachevski had behaved with that particular indiscretion which seems, unless vigorously and continuously suppressed, to be a national characteristic. He had acted at the same time the part of diplomat, roving military attaché, and socialite. He wined and dined with Madame Tabouis,[8] and she quoted him in her column; he made contact with General Miller, the head of the Tsarist officers in exile. The Germans he lectured that ". . . if it came to war, Germany would not be meeting the old Russia." Although qualifying his overtures with the formal disclaimer, "We are Communists, and you have need not to forget that we must and will remain Communist," Tukhachevski went on ". . . if Germany adopted a different position, nothing need stand in the way of further Soviet-German collaboration —if both countries enjoyed their friendship and political relations as in the past, they could dictate peace to the world."

To the French, on the other hand, Tukhachevski declared that he "would like to see an intensification of the relations between the French and the Red Army." He spent a week as guest of the French General Staff, and at the end boasted to Gamelin (apropos of ordering new equipment), "As for me, I get all I ask for."

What he was asking for, in the strictly colloquial sense, Tukhachevski was shortly about to receive. For Death was already standing at his shoulder, as it was for more than half of his senior colleagues. Less than a year after Tukhachevski's return the first cloud appeared in the sky, which immediately began to darken with nightmare speed. On 28th

[8] Madame Tabouis was a journalist who moved freely in Parisian high society in the 1930's. She appears to have combined with equal facility the role of *femme fatale,* procuress, and oracle (she also dabbled in astrology), and tidbits of information derived from the resulting intimacies were passed on in her column as revelations, or "predictions."

April, 1937, an article in *Pravda* on the necessity for the Red Army man "to master politics as well as techniques" and the assertion that the Red Army existed *"to fight the internal as well as the external enemy"* carried implications that were sinister in the extreme. Stalin had decided that the time had come when the Army was to be purged, in conformity with the ruthless pattern which had been set the previous year, when the "old guard" was driven out of the Party and shot; that the certainty of political reliability was more important than the risk of a loss of martial efficiency.

There is also some evidence that the Russian dictator had become alarmed by developments in Spain, where the Red Army contingent fighting against Franco (besides acquiring valuable tactical experience) was beginning to show its teeth in conflict with the members of the NKVD who were attached to it.

Whatever Stalin's motives, and whether or not he intended to go as far as he did, the final figures were staggering. Only Budënny and Voroshilov survived among the marshals. Out of eighty members of the 1934 Military Soviet only five were left in September 1938. All eleven Deputy Commissars for Defence were eliminated. Every commander of a military district (including replacements of the first "casualties") had been executed by the summer of 1938. Thirteen out of fifteen army commanders, fifty-seven out of eighty-five corps commanders, 110 out of 195 divisional commanders, 220 out of 406 brigade commanders, were executed. But the greatest numerical loss was borne in the Soviet officer corps from the rank of colonel downward and extending to company commander level.

Before the purge the Red Army had been a vigorous and perceptive body, abundantly equipped and alert for new ideas. Now innovation slowed down to walking pace; technique disappeared, the "Mass Army" reclaimed its position as the proletarian ideal—but the trained reflexes which can quicken a mass and make it formidable had been eliminated. Its training and indoctrination were primarily offensive. But, unlike the Germans, who were the only European army to consider the offensive concept with any optimism, the Russians had not absorbed the teachings of Liddell Hart and Fuller on the correct employment of armour. Thus, although by 1941 they had accumulated no fewer than thirty-nine armoured divisions (compared with the German strength of thirty-two) these were not grouped, as were the German, in independent corps and armies, but distributed evenly, in close support of the infantry divisions; duplicating with a heavier weight the tactical principles of close support that were indoctrinated in the tanks and artillery directly attached to the infantry.

This may be explained by three factors. During the early thirties the Russians, unlike the conservatively inclined staffs of the Western powers, had paid considerable attention to the development of tactics and design in the United States Army. The Americans, who had arrived late on the scene in World War I, at a time when the German armies were already breaking, had not the same traumatic memories of frustrated attacks on fixed defence systems as had the British and the French. In 1918 use of the tank in "packets," with groups of infantry and backed by a huge weight of artillery, had seemed the key to all fortifications, however complex, provided only that the two arms did not become separated and the tanks did not "outrun" the soldiers on foot. Since then the Americans had adopted the idea of using tanks not simply as nutcrackers, but in reconnaissance and as "cavalry." They had developed a number of lightly armoured fast tanks, and one of these, the Christie, was sold to the Russians.[9] But although they were groping in the right direction, the Americans had never really taken hold of the Panzer concept in its essence, as conceived by Liddell Hart and developed by Guderian —the heavy, balanced force, moving on tracks not to "reconnoitre" but to strike and to exploit. Consequently the Russians gradually built up a "tank park" with machines eminently suitable for mobile armoured warfare (in 1932 they had also bought from Britain the Vickers Six-Ton tank, from which they developed their own T 26 series), but they remained wedded to an offensive principle which rejected—if it ever considered—the radical notion of independent operations by a single arm.

In 1937 a number of Russian officers had been attached to the Republican forces in Spain, and here they saw these principles given practical endorsement. Except under conditions of street fighting the defence was everywhere overcome by the relentless pressure of a balanced force of tanks, infantry, and artillery. The Iron Ring of Bilbao, the Ebro Line—a system of permanent emplacements seemed capable of imposing only a delay, never a stalemate. General Pavlov, the tank expert who had gone

[9] In fact, this design was to form a working basis, through the BT series, for the famous T 34; "the best tank in any army up to 1943," Guderian called it.

to Spain (and who was to be shot in the opening weeks of the war, for incompetence) had reported to Stalin and Voroshilov, "The tank can play no independent role on the battlefield," and he recommended that the tank battalions be distributed in an infantry-support role.

Finally, as a reminder that the offensive though sound in concept must not be foolhardy in execution, came the Finnish war of the winter of 1940. Here, underestimating the courage and adaptability of the defenders, the Russians had tried to circumvent the permanent defences of Lake Ladoga by wide and deep outflanking movements in the north. The columns of the Red Army thrust deep into Finnish territory, were surrounded and annihilated. Then in the second stage of the war it was found that the permanent Finnish defences on the Karelian Isthmus could gradually be eroded by steady pressure from tanks and infantry acting in close support.

In this way, by ignoring the effect of local conditions in each case, the Russians drew on their experience to formulate a doctrine of the general offensive, an integrated "steam roller" of all arms that was nothing more than their traditional military posture dressed up with modern equipment. This attitude was firmly grounded in the personal experience of the two soldiers who would be primarily responsible for the direction of the Red Army when the German attack came. Marshal Shaposhnikov, Chief of the General Staff since 1937, had been called in to supervise the planning of the final stages of the attack on the Mannerheim Line. The Chief of the Army Staff, General Zhukov, had been appointed after the disastrous winter of 1939-40, and he, too, had moved to close quarters with the "Finnish question" at the very moment when orthodox mass tactics were finally producing results. Moreover, Zhukov's appointment owed much to his successes in the most important engagement fought by the Red Army up until the German invasion, the previous year against the Japanese in the battles of Khalkin Ghol.[10] This costly operation had been executed with competence rather than originality;

[10] This operation, coming after seven years of intermittent but bloody jockeying for position between Russia and Japan in the Soviet Far East, finally settled the issue in favour of Russia. Although over a quarter of a million men were engaged, the battle received scant attention in the West, coinciding as it did with Hitler's attack on Poland and the outbreak of World War II. But it had a profound strategic importance. The Japanese never moved—nor looked as if they would—against Russia again, even in the dark hours of November 1941. They had learned the painful lesson of underestimating the Soviets and, unlike others, had no desire to repeat it.

and although tanks had been employed extravagantly (Zhukov had disposed of nearly five hundred), the rewards seemed due chiefly to "persistence," i.e., the dismissal of subordinates who were squeamish about casualties, and rigid co-operation between all arms, especially with the artillery.

While the Soviet Union was engaged against opponents who fought along orthodox military principles, sheer weight of flesh and metal would guarantee its victory in the end. But against the fast-moving, highly trained Panzers with their tremendous volume of fire power the Russians were going to have to learn, and learn very fast, if they were to survive.

To make matters worse for the Red Army its disposition in Eastern Europe at the start of the German attack was extravagantly vulnerable. It was the compromise product of a continuing and barely articulate disagreement between some of the senior generals and Stalin, which was itself a function of the hesitant approach to tactics.

Zhukov had agreed that it was desirable to occupy the western territories in order to forestall entry by the Germans, but wished to do so with a light screen and revise Tukhachevski's plan by dividing the strategic reserve between Kiev and the Novgorod-Lake Ilmen region in the north.

During the summer and autumn of 1940 it seemed as if Zhukov were getting his way, as there were only fourteen Russian divisions in Poland and seven in Bessarabia, while the Novgorod region was becoming a substantial concentration area with upward of twenty divisions, of which eight were armoured. But following on the Vienna Award[11] and the mounting evidence of German infiltration into the Balkans this pattern of concentration altered. The shift in emphasis gathered speed and weight during the winter, after the rejection of Stalin's letter of 27th November seemed to have made conflict between the two powers inevitable; and the effect was that by the spring of 1941 the Russian dispositions resembled a caricature of Tukhachevski's old plan, with the troops bunched on the new frontier, which they had little time to prepare for defence and with their communications to base areas already stretched.

Indeed, there is a certain parallel, on a vaster scale, between the Rus-

[11] For the political background to relations between the two powers and the Berlin conference of November 1940, see Ch. 1.

DISPOSITION OF THE SOVIET ARMIES AT THE START OF *Barbarossa*, 22ND JUNE, 1941

Leningrad Military District (H.Q. Leningrad)
Commander:
Lt. Gen. M. M. Popov
Chief of Staff:
Maj. Gen. D. N. Nikishev
Commissar:
Corps Commissar N. N. Klement'ev
(Designated "Northern Front" 23rd June, 1941)

Baltic Military District (H.Q. Riga)
Commander:
Col. Gen. F. I. Kuznetsov
Chief of Staff:
Lt. Gen. P. S. Klenov
Commissar:
Corps Commissar P. A. Dibrov
(Designated "Northwestern Front" 23rd June, 1941)

Western Military District (H.Q. Minsk)
Commander:
Gen. D. G. Pavlov
Chief of Staff:
Maj. Gen. V. E. Klimovski
Commissar:
Corps Commissar A. Ya. Fominyi
Deputy Front Commander:
Lt. Gen. I. V. Boldin
(Designated "Western Front" 23rd June, 1941)

Kiev Military District (H.Q. Kiev)
Commander:
Col. Gen. M. P. Kirponos
Chief of Staff:
Lt. Gen. M. A. Purkayev
Commissar:
Div. Commissar P. E. Rykov
(Designated "Southwestern Front" 23rd June, 1941)

Army Strength	*Armour & General Reserve*
14th A Lt. Gen. V. A. Frolov (Murmansk) 7th A Lt. Gen. F. D. Gorolenko (N.E. Lake Ladoga) 23rd A Lt. Gen. P. S. Pshennikov (Karelian Isthmus)	10th Mechanised Corps (to Northwestern Front 27th June, 1941)
8th A Maj. Gen. P. P. Sobennikov (Coastal Defence & Dago & Ösel) 11th A Lt. Gen. V. I. Morozov (E. Prussian Frontier) 27th A Maj. Gen. N. Berzarin (Dvina R.)	1st Mechanised Corps (plus two cavalry divs.)
3rd A Lt. Gen. V. I. Kuznetsov (Grodno) 11th Mechanised Corps 10th A Maj. Gen. K. D. Golubev (Bialystok) 6th Mechanised Corps 13th Mechanised Corps (understrength) 4th A Maj. Gen. A. A. Korobkov (Brest-Litovsk) 14th Mechanised Corps (Pruzhany-Kobrin)	13th A Lt. Gen. P. M. Filatov (Minsk) 7th & 5th Mechanised Corps (Bobruisk) 16th, 21st, 22nd Armies (Skeletal only. Lt. Gen. F. A. Yershakov) (Vitebsk)
5th A Maj. Gen. of Tank Troops M. I. Potapov (Lutsk) 8th Mechanised Corps 6th A Lt. Gen. I. N. Muzychenko (Lvov) 6th Mechanised Corps 26th A Lt. Gen. F. Ya. Kostenko (Borislav) 15th Mechanised Corps 12th Army Maj. Gen. P. G. Ponedelin (Czernowitz) 22nd Mechanised Corps	19th & 9th Mechanised Corps (Zhitomir)

sian layout and the manner in which the French and British armies deserted their own positions and rushed headlong into Belgium to meet the invader in May 1940. In explanation, though, motives that are less high-minded than the desire to offer immediate succour to a small ally suggest themselves. During the winter of 1940-41 the strength in the Novgorod concentration area shrank again, and there was a corresponding build-up (twenty infantry divisions, two cavalry divisions, and five armoured divisions) along the Finnish frontier. Two separate army groups were formed (normally the whole area would come under one of the Leningrad army groups) under Generals Meretskov and Govorov, and this fact, together with certain remarks of Molotov's which have been recorded in the minutes of the Berlin conference, suggests that the Russians were preparing a renewal of their attack on Finland in the summer of 1941.

The even heavier concentration in the area between Lemberg (Lvov) and the upper Prut was partly an extension of Tukhachevski's original plan, partly also a means of strengthening Russia's hand in the intensified power politics that were being played out in the Balkans. For Stalin's opinion was that further Russian annexation would be possible in the Balkans if Germany became more deeply involved in the West, either by an attempted invasion of the British Isles or in the Mediterranean. When Stafford Cripps presented to Stalin comprehensive evidence of the German plan (supplied by Hess), the Russian leader thought that it was a plant, sharing the view of Voroshilov that "We have the time to play the role of gravedigger to the capitalist world—and give it the finishing blow."

The result of this divergence of opinion between the *Stavka* and GOKO was an exceedingly unwieldy and top-heavy distribution of the Russian Army. By the middle of May 1941 there were nearly 170 divisions, or over five sevenths of the country's total armed strength, outside the 1939 frontiers. They were distributed in five "military districts" running from north to south as "Leningrad," "Baltic," "Western," "Kiev," and "Odessa" commands, and under generals whose names—Popov, Tyulenev, Pavlov—were destined, if they survived the first desperate days of battle and the punitive firing squads that punctuated them, for obscurity.

But although the Red Army was at a disadvantage because of this vulnerable distribution and was to suffer fearfully from clumsy, hesitant, and incompetent leadership, it was more than the equal of the Germans in the purely logistical field of equipment and supply. There were deficiencies, notably in the field of medical services and radio communication, but in the key figures of tank strength (over seven thousand in the forward area) and field artillery the Russians were superior.

There were three types of divisions: the infantry, composed of three regiments, each of three battalions, and one reserve regiment of two battalions; the cavalry, with four regiments, each of two battalions; and the armoured division. In the later stages of the war there were separate motorised infantry divisions, but in 1941 the infantry had no motor transport and depended on horse-drawn wagons. The only motorised infantry was that attached to the armoured divisions. Each infantry division had an artillery component, and this had wheeled and tracked vehicles for drawing the guns and carrying the ammunition. The infantry divisions also had a tank strength attached to them, but this was made up mostly of old French designs of the twenties. Output of the T 34 was restricted to the armoured divisions.

The cavalry, far from being an anachronism, was of immense value. Recruited from Cossacks and Kalmuks—peoples who spent their lives in the saddle—it had an extraordinary mobility. Its men were trained to fight as infantry, but would use the horses to cover huge distances over bad ground, and to tow their light artillery and mortar limbers. They were adept at the art of concealment and dispersion. "A Soviet cavalry division," Manstein grumbled, "can move, in its entirety, a hundred kilometres in a night—and that at a tangent to the axis of communication." They were invaluable under conditions of fluid fighting, and their horses, shaggy little Kirkhil ponies from Siberia, could stand temperatures of 30 degrees below zero.

The importance of the cavalry divisions was heightened by their status as the only mobile units capable of operating with any degree of independence. For following on Pavlov's recommendations in 1939 the armoured divisions had been broken up and their strength distributed as "brigades" throughout the infantry armies. Although the divisional organisation was retained in a number of cases, the breakdown of the brigades into "heavy," "medium," and "reconnaissance" spelled the end of the tank force as an independent arm.

Then, following the success of the Panzer divisions in Poland and

France, efforts, first lethargic, then frantic, had been made to start the regrouping of the tank brigades back into armoured divisions. But this process was just beginning by the summer of 1941, and the Russian commanders had had no time to acquaint themselves with the problems—much less the solutions—of handling large tank forces. Nonetheless, the actual weight of the armour deployed was, in the aggregate, very formidable (some authorities have put the total number of tanks in the Soviet Army at the start of the campaign as high as twenty thousand), and its even distribution endowed the regular infantry divisions with a fire power that was at least the equal of their German equivalent.

Mass, then, the Russian Army possessed in abundance—as always in its history. In equipment, too, it was better off than any of the Wehrmacht's earlier victims. The key question remained, what of its morale and its leadership?

In Russia, as in Germany, the relationship between Army and state was a delicate one. In both countries a personal dictatorship and a "Party" organisation had been faced with the problem of disciplining the military and subordinating it to their own political purpose. In both this had been achieved, but by completely different approaches, which in turn left residual influences of profound importance. Hitler had outmanoeuvred his generals and, within a few years, achieved their exclusion from the field of politics, where for half a century they had ruled as arbiters. Then with bribes, cajolery, and browbeating he canalised their energies and their expertise into one field, the pursuit of pure military efficiency.

But the Russian officer corps was not isolated, it was crushed. When the purges were over, the Red Army was obedient to the point of witlessness; dutiful but without experience; stripped of political weight or ambition, at the expense of initiative, experiment, or the desire to innovate. The question remained, had their native patriotism, the primaeval love of "Mother Russia" which had quickened ancestors suffering under regimes more barbarous and tyrannical even than Stalinism, to rise and reject an alien invader, also been eradicated? For this, and will power, and fatalism, and that readiness to accept terrible sufferings that are essentially Russian qualities, would all be needed to the full in the first dreadful weeks of the German assault.

At the beginning of 1941 the OKW intelligence branch had estimated Russian strength at "not more than" two hundred effective divisions.

Since the war Halder has said, "This was a gross underestimate, the figure was more like three hundred and sixty." In actual fact, the original figure was probably much nearer the truth, but the Soviet mobilisation machinery was highly efficient, succeeding in putting over a million men under arms before the end of July. In this prodigious feat the Russians were greatly helped by Osoaviakhim, which had thirty-six million members, of whom 30 percent were women. It was a nationwide paramilitary organisation which "implanted in them the rudiments of civil defence and close fighting. Its clubs were formed of units to defend local areas, units of pilots, of parachutists, of Partisan cadres and even for the use of dogs in warfare. It was entrusted with the neutralisation of mine fields and the recovery of equipment in the rear of armies . . ."

Hitler dismissed the latent strength of such an organisation. He believed that the Soviet military machine was so riddled with Communism, insecurity, suspicion, and informers, and so demoralized by the purges that it could not function properly. Intelligence had drawn up a clear picture of the Russian Army in Poland and of the vulnerability of its disposition.

"You have only to kick in the door," he told Rundstedt, "and the whole rotten structure will come crashing down."

It is certainly paradoxical to find Hitler, whose own contempt for the professional soldier was unbounded, and who never ceased to exalt the ties of Party over the scruples of caste, expressing so orthodox a view on the corrupting effect of politics on a military system. But whatever his reasoning, he had, in his estimate of the Russian potential, overlooked one very important factor. The Wehrmacht was now confronted by an opponent of a completely different kind from the soft nations of the West. "The Russian soldier," Krylov has said, "loves a fight and scorns death. He was given the order: 'If you are wounded, pretend to be dead; wait until the Germans come up; then select one of them and kill him! Kill him with gun, bayonet, or knife. Tear his throat with your teeth. Do not die without leaving behind you a German corpse.' "[12]

[12] A more formal rendering of these instructions can be found in par. 2 of the "General Principles" to the Provisional Field Service Regulations of the Workers' and Peasants' Red Army (People's Commissariat for Defence 1937):

"The constant urge to get to grips with the enemy, with the aim of destroying him, must lie at the basis of the training and activity of every commander and soldier of the Red Army. *Without special orders to this effect* the enemy must be attacked boldly and with dash wherever he is discovered." My italics.

three | THE CLASH OF ARMS

> Weighted down with heavy cares, condemned to months of silence, I can at last speak freely—German people! At this moment a march is taking place that, for its extent, compares with the greatest the world has ever seen. I have decided again today to place the fate and future of the Reich and our people in the hands of our soldiers. May God aid us, especially in this fight.

Hitler's proclamation was read by Goebbels over the radio to the whole nation at seven o'clock on the morning of 22nd June. Four hours earlier the glare from six thousand gun flashes had lit the eastern dawn, overwhelming the bewildered Russians in a tumult of fire and destruction. The frontier guards, awakened by the squeal and clatter of tank tracks, were shot down as they emerged from their barracks, running half dressed through the smoke. From gun positions in the line the Germans intercepted again and again the same message: "We are being fired on; what shall we do?" [1]

What an appalling moment in time this is! The head-on crash of the two greatest armies, the two most absolute systems, in the world. No battle in history compares with it. Not even that first ponderous heave of August 1914, when all the railway engines in Europe sped the mobilisation, or the final exhausted lunge against the Hindenburg Line four years later. In terms of numbers of men, weight of ammunition,

[1] General Günther Blumentritt, Chief of Staff to Field Marshal von Kluge, quotes the retort of Russian headquarters: "'You must be insane. And why is your signal not in code?'"

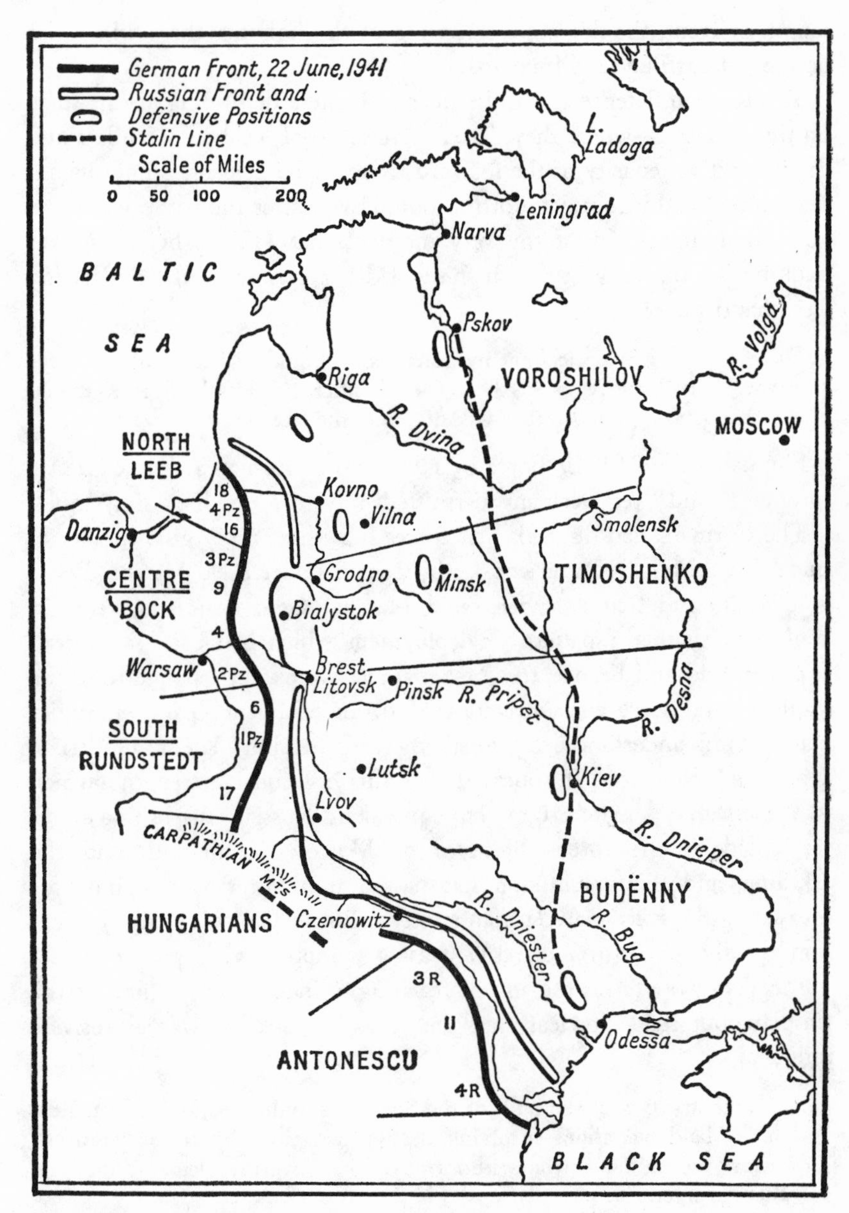

The Eastern Front on 22nd June, 1941

length of front, the desperate crescendo of the fighting, there will never be another day like 22nd June, 1941.

The Russian defence was quite unco-ordinated, depending at this stage on the initiative—where they dared exercise it—of local commanders and the instinctive tenacity of the forward troops, who held on grimly in undermanned and incomplete fortifications. Even after the battle was three and a half hours old, at the very moment that Hitler's broadcast was exulting in "the greatest march the world has ever seen" the Red Army Command was ordering:

> . . . troops will attack enemy forces and liquidate them in the areas where they have violated the Soviet frontier [but] unless given special authorisation ground troops will not cross the frontier.

Flights by the Red Air Force over Finland or Rumania were expressly forbidden, and over Germany permitted only to a depth of sixty miles.

The Germans had divided their forces into three army groups; North, under Field Marshal Ritter von Leeb; Centre, under Field Marshal Fedor von Bock; and South, under Field Marshal Gerd von Rundstedt. In conformity with the pattern of deployment which had been so successful in Poland and France, the Panzer forces were kept separate from the infantry, and were concentrated in four independent *Gruppen,* under young commanders of exceptional vigour and skill—Kleist, Guderian, Hoth, and Hoepner.[2] It appeared that this division of strength (which was soon to be matched by an equivalent Russian disposition) corresponded to the three objectives of Moscow, Leningrad, and the Ukraine, and this assumption has passed into history as a guide for measuring the success of German strategy. But in fact the "general intention" of the *Barbarossa* directive was, geographically, imprecise. It set out in very loose terms the aim of reaching a line from Archangel to the Caspian, but made it clear that the primary objective was exclusively military:

> . . . Destruction of the bulk of the Soviet Army located in Western Russia by bold operations involving deep penetrations by armoured spearheads; prevention of the withdrawal of battleworthy elements into the Russian interior . . .

The Panzer forces were to carve up the Soviet Army, the slower-moving infantry and artillery were to force their surrender. Hitler had

[2] For opposing orders of battle see charts in this chapter.

no desire to fight for, or in, the cities of the Soviet Union, and many of the generals on the staff agreed with him. The battle of France had been won by striking for the Channel—not for Paris.

This formula, as will be seen, carried in itself the seeds of trouble. There was often friction between the Panzer commanders, who believed that they had the whole of Russia at their feet and longed to be sent after the glittering spires of the capital cities, and the infantry locked in combat with the stubborn Russian masses in the rear, who felt that the tanks should be held back to help them. This friction was to cause a number of local tactical errors, and gradually came to infect the whole High Command with indecision, leading to a succession of command crises in the early autumn. But in June it certainly seemed as if the terms of the directive were being followed to the letter.

In the vital central sector, where the eight hundred tanks of the 2nd *Panzergruppe* were piled up behind the Bug, both bridges over the river to the south of Brest-Litovsk fell, intact and undefended, at the first rush. North of the town the 18th Panzer Division, using tanks which had been specially waterproofed for "Sea Lion" (the projected invasion of England), forded the river and struck across the marshy ground to the main Russian defences on the left bank of the Lesna, reaching them three hours later and there also capturing the bridge intact. Minute by minute, as the tanks probed deeper and the German guns lengthened their range, the tremors which shook the Russian front magnified in strength and frequency. A few fragments, a gradual crumbling—by midday the vital sectors were already in landslide.

During the afternoon, when the first positive orders began to percolate to the defenders, a gradual stirring at corps and divisional levels took place. But there was no real effort at concentration—not, at least, in any co-ordinated sense. It is simply that all the units grouped behind the frontier seem to have packed up as best they could and moved off to encounter the Germans head on. And in that time the Luftwaffe had finished its work on the forward Soviet airfield network, and these approach marches led straight into the German bombsights. Roads were smashed and raked with machine-gun fire; tank parks were blasted; fuel stores set alight; thousands of horses were scattered, wounded and in terror, across the countryside. It was the classic stencil of *Blitzkrieg,* imprinted now on the broadest canvas.

In addition to the advantage of surprise the Germans had secured a devastating superiority of numbers and firepower at the points selected for their armoured penetration. Halder's plan had put the entire tank strength of the German Army into these opening attacks, dividing it into four *Panzergruppen* whose purpose was to perforate the Russian defensive membrane at the first blow, then to wheel inward, isolate, and cut to pieces the mass of Soviet army as it stood on the frontier. The map will show how effective a degree of concentration was achieved.

In the north three Panzer divisions (over six hundred tanks) and two infantry divisions had an attack frontage of less than twenty-five miles. Opposite them stood one weak Russian unit, the 125th Rifle Division. In the centre, where Bock's army group carried the *Schwerpunkt*[3] of the opening days, the two Panzer groups, under Hoth and Guderian, comprised seven divisions with nearly fifteen hundred tanks between them. They were opposed by one complete rifle division (the 128th), regiments from four others, and a tank division (the 22nd) which was understrength and in process of reorganisation.

On the southern front two Soviet rifle divisions faced six infantry divisions with about six hundred tanks distributed among them in close support. Small wonder, then, that the comment of a German lieutenant of the 29th Motorised Division was, ". . . the Russian defences might have been a row of glass houses," and that by the afternoon of 22nd June the leading elements of all four German armoured groups were motoring fast along dry, undamaged roads, with the sound of gunfire fading in their rear.

These "reconnaissance detachments" were mixed groups of motorcyclists with armoured cars and half-track infantry carriers towing anti-tank guns; sometimes they were supported by a sprinkling of light or PzKw III medium tanks. On the road they moved at about twenty-five mph. Immediately behind them travelled the mass of the tank strength, in continuous radio contact with the leaders and ready to deploy into attack formation if the head of the column should get held up. Still farther to the rear was a "sandwich" of mechanised infantry, divisional artillery, and more infantry. The whole column, deployed in extended order of advance, stretched over a distance of from seven to ten miles, yet by the evening of 22nd June all the leading Panzer divisions were well clear of the fighting zone and had penetrated to nearly twice their own length.

[3] A *Blitzkrieg* term, meaning spearhead, point of maximum concentration.

The deepest advance had been made by Manstein's 56th Corps in the north, which had crossed the East Prussian frontier at dawn and captured the bridge at Airogola, over the Dubisa gorge, before sundown—a forward leap of over fifty miles! In the centre Guderian's columns had joined up on either side of Brest-Litovsk, captured Kobrin and Pruzhany, and crossed the line of the Krolewski Canal.

But even before dusk on the 22nd certain differences from previous campaigns were apparent. Like some prehistoric monster caught in a net, the Red Army struggled desperately and, as reflexes gradually activated the remoter parts of its body, with mounting effect. Until that day the Germans had always found that bodies of surrounded enemy lay down and died. There would be a contracting of perimeters, a drawing in of "flanks," perhaps some perfunctory efforts to break out or counterattack, and then—surrender. The speed and depth of a Panzer thrust; the tireless ubiquity of the Luftwaffe; above all, the brilliant coordination of all arms, had given to the Germans an aura of invincibility that had not been enjoyed by any other army since the time of Napoleon. Yet the Russians seemed as ignorant of this as they were of the rules of the military textbook.

The reaction of the surrounded formations was in every case vigorous and aggressive. Their very lack of co-ordination bewildered the Germans and hampered the plans for containing the various pockets. Whole divisions would assemble and move straight into the attack, "marching towards the sound of the guns." During the day the tank parks emptied as one brigade after another took on fuel and ammunition and clattered off to be destroyed piecemeal in the sights of the German artillery. By the afternoon fresh masses of aircraft, summoned with desperate urgency from the flying fields of central Russia, began to appear over the battlefield, though "It was infanticide, they were floundering in tactically impossible formations." By that time Stalin's restriction against sorties over German territory had been lifted, and the Russian bomber force (which had largely escaped the first Luftwaffe strike, owing to its bases being farther from the frontier) took off obediently in accordance with an already outdated operational plan. Over five hundred were shot down. On 23rd June, Lieutenant General Kopets, commander of the bomber group, committed suicide, and within a week General Rychagov, the commander of aviation on the northwestern front, was under sentence of death for "treasonable activity" (that is to say, having been defeated). In the first two days the Russians lost over two thousand aircraft—a casualty rate without precedent. The (numerically) strongest air force in the world had been virtually eliminated in forty-eight hours.

The effect of being thus completely deprived of air cover was, on the frontier armies, disastrous. For the rest of the year the Russians were to fight with only minimal support from their Air Force, and were quick to adjust themselves to the operational limitations this imposed. But in those first hectic days of confusion and encirclement, when there were no orders, when there was no central direction, nothing more specific than the standing instructions, ". . . attack the invader whenever and wherever he be encountered," casualties were increased tenfold by this blindness in reconnaissance and vulnerability on the march.

While the Panzers streaked across the plain, toward objectives seventy miles distant, a slow polarisation took place among the Russian armies left standing in Poland. Like giant cedars, which remain erect after their roots have been cut, they stood up to assaults whose result was certain before crashing down to disappear forever under the saw. In the first week of the campaign four major "battles of annihilation" cleared the way for the German Army to step bodily into European Russia as far as the line of the Dnieper.

The idiotic disposition of the frontier armies[4] had left Pavlov with a weak centre (known in the first ten days of the campaign by its peacetime designation, "Western Military District") and a bare numerical parity with the Germans opposite him in terms of infantry. In tanks Pavlov was completely outclassed, for he faced nearly 80 percent of the German strength, including the *Panzergruppen* of Hoepner, Hoth, and Guderian.

Pavlov had three armies, the 3rd, 10th, and 4th, drawn up in a line running south from the Latvian frontier to Wlodawa, on the fringe of the Pripet Marshes. In close reserve there were five mechanised corps (little bigger than a division, in reality), which were evenly distributed and fully occupied in training to assimilate the *volte-face* which had come over the Red Army Command's attitude to the employment of armoured forces.[5]

Hoepner brushed his sleeve against the right wing of the Russian

[4] Refer to disposition chart.

[5] See Ch. 2.

3rd Army on the first day, tearing a wide gap between it and the edge of the Baltic Military District area, and through this Manstein's 56th Panzer Corps flowed at breakneck speed. Russian counterattacks during the afternoon had run into the full strength of the 4th Panzer Army, now fast eroding the walls of the breach, and withered under its fire. By nightfall three Russian infantry divisions had gone completely—men, guns, staff organisation, transport, everything—and another five were licking their wounds. More serious, half of Pavlov's tank strength was lost in the desperate confusion of that first afternoon's encounter. The 14th Mechanised Corps, assembling in the Pruzhany-Kobrin area, had been so badly punished by German bombers that it never got under way; the 13th, being nearer the point of impact, was in action by six o'clock in the evening, but shortage of fuel, mechanical failures, and unsuitable ammunition[6] dissipated its effect, for the brigades went into battle singly, often following their predecessor's tracks and repeating his mistakes.

During the night Pavlov attempted to draw off the remainder of his tank strength from the 10th Army, forming the 6th and 11th Mechanised Corps and 6th Cavalry Corps into a special "shock force" under his deputy, Lieutenant General I. V. Boldin, with instructions to attack the southern flank of the German penetration on the 23rd. It is probable that these orders were not evenly disseminated during that first hectic night; likely also that the 10th Army commander, Major General K. D. Golubev, was not overanxious to hear them *en clair,* as his own front was under mounting pressure. At all events, only the 11th Mechanised Corps was in position the following morning. Both the 6th and the cavalry were still on the road, strung out in all directions, vulnerable and understrength. During the morning all were visited by the Luftwaffe, and the cavalry, in particular, paid a terrible price for their delay. The result was that no move was made by Pavlov's armies to close the gap during the 24th.

In the meantime the commander of the Baltic Military District (now redesignated the "northwestern front") had been assembling such tank strength as remained to him, and during the afternoon of the 23rd it was all (about the equivalent of three divisions in strength) committed in an attack southwestward from Shaulyai. It is highly doubtful that the

[6] The majority of Russian tanks (mainly T 28 and the T 50-60 light) were fitted out with high-explosive ammunition for "close support," and altering the ration in favour of armour-piercing shot, for anti-tank work, was just beginning.

gap could have been closed even had this attack been simultaneous with that of Boldin's group. With Boldin inactive, it was doomed to failure, running straight into the concentrated strength of Reinhardt's 41st Panzer Corps, which was deploying to attack Kovno (Kaunas). The following day, 24th June, Boldin at last put in his attack, but punishment on the march and the isolated character of the operation made it, too, a failure. By now the northwestern front, denuded of its armour, was disintegrating fast, with the surviving armies falling back on Riga and uncovering the approaches to Dvinsk. By 24th June, Manstein had penetrated over a hundred miles, as far as Wilkomierzi; on the 25th he was in sight of the town; on the 26th he entered it, the motorcyclists of the 8th Panzer capturing the huge road bridge over the Dvina at the very moment that the sentries were fumbling with the demolition charges.

Now a corridor, almost a hundred miles wide at its entry, was leading directly toward Leningrad. In five days the Germans had halved the distance which separated them from the "Cradle of the Revolution."

Frantic to close this gap and to regain contact with the disintegrating northwestern front, Pavlov continued to shift divisions pell-mell out of the 10th Army area northward to stiffen the shaky 3rd Army. This uncovered Minsk and left the luckless 4th Army commander, Major General A. A. Korobkov, without support on either flank. Had the Russians but known it, the threat to Leningrad was as nothing beside the menace bearing down on the 4th Army. With his centre under pressure from Kluge, Korobkov was isolated to the north by Hoth's 3rd *Panzergruppe* and his left flank driven in by Guderian's 2nd *Panzergruppe*. In three days Guderian had driven a hundred miles northeast to Slonim, drawing, with Hoth, a noose around the bulk of the Soviet infantry and the remaining armour, which Pavlov had left in position. On 25th June, the 26th Panzer Corps took Lesna and advanced fifty miles toward Slutsk; on the 26th, the 66th Panzer Corps captured Baranovichi in the morning and drove nearly sixty miles during the day to enter Stolpce at nightfall. On the 27th this corps covered the remaining fifty miles to Minsk, where it joined up with the southern arm of Hoth's pincer, putting a "long-stop" behind the Slonim pocket and achieving one of the most spectacular marches in the history of armoured warfare.

In the south the Red Army held its ground better, though at a fearful price in men and equipment. The front commander was Colonel General

M. P. Kirponos (commander of the Kiev Military District), and the forces of which he disposed were substantially stronger both than those of his colleague to the north, the unfortunate Pavlov, and of the Germans opposite him.

The main German thrust was directed down the relatively narrow gap between the southern edge of the Pripet Marshes and the foothills of the Carpathian range. Here Rundstedt, the commander of Army Group South, had concentrated the whole of the 1st Panzer Army (Colonel General von Kleist) and the 6th Army (Field Marshal von Reichenau) and the 17th Army (Colonel General von Stülpnagel). The longer front along the Prut and down to the shore of the Black Sea had only one German army, the 11th (General von Schobert), to stiffen a large mixed group of Hungarians and Rumanians. These last were slow in getting off the mark, and being fitted out with French equipment, were not formidable.

Kirponos therefore was free to concentrate against Kleist and Reichenau. He had four infantry armies,[7] three mechanised corps in close support (the 22nd, 4th, and 15th), one (the 8th) in reserve, about 250 miles inland, and two in "strategic reserve" at Zhitomir (the 19th and 9th). But this powerful force was dissipated in a sequence of piecemeal counterattacks, and due largely to command difficulties and the inexperience of the senior officers of the Red Army in handling masses of armour, the strongest concentration of Russian tank strength in the east lost its cutting edge before the really critical phase of the southern battles developed. On 22nd June, Kirponos had ordered up all three mechanised corps from the reserve with the intention of concentrating them northeast of Rovno and staging an attack, together with the 22nd (which was already in position there), against Kleist's left flank. In fact, the 22nd Mechanised Corps was drawn into battle on the first day and cut to pieces. The 15th Mechanised Corps, attacking from the south, was likewise fought to a standstill in front of the German anti-tank screen. With his tank strength seriously diminished, Kirponos held on grimly, but by the time the 8th Mechanised Corps had completed its forced march the situation had become so bad that it was sent straight into action alone. Once again the Russian tanks took a severe mauling, though better

[7] These were the 5th (Major General of Tank Troops M. I. Potapov), the 6th (Lieutenant General I. N. Muzychenko), the 26th (Lieutenant General F. Kostenko), and the 12th (Major General P. G. Ponedelin).

combat discipline and more up-to-date equipment (some regiments had just been refitted with the T 34)[8] helped the corps preserve its cohesion. When finally the 9th and 19th corps arrived from Zhitomir, things were so critical that they, too, had to go straight into action—at half the strength originally planned. The inexperienced Russian tank crews, exhausted by four days on the march and round-the-clock hammering by the Luftwaffe, were no match for the confident veterans of the 1st Panzer Army, who knew how to concentrate, when to disperse, the secrets of holding fire and picking ground. Once again many of the Russian tanks broke down, others floundered into German ambushes or lost their way. One division followed its corps commissar into a swamp, and all the tanks had to be abandoned.

Yet although the situation seemed desperate from the Russian side, the Germans found their opponent's strength highly perplexing. "The enemy leadership in front of A. G. South," Halder grumbled, "is remarkably energetic, his endless flank and frontal attacks are causing us heavy losses." Again, on the following day, "One has to admit that the Russian leadership on this front is doing a pretty good job."

At least, by his lavish expenditure of lives and machinery, Kirponos was holding the southern front in being. But its days were numbered, for north of the Pripet Marshes the Russian armies of the centre were fast breaking to pieces. A general breakdown in communications aggravated the fragmentation of the various commands. Signals, radio, telephones, nothing functioned properly. Roads and railways were raked by the Luftwaffe; some units had their effectiveness reduced by as much as half while on the march.

Only the regional machinery of the *Osoaviakhim* functioned with efficiency, continuing to churn out a mass of conscripts under its mobilisation decrees. These wretched fellows, the cadres of 1919, 1920, 1921, with those of other years following on their heels, were brought from all over Russia in slow-moving freight trains and dumped as near the front as the Luftwaffe allowed. Out they clambered, in their civilian clothes, holding their cardboard suitcases, and set off on foot, toward mobilisation centres long since overrun.

[8] On 24th June, Halder, besides noting, "Interesting historical coincidence that Napoleon also took Vilna on 24th June," also wrote (underlining the sentence), "New enemy heavy tank!" The T 34 was issued to some armoured brigades in May 1941 and went into action in the first week of the campaign. Not, as is sometimes claimed, at the "relief of Moscow."

In the huge no man's land of White Russia, which had a week, some parts only a few days, of grace before falling to the enemy, those fittest to command survived. A few commissars together with some Red Army officers of courage and foresight struggled day and night to form fresh units out of the unarmed reservists, wandering stragglers, men on leave, and garrison brigades which littered the area. Installations were demolished, dumps set ablaze, extempore fieldworks thrown up, cattle and fowl slaughtered or driven east. Over the whole scene brooded the "rear security detachments" of the NKVD, machine gunners held ready "to check panic . . . and prevent unauthorised withdrawal." On 28th June, Korobkov had been taken back to Moscow and shot for cowardice. Pavlov was to follow him, together with his Chief of Staff, Klimovski, and his signals commander, Grigoriev.

As the frontier force withered in battle, new armies, under new commanders, took shape in the interior. To speed their concentrations the Russians made all the major rail lines west of the Dnieper one-way traffic; only the engines went careering back to collect their loads. This puzzled German intelligence.

> Air reconnaissance shows enormous mass of rolling stock accumulating in marshalling yards. Appears to be empty. Is this a bluff?

Halder's reaction was typical of that of all Germans who came face to face with the extraordinary Russian profligacy in battle. First, exultation: the Germans counted heads, measured the miles of their advance, compared it with their achievements in the West, and concluded that victory was around the corner. Then, disbelief: such reckless expenditure could not go on, the Russians *must* be bluffing, in a matter of days they would exhaust themselves. Then, a certain haunting disquiet: the endless, aimless succession of counterattacks, the eagerness to trade ten Russian lives for one German, the vastness of the territory, and its bleak horizon. A German Colonel Bernd von Kleist, wrote:

> The German Army in fighting Russia is like an elephant attacking a host of ants. The elephant will kill thousands, perhaps even millions, of ants, but in the end their numbers will overcome him, and he will be eaten to the bone.

There were differences, too, in the manner of the fighting. Manstein has described how, on the very first day, he was shown the bodies of a

German patrol which had been cut off, and "gruesomely mutilated," and the Soviets' practice of "throwing up their hands as if to surrender and reaching for their arms as soon as our infantry came near enough, or . . . feigning death and then firing on our troops when their backs were turned." As early as 23rd June, Halder had been complaining of the "absence of any large take of prisoners," on the 24th that "the stubborn resistance of individual Russian units is remarkable," on the 27th, again, dissatisfaction at "the singularly small number of prisoners." The fissures in Russian morale which were to open that autumn (and as suddenly to be closed by German brutality and miscalculation) were still far below the surface.

All this had been immediately apparent to the German infantry, which was fighting at close quarters. But on the Panzer crews, riding out on the armoured decks of their vehicles, the sun shone. For the first few days it seemed almost like the summer campaign in the West, as the undamaged villages slid beneath their tracks, the bewildered population peering from windows and doorways. Soon, though, this similarity began to fade. The first effects of the distance they were travelling began to be felt. Many of the motorised divisions had been re-equipped with captured French trucks, and these were starting to break down on the poor roads. Spare parts had to be flown in as the long trails that stretched west behind the armoured spearheads were dangerously vulnerable to wandering bodies of "surrounded" Russians. "In spite of the distances we were advancing," wrote a captain in the 18th Panzer Division, ". . . there was no feeling, as there had been in France, of entry into a defeated nation. Instead there was resistance, always resistance, however hopeless. A single gun, a group of men with rifles . . . once a chap ran out of a cottage by the roadside with a grenade in each hand . . ."

On 29th June, Halder, after summarising the day's progress in his diary, concluded:

> Now, for once, our troops are compelled to fight according to their combat manuals. In Poland and in the West they could take liberties, but here they cannot get away with it.

There is a note almost of smugness about about this entry. It is as if the dedicated graduate of the General Staff College was gratified to see the rules of war beginning to assert themselves. But, "for once . . ." For always. Had the Germans but known it, the first (and, for their arms, the most spectacular) phase of the Eastern campaign was already fading into memory.

The 30th of June was Halder's birthday, and at OKH the anniversary was a happy occasion. On coming down to the breakfast room the Chief of the General Staff found that it had been specially decorated. The junior officers stood in a line and presented their compliments, preceded by "the H.Q. Commandant, accompanied by a man from the guard unit who brings a bunch of wild flowers." Halder read the teleprints from army group headquarters and pronounced the news satisfactory. The Russians were in full retreat, and Luftwaffe reports from the southern front told of disorganised columns three and four abreast. Of the total of two hundred aircraft shot down the day before, the majority had been old types, TB 3 high-wing bombers dragged up from the training airfields of central Russia. It was evident that the enemy was scraping the barrel.

It is nothing if not paradoxical to think of these precise and immaculate staff officers, dressed this day in their best uniforms, seated at a table with a clean cloth, exchanging formal pleasantries with one another. These men were at the nerve centre of the German war machine in the East. Each day they sifted reports which expressed in cold print a fresh and enormous sum of human agony—men dying of wounds and thirst, villages smashed and burning, animals slaughtered, families separated and sent into captivity. They had heard Hitler speak of his intentions toward the Russian people, his rejection of the Geneva convention on prisoners of war, of the "Commissar order," of his wish to "level" Leningrad in order not to be embarrassed by the size of its population. They knew, too, what Nazi occupation meant: they had all fought in Poland and seen the revolting behaviour of the SD detachments at close quarters; and there, no farther than the ration-strength sheet on the wall, the movement orders in the daily file, was confirmation that these same criminals were operating close up behind their own soldiers. Yet such is the schizophrenic capacity of the human mind that all this could be submerged with facility, and like schoolboys, they set out to enjoy themselves at their housemaster's birthday party.

Brauchitsch, or "ObdH," as he was affectionately called by Halder,[9]

[9] Not to his face, of course. It is an abbreviation for Oberstdas Heeres, Commander in Chief of the Army, under which Brauchitsch's name always appears in Halder's diary.

punctilious as ever, had sent red roses and strawberries for the table. When Halder telephoned to thank him, the Commander in Chief revealed some exciting news. Hitler had decided to visit OKH headquarters in person. He would be arriving for tea. Overcome by the atmosphere of good feeling which Halder's birthday celebrations had generated, Brauchitsch went on to say (quite mendaciously) that the Führer's visit "is primarily on your account." Other "well-wishers" then took the telephone, ending with the fanatically Nazi Frau Brauchitsch, who rang off with a strident "Heil Hitler!"

During the day the collapse of the Russian front went several stages further. In Kirponos' command, the only area where the defence still held a certain degree of cohesion, the valiant 8th Mechanised Corps had fought itself to a standstill, and with his tank arm almost eliminated, Kirponos ordered a retreat to the positions on the old Soviet-Polish frontier. In the north Pavlov's forces were in a state of complete disintegration, their strength broken by a sequence of counterattacks which for clumsiness and extravagance were to be rivalled only by Budënny's later performance in the Ukraine. In the centre the Soviet mass was now enclosed in two pockets, at Slonim and Minsk, and the way seemed clear for the Panzers to roam undisturbed. After eight days' fighting the bulk of the Soviet forces standing on the frontier had been splintered, and accordingly, within the terms of the *Barbarossa* directive, OKH now ordered that the crossings over the Dnieper be seized.

Hitler arrived at teatime, and an SS adjutant brought a large silver flagon of cream. After a tour of the wall maps the Führer sat down, and the conversation—if such a term may be used of the discreet assent with which Hitler's rambling monologues were received—turned to "global subjects."

After some grumbles about Germany's African colonies (the return of Togo was "not essential") Hitler began to develop, with an uncharacteristic benevolence, the theme of "European unity after the war." From England there was still some hope, "Especially," Halder records, "the possibility of Churchill's overthrow by Conservatives with a view to forestalling a Socialist-Communist revolution in the country." The Führer was in excellent spirits. Some of those present may have been reminded of the occasion, almost exactly a year before, when he had danced a victory jig in the Forest of Compiègne.

During these first halcyon days of victory, when the campaign seemed almost to be running itself, Hitler relaxed happily into dreams of a colonial East. Now, truly, it seemed as if that most fantastic of the Nazi visions—a million square miles of Slavic helots, ruled by a handful of *Herrenvolk*—were on the point of realisation. Hitler envisaged a mixture of British India and the Roman Empire: "A new type of man will take shape, real masters . . . viceroys."

But reality, though maturing with delirious speed in the field of military achievement, lagged sadly in that of administration. The quality of the "viceroys" was far from uniform, for

> When ministries were summoned to supply their quotas of civil servants for the new *Führerkorps Ost* . . . [they had seen] in this call a welcome opportunity to rid themselves of personal enemies, obnoxious meddlers and incompetent chair-warmers.

The result was

> A colourful and accidental conglomeration of Gauleiters, Kreisleiters, Labour Front officials, and a great number of SA leaders of all ranks, who assumed high positions in the civil administration after listening to a few introductory lectures delivered by Rosenberg's staff at the Nazi training school at the Croessinsee.

This motley crew owed a nominal loyalty to their chief, Rosenberg. In fact, they were infiltrated, particularly in the higher echelons, by the personal representatives of other Nazis who were determined to carve their own empires out of the Eastern territory while the going was good. Besides Rosenberg, the two most persistent and avaricious rivals were Bormann and Himmler, with occasional (and waning) intervention by the Reichsmarschall, Goering, who based his claims on his responsibility for the "Four-Year Plan."

Rosenberg's own views had been set out in a long memorandum in April. Part of this document is unintelligible rambling, but its essence may be found in the following paragraph:

> The aim of our policy to me, therefore, appears to lie in this direction: to resume in an intelligent manner and sure of our aim, the aspirations to liberation of all these peoples [the "imprisoned nationalities" of the Soviet Union] and to give them shape in certain forms of states, i.e., to cut state formations out of the giant territory . . . and to build them up against Moscow, so as to free the German Reich of the Eastern nightmare for centuries to come.

This plan—the "Wall against Muscovy"—may have had a certain romantic appeal for Hitler, with its suggestion of the legions standing guard on the Barbary frontier, but privately the Führer rejected Rosenberg's principles—at least on a political level. With characteristic brutality of logic Hitler declared:

> Small sovereign states no longer have a right to exist . . . the road to self-government leads to independence. One cannot keep by democratic institutions what one has acquired by force.

His own view, which he was to express at the notorious 16th July conference[10] on the future of the occupied East, was:

> While German goals and methods must be concealed from the world at large, all the necessary measures—shooting, exiling, etc.—we *shall* take and we *can* take anyway. The order of the day is
> first: conquer
> second: rule
> third: exploit.

Sometimes it is hard to understand why Hitler ever installed Rosenberg as chief of the *Ostministerium* or gave even qualified endorsement to his schemes. But it must be seen in a context separate from Reich foreign policy and in relation to the personal power struggles that cut fissures across the Nazi hierarchy. Pursuing the analogy of the Roman Empire, Hitler must have seen that the only threat to his own position in the future—a future of German domination, actual and undisturbed, over half the globe—would come from the provincial governors, "overmighty subjects" who were allowed an excess of freedom in building up their private empires. Indeed, Ovens' assessment of Bormann can be applied, *a fortiori,* to Hitler.

> He preferred a crack-brained Ostminister to a clever one; a blockheaded foreign minister to an adroit one; a wishy-washy Reichsmarschall to one hard as iron.

After Hitler the two most powerful figures in the Reich were Himmler and Bormann. Each was a direct claimant to his succession and each saw in the limitless potentialities of an Eastern empire the means to tip the balance in his own favour. Their rivalry and their mutual personal dislike lie at the root of all the inconsistencies in German *Ostpolitik,* for first one, then the other would use the bewildered Rosenberg as an indig-

[10] See Ch. 1.

nant pig in the middle, blocking, perverting, or exploiting his policies to achieve their own long-term ambitions.

Rosenberg's great weakness was that he had no personal *corps d'élite,* and the quality of the material from which he was compelled to draw to staff his Ministry and execute its policy has already been the subject of remark. Bormann, on the other hand, had at his disposal the mass of the SA, decapitated by the purge of 1934, but still substantial, frustrated, and experienced in politics and administration. From the very day when the *Ostministerium* was incorporated, it was subject to a double stress—from Himmler, who wished to sterilise it completely, and from Bormann, who tried to staff its higher posts with his own nominees.

As early as April 1941 talks had begun between the SS and OKW concerning the operation of the SD detachments in the rear of the advancing troops. Himmler rapidly forced the pace and tried to extend the "talks" into a general agreement that the Army would be left as undisturbed master of the forward zone, "with the SS as a free corps in effect responsible for the New Order in the East . . . the SD would be advance teams of the future commissariats." At the last moment the Army took fright and started to back away— ". . . these demands must be refused," Halder noted grimly in his diary. Bormann, who had got wind of the scheme, persuaded Hitler to "discuss the affair with everyone concerned," not in conference, but one by one.

When his turn came, Bormann had warned Hitler that an accommodation between the SS and the Army would result in "a measure of power which was inconvenient, perhaps even dangerous, to the Party." Rosenberg put things more formally, and unlike Bormann, was not reticent about declaring his views to anyone who would listen. Hitler threw the scheme out, although he reserved "police matters" to the SS, and Himmler blamed his defeat on Rosenberg's duplicity.

In a state of pique, Himmler complained innocently to Bormann:

> The manner in which Rosenberg approaches this question once again makes it endlessly difficult to work with him, man to man . . . to work with, let alone under, Rosenberg is surely the most difficult thing in the Nazi Party.

Inflamed by his "victory" and rampant with megalomania, Rosenberg now proceeded to claim the right "to approve all assignments of SS personnel to the East." Once the campaign had begun and conquered territory began to accumulate, relations between the various agencies deteriorated to such an extent that Hitler was obliged to call another conference (on 16th July). Himmler was not present, but Goering, Rosenberg, and Bormann all took part with vigour, and there were some undignified scenes—particularly when it came to selecting the names for the actual commissariats, or regional governorships. At the end a Führer directive promulgated that the conquered regions should pass from military to civilian administration "once they had been pacified." The authority of the Army, the SS, and the Four-Year Plan were to be defined under separate agreements, and it was to be hoped that ". . . in practice the conflict [between the different bodies] would very soon be settled."

In practice, however, nothing had been settled except the names of the commissars. Each of the separate directives, being negotiated separately and under the pressure of its own particular lobby, granted a measure of overlapping authority to that agency with which it was concerned. For example, the SS was specifically delegated responsibility for "police security" in the East, and by Article II the Reichsführer (Himmler) was empowered "to issue directives on security matters" to Rosenberg's subordinates. To ensure that his privileges would be enforced, and that he would be kept informed of any opportunity for their extension, Himmler appointed as "liaison officer" to the *Ostministerium* Reinhard Heydrich, his most trusted deputy and one of the most evil figures in the Nazi Party.

The effect of this squabbling was that the Nazi machine was to administer Russia on a basis of almost complete fragmentation—at the levels of both policy and personality. The only sentiment which may have united them was Backe's[11] when he spoke of "The Russian . . . who . . . has stood poverty, hunger, and austerity for centuries. His stomach is flexible; hence, no false pity!"

The General Commissar for Belorussia (the vital central sector of the front, with civil responsibility behind Bock's army group) was Wilhelm Kube, a former Nazi member of the Reichstag who had been duly promoted in the West Prussian administration on Hitler's accession to power, but whose subsequent behaviour had been so scandalous that he had been "retired" before the outbreak of war. By the end of June, how-

[11] Herbert Backe, German Minister for Food and Agriculture.

ever, he was installed in Minsk and making the most of his vice-regal powers. Kube was delighted to find that many of the Belorussians were "blondies and blue-eyed Aryans." He also spoke highly of the vodka and the beer. He found himself a magnificent building for his commissariat and embraced a number of peasant girls in his domestic service.[12]

The administrative staff, in contrast, was far from decorative. It

> consisted of woefully unprepared personnel . . . Nazi waiters and dairy men, yesterday's clerks and superintendents, graduates of quick training courses . . . dizzy with power, yet quite unfit for their jobs. In practice Kube's instructions were often disregarded by his subordinates . . .

Another factor which irritated Kube was the constant encroachment by the SS upon his jurisdiction and the manner in which its members held themselves above either civil or military law. They were particularly prone to "sequestrate" gold and silver in any form, and their indiscriminate violence against the civilian population was already having effect. A typical day in Slutsk saw the arrival of a black-uniformed SD detachment which

> fetched and carted off all the Jews . . . with indescribable brutality they were brought together from their apartments. There was shooting from all over the town, and corpses of dead Jews [and Belorussians, too] were piled up in several streets. Besides the fact that the Jews . . . and the Belorussians . . . were mistreated with frightful roughness before the eyes [of spectators] and "worked over" with rubber belts and rifle butts. There is no more question of a Jewish action. Much rather it looked like a revolution.

On another occasion, in Minsk itself,

> the SD one day took about 280 civilian prisoners from the gaol, led them to a ditch, and shot them. *Since the capacity of the ditch was not exhausted,* another thirty prisoners were pulled out and also shot . . . including a Belorussian who had been turned in to the police for violating the curfew by fifteen minutes . . . and twenty-three skilled Polish workers who were quite innocent but had been sent up to Minsk from the Government-General [i.e., Poland] to relieve the shortage of specialists and had been billeted in the jail . . . because there were no other billeting facilities.

[12] The harem, as oriental scholars know, has its own perils, and these are not exclusively venereal. Eventually one of the "blondies" put an anti-personnel mine in Kube's bed and he was blown to pieces. See Ch. 15.

In this particular case Kube's protest got as far as Rosenberg, and in due course wound its way from the *Ostministerium* to Lammers,[13] who presided over the withered rump of the German judiciary. The essence of Rosenberg's case was not (need it be said) humanity, but administrative protocol:

> It impinges most emphatically upon the responsibility entrusted to me by the Führer for the administration of the occupied Eastern territories.

But when Lammers' reproof finally reached Heydrich, the SS liaison officer brushed it off. "The executions were due to a danger of epidemics."

All the same, Kube continued to grumble. Not only was the SS competing in its efforts to run the territory by issuing its own decrees, but the wholesale slaughter which it practiced daily was already having its effect on the economy:

> Jewish artisans simply cannot be spared because they are indispensable to the maintenance of the economy.

The confusion was being aggravated by Goering, who was eagerly extending the net of his own administrative machine and finding to his annoyance that he was being forestalled by Himmler. Throughout European Russia the SS was "requisitioning various industrial and commercial enterprises." Forced to operate through the corrupt and rickety machinery of the Reich commissars and without a private army of his own (a state of affairs which he was soon to remedy), Goering was compelled to bow out with as good a grace as he could muster,[14] but the effect of this triangular rivalry in the fields of murder, plunder, and administration on the smooth running of the occupied territory needs no emphasis.

In the Ukraine the Reichsmarschall was better served, for at the 16th July conference his own nominee, Erich Koch, had been chosen for the commissariat. Rosenberg had protested vigorously against this choice, believing with some reason that the whole of his delicate and crack-brained scheme for racial discrimination would be placed in jeopardy by

[13] Hans Lammers, Chief of the Reich Chancellery 1933-45.

[14] On 26th August, 1941, Goering wrote to Himmler (Doc. NO-1019), "I have asked the Reichskommissar for Ostland to handle your requests for the supply and disposal of service and consumer goods with the requisite understanding . . ."

a man who was already notorious for sadistic taste and corrupt practices.[15] The Ostminister had also considered the close personal friendship between Koch, Bormann, and Goering, and the direct link which his (nominal) subordinate would thereby enjoy with the Führer.

In fact, Koch agreed with Goering that "The best thing would be to kill all men in the Ukraine over fifteen years of age, and then to send in the SS stallions," and the two of them made an informal deal with Himmler that the SS would be allowed a free hand in its extermination program in return for allocation of the economic resources and "general loot" to Goering.

Koch had begun as a railway official in the Rhineland (and his subsequent career may be studied with some misgivings by persons who have had the misfortune to attempt travel in Germany or Switzerland with tickets that are out of order). Under Goering's patronage he had risen to be Gauleiter of East Prussia, and he retained this title even after being "given" the Ukraine. He had his own notions of colonial-style government, and liked to carry a stock whip. He persuaded Goering to extract certain provinces of Belorussia and the Bialystok forests from the general carve-up that took place in the first weeks of the German advance, and added these to his dominion, frequently boasting that he was the "first Aryan to hold sway over an empire from the Black Sea to the Baltic." The essence of Koch's theme was that propounded to him by Himmler:

> Like the skimmed fat at the top of a pot of bouillon, there is a thin intellectual layer on the surface of the Ukrainian people; do away with it and the leaderless mass will become an obedient and helpless herd.

Against this attitude Rosenberg kept up a running fight, handicapped by the disloyalty and incompetence of his own staff, and by his periodic tiffs with Hitler. After one such scene, at which Rosenberg complained:

> Koch, through various remarks to officers of the OKW, has given the impression that he has the privilege of reporting directly to the Führer and, in general, that he intends to reign without reference to Berlin [i.e., to the *Ostministerium*] . . .

[15] Gisevius, in *To the Bitter End,* London, 1948, has described how, while Gauleiter of East Prussia, Koch had established the "Erich Koch Institute," and "cheerfully watered the stock whenever he needed money for his palaces or similar amusements . . ."

> Similar remarks to the effect that *he* made policy have been made to my associates . . . I have made it clear to him that a distinct relationship of subordination exists . . .

Hitler agreed to receive Koch "only in my [Rosenberg's] presence."

This, however, was a meaningless concession, for Koch could always obtain access at the shortest notice through Bormann, who himself nurtured personal schemes of empire building through "nominees." Bormann encouraged Koch to issue a proclamation to the effect that the Reichskommissar was

> the sole representative of the Führer and the Reich Government in the territory entrusted to him. All official agencies of the Reich must therefore . . . be subordinated to the Reichskommissar.

Poor Rosenberg! At the very moment when he was locked in combat with Koch he was distracted by interference from a new and unexpected quarter. For he found that his principles—or a rationalisation of his principles—were being taken up and pushed hard by yet another organisation, which, although the last to climb on the bandwagon, was nonetheless determined to take its share of the spoils and the power.

This latest intruder was none other than the Reich's Foreign Minister, Joachim von Ribbentrop. In the weeks immediately preceding the start of *Barbarossa,* Ribbentrop had been hastily accumulating a diversity of "experts" and émigré leaders within the confines of his offices at the Wilhelmstrasse. Their purpose was to identify and encourage separatist movements in Russia, whether they existed on a basis of nationality (Balts, White Russians, Galicians, and so forth) or simple "anti-Bolshevism." The most civilised of these "experts" was the former German Ambassador in Moscow, Count Werner von der Schulenburg,[16] who believed

> the definitive status of the Ukraine can only be settled after the conclusion of the war. As possible solution [I] envisage a strong autonomy of the Ukraine within a Russian confederation, or under certain circumstances an independent Ukraine within a confederation of European states.

This, of course, was the only policy which could, in the fullest sense, solve the problem of "pacification" in the rear areas and bring the oc-

[16] Not to be confused with his kinsman Count Fritz von der Schulenburg, the Deputy Police President of Berlin. Both men were subsequently members of the 20th July plot.

cupied territories solidly into the German war effort. Ribbentrop was pressing it, not because of its obvious justice and humanity but because he thought that within weeks the war would be over and within months the whole world would be at Hitler's feet. Then the only *raison d'être* for the Foreign Office would be as the apparatus which continued to dabble in nationalities and play with countries in a world of make-believe diplomacy, where the Reich Foreign Minister would always have the last word.[17]

It is this feeling, amounting to a conviction, that the war would be over in a week or so, that conditioned the attitude of every person concerned with the administration of occupied Russia in 1941. There was no cause to fear retribution, no restraint on the grossest indulgence of personal greed and lust—whether for blood, torture, or "blondies." Only Rosenberg, half mad with vanity, continued with his plans to separate and purify the racial strains in his kingdom, and it is precisely because the theories of Ribbentrop and Schulenburg came nearest to, and thereby carried a direct threat to supplant, his own schemes that Rosenberg opposed them with all his resources.

After several months of correspondence, urgent and clandestine approaches to the Führer, complicated and at times farcical manoeuvres[18] in a steadily rising temperature, Rosenberg got his way. Hitler sent for Ribbentrop and put him straight in a "down-to-earth" talk. The Foreign Minister returned to Berlin and declared to his bewildered aides, "It is all nonsense, gentlemen! In wartime nothing can be achieved with your sentimental scruples."

The decision was codified by a Führer directive to the effect that "The Foreign Office is not to concern itself with countries with which we are at war." The files on all the émigrés in Berlin were turned over to Rosenberg and in due course fell into the hands of Himmler, who

[17] A week before the invasion Ribbentrop had addressed a pompous note to Lammers:

> The territory to be occupied by German troops will on many sides border foreign states, whose interests will thereby be most strongly affected. . . . The Foreign Office cannot acquiesce in the absence on the spot of representatives schooled in matters of foreign policy and versed in local conditions. (NMT, NG-1691, xiii 1277-79)

[18] In April 1942 a "conference" of émigré leaders was arranged by the Foreign Office at the Hotel Adlon. Some forty persons attended, drawn from governments in exile as far away as Ankara, and including Count Heracles Bagration, Pretender to the throne of Georgia, and the grandson of the Caucasian bandit, Said Shamil.

threw most of the persons named there into concentration camps. (Much later, as will be seen,[19] some of the survivors were hauled out and allowed to restart their movements, but by then there was little incentive for them to do so, as the probability of German defeat loomed large.)

This, then, was the brief tale of the only policy that might have achieved solid gains for the Germans in the occupied East. It had originated in grounds not of justice but of expediency, and was rejected because, on the shortest possible view, it was not so much inexpedient as inconvenient. Rosenberg regarded its rejection as a personal victory, and if it was such, it was certainly his last. But even then he would scarcely have been reassured to hear Hitler's private opinion:

> Anyone who talks about cherishing the local inhabitant and civilising him goes straight off into a concentration camp . . . my one fear is that the *Ostministerium* will try to civilise the Ukrainian women.

While the *Ostministerium* was occupied in repelling the usurpers from the Foreign Ministry, Koch tightened his hold on the Ukraine. Daily there were executions—if that term, with its overtones of judicial rectitude, can still be used of the rattling machine guns and haphazard mass graves that characterised the terror—and nightly the trucks of the SS rumbled through the streets, collecting "suspects." Whippings (usually to death) were a feature of Koch's regime, and they were conducted, for "exemplary" reasons, in public places such as squares and parks. In these first weeks of the occupation there was no systematic plan of exploitation. It was pure recreation for the Germans, "scraping the icing off the cake." Nor was there any resistance worthy of the name on the part of the local population. Yet in this orgy of sadism and misgovernment it required no gift of prescience to see, as Rosenberg explained in one of his many letters of reproof to Koch, that

> There exists a direct danger that if the population should come to believe that the rule of National Socialism would have even worse effects than Bolshevik policy, the necessary consequence would be the occurrence of acts of sabotage and the formation of partisan bands. The Slavs are conspiratorial in such matters. . . .

In contrast to the regime in the Ukraine and Belorussia, that which prevailed in the Baltic provinces at the northern end of the front seems

[19] See p. 408.

easygoing. Lohse, the Commissar, was the German bureaucrat par excellence. He liked good food, and overindulgence of this taste compelled him to take frequent leaves of absence at curative spas. But when he was at his office his appetite for detail was insatiable. He sent out "A flood of decrees, instructions, and directives which covered thousands of pages. Lengthy correspondence took place between Riga [Lohse's headquarters] and the four general commissariats on the most trivial administrative problems. Price control was established for [*inter alia*] metal wreaths for geese 'with' and 'without' heads, alive and dead. A decree of 'maximum prices for rags' was promulgated, with differences of ten pfennigs per kilogram between light brown and dark brown rayon rags in Latvia. Even NO SMOKING signs had to be signed by Lohse personally."

The commissar's attitude to the "subjects of Ostland" was summarised in an address to his staff the following year:

> So long as a people is peaceful, one should treat it decently. To make political mistakes and to hit people over the head—anyone can do that.

Moreover, Lohse cherished the hereditary principle:

> I am not working for myself. I work so that my son, who has just been born, can some day put the hereditary ducal crown on his head.

The result of this policy was twofold. First, the industrial capacity of Ostland contributed far more to the war effort of the Reich than that of other, potentially richer areas where the administration was needlessly harsh and oppressive. Although even here the effect was diminished by a ludicrous corruption and inefficiency—it was open season for German businessmen, large and small, who built up private industrial empires by confiscation and "licensing" and then used them to manufacture and market luxuries (such as perambulators) which were banned in the more tightly organised economy of the Reich.

The second result was that the Partisan movement never became a major adjunct of the Red Army, as it was to become in the other parts of Russia, and indeed, as it became in Ostland itself once the frontiers of the Baltic states had been passed and the dark forests of the Narva gave cover. In Estonia, Latvia, and Lithuania the population lived through the war in a state of dumb resignation, tempered, during its later years, by a fearful apprehension of the revenge the Red Army would exact on its return.

While the rear areas succumbed to a hodgepodge regime of terror, bungling, and exploitation, the German Army continued with its gallop across the steppe. On 1st July, Guderian had crossed the Berezina at Svisloch with the 4th Panzer, and the following day the 18th Panzer won a bridgehead higher up the river, at Novy Borisov, entering the town simultaneously with tanks from the 14th Panzer on Hoth's right wing. But the clash of personalities, which compounded the incompetence of the civilian administration, was by no means absent from the conduct of military operations, even at this early stage. In particular, the activities of Army Group Centre suffered from this affliction.

The primary source of this trouble was friction between Guderian and Kluge. Kluge had commanded the 4th Army in Poland (when Guderian had a corps), and in France (and again Guderian was only a corps commander). Now Kluge was still commanding the 4th Army and his erstwhile subordinate had a whole *Panzergruppe*—the very tip of the spear which, wielded not by Kluge but by the army group commander, Bock, was to lance the centre of European Russia. Kluge, it was true, had been promoted to Field Marshal for his achievements in France, and the 4th Army was the most powerful of the eleven which the Germans deployed at the start of *Barbarossa;* but though these things may have diminished, they could not eradicate Kluge's chagrin at being denied a *Gruppe*.

Partly out of deference to the Field Marshal's feelings, partly from a supposed administrative convenience, the chain of command at Army Group Centre had been altered so as to make Guderian subordinate to Kluge and not (as would have been the normal arrangement) directly to Bock. The circumstances under which this had arisen were as follows. The fortress of Brest-Litovsk had stood directly in the path allotted to the 2nd Panzer Army. To subdue it Guderian had asked for an infantry corps from the 4th Army and had divided his tanks, sending a column to the north and one to the south. Guderian's version of the arrangement is that

> In order to ensure unity of command, I asked that these troops be temporarily subordinated to me, and expressed my willingness to place myself under the command of F-M von Kluge during this time. [These arrangements] . . . *involved a sacrifice on my part; Field-Marshal von Kluge was a hard man to work under.*

Whether it is true that the subordination was done at Guderian's request or (as seems more likely) was decided at OKH, the operational phrase was "during this time," i.e., during the siege of Brest-Litovsk. But in fact, Kluge continued to interpose himself between Guderian and Bock long after the river Bug had been crossed and the Panzers were in open country. And at this point personal animosity began to aggravate the very real differences in tactical approach which characterised the two men.

Guderian was the tank general par excellence; more than Manstein, O'Connor, Model, with a cooler nerve than Rommel or Patton, he knew how to handle an armoured division. He was one of the very few men who had really absorbed the teaching of Liddell Hart on the importance of speed, mobility, and firepower; who understood the tank as an independent arm, not simply as an adjunct to orthodox deployment. Kluge, on the other hand, disliked seeing the Panzers rushing too far ahead, and claimed that their extra weight was required to keep the Russian pockets "contained." Guderian believed that this was a task for infantry, that the tanks should keep moving, and were vulnerable only when they stopped. Bock, commander of the entire army group, privately backed Guderian. But he could see the risks, risks which in those first hectic weeks were more apparent at headquarters than through the observation slit of an armoured command car. Kluge's policy was the safe one, and the prime concern of Bock was that nothing should prejudice his own prospects of acclaim as the captor of Moscow and the hammer of the Soviets. He tried to compromise between Kluge's restraint and Guderian's audacity—and always to keep himself covered in case anything should go wrong.

> It is characteristic of this officer [wrote Halder testily] that he should demand a written confirmation of an order from my headquarters simply because he does not agree with it.

Halder himself, who could have given a lead, shirked the responsibility. From his desk at OKH he could see the opportunities, but his strict professional training urged caution. It is also the case that he was subject to a steady stream of telephone calls from the nervous ObdH, who himself was under more or less constant cross-examination by Hitler. Was the pocket at Slonim holding fast? Was the 292nd Division across the Desna yet? Was it true that two Russian corps had been identified in the Naliboki forest? How many serviceable PzKw III's were left in the 29th Motorised? "There they go, fussing again." One can imagine Halder's irritation as he scribbled down this phrase before leaving his evening journal to answer, once again, a telephone call from Rastenburg.

Drawn in two directions at once, Halder committed himself on 29th June to one of the most craven admissions of executive impotence that can ever have been uttered by a member of the General Staff:

> Let us hope that the Commanding Generals of Corps and Armies *will do the right thing without express orders,* which we are not allowed to issue because of the Führer's instructions to ObdH.

Yet at this point Halder himself does not seem to have been clear as to what "the right thing" was. For the very next day he was echoing Kluge's complaints about Guderian's headstrong advance:

> . . . in disregard of its orders [the Panzer group] has neglected to attend to the mopping up of the territory traversed by it, and now has its hands full with local enemy break-throughs.

Among these five generals a multilateral dispute (which, with Hitler's intervention, was to blow up into a major crisis by the end of July) was already in the making.

On 1st July a sharp Russian attack against the east side of the Slonim pocket had penetrated the German screen and allowed the remnants of two tank brigades to escape into the marsh and forest area between the 47th and 24th Panzer corps. This setback had occurred almost simultaneously with the forcing of the Berezina by the 18th Panzer, over sixty miles to the northeast. The 18th Panzer was stretched to the limit, and with a hostile (if somewhat ragged) Russian brigade group straggling its lines of communication, the question of its reinforcement was urgent. Guderian had ordered the 17th Panzer, on that day in leaguer to the south of Minsk, to drive at once to Borisov. But Kluge countermanded this, personally communicating directly with Weber, the divisional commander, instead of through Guderian.

Thus far, although Kluge had not been tactful in his approach, the incident was unremarkable. But now an element of mystery intrudes. Guderian had been touring the forward units throughout the day, and learned of the 4th Army order only when he arrived at Weber's headquarters in the afternoon. He does not mention any conversation between the two, and no other source is available concerning their exchanges as

Weber was mortally wounded a week later. However, when Guderian finally arrived at his own command post that night he "immediately despatched a signal to Fourth Army, informing them that *a mishap had occurred in the transmission of orders to 17th Panzer;* part of the division had not received the order to remain on the encirclement front and had, therefore, set off for Borisov. . . . It was too late to do anything about it."

The reply from Kluge's headquarters was instantaneous—a summons to report there in person at eight o'clock the following morning. Guderian says that he was "strongly taken to task," and in view of the fact that Kluge raged about a "generals' conspiracy" (the same sort of "muddle" had occurred earlier in Hoth's *Gruppe*) and threatened him with court-martial, this can hardly be called overstatement.

In the result the forward dispositions were not affected, and neither the strength around the Slonim pocket nor the striking power of Guderian's left wing was adequate. On 3rd July it rained all day, and forward movement stopped. Evidence that the Russians were going to make a fight for the Dnieper soon began to accumulate. On 6th July a strong Russian force drove back the 10th Motorised and the cavalry division from Shlobin, and an attempt to storm Rogachev by the 3rd Panzer was repulsed. Then the following day came violent pressure against the left wing; the 17th Panzer was ejected from Senna.

Guderian, however, was undeterred. The revival of Russian strength made it all the more urgent, he believed, to force the crossing of the Dnieper at the earliest possible moment. Instead of clearing his flanks he contracted them. The 17th Panzer and 10th Motorised were ordered to "break off the engagement," and to content themselves with keeping the enemy "under observation." Guderian's plans suffered another setback when SS *Das Reich* was bloodily defeated in an attempt to capture the bridges at Mogilev—right in the centre of the Panzer group's front, but even this failed to deter Guderian, and finding weak spots at Kopys and Shklov, he prepared to force both the 47th and 46th Panzer corps across the river.

By this time Kluge was not the only one to be alarmed. Halder recorded that "Everyone [at OKH] is vying for the honour of telling the most hair-raising tales about the strength of the Russian forces [behind the Panzer group in the Pripet Marshes]. Foremost are the radio intelligence people who claim three Armoured Corps, and two

Infantry Corps." Another disturbing feature was the evidence of mounting concentration at Bryansk and Orel, and the fact that the remaining Soviet fighter aircraft seemed to be devoted to protecting the railway stations there.

July 9th was marked by "exceptionally heated conversations." Kluge flew to Guderian's headquarters at first light and "ordered that the operation [i.e., crossing the Dnieper] be broken off and the troops halted to await the arrival of the infantry." Guderian claimed that his preparations had "already gone too far to be cancelled." Guderian continued by asserting, ". . . this operation would decide the Russian campaign in this very year if such a decision were at all possible."

After a good deal of argument Kluge was convinced, and gave his approval. But there is no doubt that he impressed his subordinate with the verdict that it was "now or never"; the headstrong General could not be allowed another such chance. At the close of their meeting he passed his celebrated judgment on Guderian's tactics:

"Your operations always hang by a thread!"

On the northern front, too, great prizes beckoned, but the generalship faltered before the diffusion and tenacity of Russian resistance. Hoepner, the commander of the 4th Panzer Army, has been widely criticised by his colleagues since the war.[20] But the fact remains that of all the Panzer armies his was the weakest, and had been given the most ambitious objectives. Hoepner was expected to strike directly for Leningrad, yet at the same time he had to protect his right wing and that of the whole of Leeb's army group from the Russian armies along the Lovat—an open flank of over two hundred miles. Hoepner's task was further complicated by the fact that his neighbour, Hoth, had an axis of advance due east, and was in fact repeatedly turning his forces *inward* to meet Guderian in their succession of encirclement battles.

Manstein has described how, after his corps had been waiting for two days at Dvinsk, Hoepner arrived in a Fieseler Storch but "could tell us nothing" except "to widen the bridgehead, and keep the crossings open." The 56th Corps commander goes on to complain, ". . . one might reasonably have expected the commander of a whole Panzer Group to be in the picture about future objectives, but this was obviously not the

[20] Treatment which cannot be divorced from his own career following his dismissal in 1941, and his execution in 1944.

case." But how could Hoepner possibly have allowed Manstein to press on with only two divisions (the 8th Panzer and 3rd Motorised)—which was what Manstein was wanting—while his sister corps was not yet abreast of the Dvina, much less across it?

And so another five days passed while SS *Totenkopf* moved up and the 41st Panzer Corps forced a crossing at Jacobstadt. The Russians meanwhile were frenziedly redeploying their forces, taking men, tanks, and airplanes from the Finnish front to bolster the crumbling armies of Popov and Kuznetsov. Instead of being husbanded for the next stage, these regular troops were used to stiffen the masses of conscripts, workers, and militia units that were beginning to take shape, and thrown into a succession of savage counterattacks, so that ". . . at a number of points the German situation became quite critical." As for the Red Air Force, ". . . with an almost mulish obstinacy one squadron after another flew in at treetop level, only to be shot down . . . one day alone they lost sixty-four aircraft."

As on so many occasions, this reckless profligacy in lives and equipment had an unnerving effect on the Germans, and Leeb, overestimating his enemies' strength and the cohesion of the Russian Command, made his first tactical mistake. When the Panzer army resumed its march on 2nd July, the axis of the two corps was separated, Reinhardt being directed on Ostrov and Manstein into the yawning void on the right flank—toward Opochka and the Lovat.

Within a few days both the 8th Panzer and the motorised division were stuck fast in swampy ground. SS *Totenkopf* made better progress, but then ran into the concrete fortifications of the "Stalin Line," where "their losses and lack of experience led them . . . to miss favourable opportunities, and this . . . caused unnecessary actions to be fought." None of the three divisions of the 56th Panzer Corps were able to give the others support, and after a week of inconclusive fighting the 8th Panzer and the motorised division were pulled back and sent in behind Reinhardt. *Totenkopf,* after this brief and violent experience of real combat, was returned to "Reserve," where it could vent its spite on the civilian population.

Reinhardt, in the meantime, had captured Ostrov, but had not the strength to press on past Pskov and along the eastern shore of Lake Peipus. And once again Manstein was prevented from lending the weight of his own corps to the main thrust; he was directed due east—now with only two divisions—with the grotesquely vague and ambitious objective of "breaking communications between Leningrad and Moscow at the earliest possible date." This bifurcation of the two weak Panzer corps was soon to have serious results.

These cumulative errors of decision on the northern and central fronts can be (and have been) attributed to many things: timidity at OKW, the conflict of personalities, the absence of a long-term strategic plan, and so on. But the hard fact remains that the Germans, even at this early stage, were attempting too much. Their mobile forces were not strong enough,[21] or numerous enough, to support three simultaneous thrusts.

Few of the German commanders realised this at the time. Each attributed other, local causes to his own (qualified) failures. On the wall maps at Hitler's headquarters the territorial gains looked enormous—and the more impressive in relation to the few weeks they had taken to acquire.

"No *Schweinhund* will ever eject me from here," said Hitler to General Köstring when the latter visited him at Rastenburg.

What Köstring, the last military attaché at Moscow, and one who knew more about the Red Army than anyone else in the room, thought may perhaps be gauged from his laconic reply.

All he could manage was, "I hope not."

[21] Strength, that is to say, in a qualitative sense. The Panzers had only a limited cross-country ability as nearly all their supply vehicles were not tracked, but wheeled, and had difficulty over bad going. The reduction in the tank quota (see p. 17) had reduced not only the number of tracked vehicles but the over-all firepower of the division.

four | THE FIRST CRISIS

At this point in the campaign there occurs a major break in the continuity of its direction. The latent conflict of attitude between Hitler and his generals, rooted in their training, instinct, and technique, assumes hereafter a gradually increasing importance. It gathers momentum until by 1944 the changed balance in the field between the Wehrmacht and the Red Army coincides with a final subordination of the professional to the amateur on the German side. This conflict had been a factor of primary importance during the political evolution of the Third Reich,[1] and its residue was to foment bitterly as the prospect of victory faded. Now for the first time it began to assume positive military significance.

To say that Hitler was an amateur is not intended as an unqualified derogation. He was a brave man who had won the Iron Cross in the field. Throughout his life he had studied military affairs. His ability to gauge the feeling of the common soldier and to inspire him is unquestioned. All these are vital ingredients in the essence of successful command. And in the early months of the war his élan, his propensity to take risks, his "intuition," had reaped a tremendous harvest.

But eight weeks after the start of the campaign in the East, these roles were reversed. The General Staff was to become virtually unanimous in its desire to reinforce Bock and strike directly on a narrow front toward Moscow. Hitler insisted on the orthodox solution after Clausewitz—the methodical destruction of the enemy's forces in the field, regardless of geographical or political objectives. As early as 13th July he had told

[1] P. 7, *et seq.*

Brauchitsch, "It is of less importance to advance rapidly to the eastward than to destroy the living forces of the enemy," and this attitude, to which he clung for the ensuing two months, was no more than consistent with the original terms of the *Barbarossa* directive, which laid down that the purpose of the operations was "to destroy the Russian forces deployed in the West and to prevent their escape into the wide-open spaces of Russia."

The problem was simple in outline, vastly complex and elusive in substance. After the first flush of success the Wehrmacht was losing momentum. Partly this was a question of supply. Food and ammunition, auxiliary services, the maintenance of machinery, all these became progressively more difficult as the front broadened and the divisions fanned out. But there was also a tactical aspect. The detailed plans worked out under Halder and Warlimont were already surpassed, and the dispersal of the armies widened daily as each fought deeper along its own prescribed axis, bypassing resistance and exploiting weakness. At this distance from headquarters army and even divisional commanders were acting more and more on their own initiative, the more adventurous fighting a series of interlocking (but not necessarily co-ordinated) local actions deep in the Russian rear, while their more placid—or less mobile—colleagues sat patiently in rings around those portions of the Soviet Army that had been cut off.

In the middle of July the German front ran true along a north-south line from the mouth of the Dniester, on the Black Sea, to Narva, on the Estonian frontier. But in the centre the reversed *S* of two gigantic salients bulged ominously. The Panzer groups of Army Group Centre, advancing on Moscow to the north and south of the Minsk highway, had already passed the longitude of Smolensk. But to their right the Russian 5th Army still held its forward positions in the Pripet Marshes. In this way there was an extra "front" of over 150 miles which lay against the exposed flanks of Army Group Centre, and of Rundstedt's left wing as it approached Kiev. The Russian salient, although giving the appearance of mass, was in reality a fragmented hodgepodge of defeated units, stragglers, men without equipment, tanks without fuel, guns without ammunition. But this was not apparent from the large-scale war map at Rastenburg, and the Germans simply did not dispose of the men to probe the area in sufficient strength to find out. And so the Russian presence, poised (as it seemed) over its supply routes, acted as a brake on the freedom of the army groups to either side. Meanwhile, as the days passed with them undisturbed, the Russians were exploiting to the full that extraordinary gift of improvisation which was to succour them on so many occasions during the campaign.

Under Potapov[2] they were busy restoring cohesion to their shattered brigades, laying the foundations of the Partisan movement, and operating vigorously with their cavalry—the only mobile arm left to them in any strength. The 5th Army and the units gathered around it were the largest concentration operating in the German rear, but there were many others still in vigorous action, even though (unlike the 5th Army) they were completely out off from the main front. The garrisons at Orsha and Mogilev, great numbers of wandering infantry—some as far west as Minsk and Vilna—the whole stretch of the Baltic coastline up to the west of Tallinn, the continued resistance of all these "pockets," lent force to the arguments of those who believed that the Wehrmacht was being dangerously overextended.

With the intention of restoring concentration and asserting at the same time a strict priority of objectives, OKW had issued, on 19th July, Directive No. 33. This opened with a reminder that although the Stalin Line had been pierced along its whole front, ". . . the liquidation of important enemy contingents caught between the mobile elements of the Centre will take a certain amount of time." The directive went on to complain that Army Group South had its northern wing immobilised by the continued resistance of the Soviet 5th Army and by the defence of Kiev. Therefore ". . . the object of the immediate operation is to prevent the enemy from withdrawing important forces beyond the Dnieper and to destroy them."

To this end:

(a) The Soviet 12th and 6th armies are to be crushed by a concentric attack of Army Group South;

(b) The inner wings of an Army Groups South and Centre are to inflict the same treatment on the Soviet 5th Army;

(c) Army Group Centre is to push only its infantry toward Moscow. Its mobile elements which are not engaged to the east of the Dnieper [i.e., against the Soviet 5th Army] are to assist the advance of Army Group North against Leningrad by covering the right flank and destroying communications with Moscow;

(d) Army Group North is to continue its advance on Leningrad

[2] Major General of Tank Troops Potapov, commander of the 5th Army.

> when the 18th Army has established contact with the 4th Panzer Army, and when the right flank of the latter has been made completely secure by the 16th Army. The Estonian naval bases are to be seized, and the enemy is to be prevented from withdrawing his forces from there into Leningrad.

This was clear enough. What the directive amounted to was a halt order to Army Group Centre (to "push on" with infantry meant nothing over these distances) while the flanks were secured.

The fact was that OKH and OKW alike were taken aback by the continued strength of the Russian armies. From the distance of their headquarters the weird convolutions of the front line, the reports of resistance so far behind their own deep salients, the mounting activity of the Partisans, all this had an air not simply of unorthodoxy but of menace. Army Group Centre was by far the strongest of the four, and with it the Russian front was to have been rent in twain. Yet in spite of its headlong advance and the brilliant victories of encirclement it had achieved, the enemy had maintained co-ordination, and his resistance seemed as tough as it had been at the start of the campaign. There is a uniformity about their accounts of the fighting at this time which illustrates the Germans' surprise at finding an enemy who continued to resist long after he had been surrounded.

> The Russians did not confine themselves to opposing the frontal advance of our Panzer divisions. They further attempted to find every suitable occasion to operate against the flanks of the wedges driven in by our motorised elements, which, of necessity, had become extended and relatively weak. For this purpose they used their tanks, which were as numerous as our own. They tried especially to separate the armoured elements from the infantry which was following them. Often they found that they in their turn were caught in a trap and encircled. Situations were sometimes so confused that we, on our side, wondered if we were outflanking the enemy or whether he had outflanked us.

The extent to which the Panzer armies of Hoth and Guderian were outrunning their supporting infantry was a constant source of worry to OKH. The Germans were very short of motorised infantry units, and those that they had operated close up with the tanks as part of the armoured spearhead. The mass of the army marched on foot, with its supplies drawn by horses and mules, and its pace was limited. On 17th July the leading elements of the 4th Army was still at Vitebsk and the 9th Army had not even crossed the Duna. But Hoth's tanks were already northeast of Smolensk, and this situation was duplicated by Guderian, who had crossed the Desna with the spearpoint of the 10th Panzer and SS *Das Reich* divisions, but whose situation map for that day shows no infantry east of the Dnieper—a distance of over a hundred miles.

The staffs had learned much about the handling of armoured forces since May 1940, when the British counterattack at Arras had caused them to halt Bock[3] before Gravelines for two critical days. But this was not northern France, with the depots of the Siegfried Line eight hours away through an intact railway system. Between the Panzers and their depots lay two hundred miles of White Russia and the whole of eastern Poland. A territory ravaged and demolished, "served" by dirt roads and single-track, broad-gauge railways, sprinkled with Partisans, and with whole areas controlled by groups of bypassed enemy.

This monstrous awe-inspiring war of movement was but a few weeks old. To say that the General Staff was bewildered would be to overstate, but it is certainly true that its meticulous professional competence left little room for the imagination required to cope with such a gigantic opportunity. Besides Guderian and Hoth there were many others who could have taken the Panzers through to Moscow. Some, like Model, were in relatively junior positions; others, like Rommel and Student, were elsewhere. Bock himself had the inclination, but not the persuasive ability, to defy his colleagues and Führer. But knowing what we do now of the strength of the Russian armies, even at that time, and of the plans they themselves had prepared for a riposte, it is by no means certain that such a thrust would have succeeded. It would have been a gigantic gamble, about which the only certain thing that can be said is that it would have ended the war—one way or another.

Hitler's own attitude was, as so often, ambivalent. Before the invasion he had assured Rundstedt that the Russian armies would be destroyed west of the Dnieper. There is no doubt that he welcomed the support of conventional opinion in his desire to halt the deep thrust to the enemy capital insofar as this involved dividing, and thus weakening, the counsels of the generals. But he seems to have had no intention of accepting the reservations they advised. Rundstedt now recommended a slowing down both in the centre and the Ukraine, and a concentration against Leningrad, with the object of freeing the Baltic and linking up with the Finns before winter. But at this stage Hitler wished to avoid the "encum-

[3] Field Marshal Fedor von Bock.

brance" of the capital cities, with their huge and hungry populations, and believed that by dividing his strength between the north, where Leningrad was to be "isolated" and "bypassed," and the south, the Panzer armies would sweep around behind Moscow, cutting off the city and the whole of Timoshenko's stubborn army. It was to be "Super Cannae," the greatest battle of annihilation that the world had ever seen. In this way, during the last week in July, both at OKH and OKW, opinions were united in the view that the advance of Army Group Centre should be slowed down. These opinions were not all arrived at by the same process nor based on the same assumptions, but in unison they represented a formidable weight. Nevertheless, within days of the publication of Directive 33 the pressure of events at the front was to make its appreciations obsolete.

To understand the fluctuations of opinion at OKW during the following weeks, and the delays in execution which resulted therefrom, it is necessary to study the course of the fighting in detail. For at no time in the campaign is the interaction of hesitation and lack of purpose at headquarters and opportunism at the front so pronounced or its strategic consequences so important.

Guderian's army was advancing more or less due east, along three separate axes. The most northerly of these ran from the Dnieper crossings below Orsha, along the line Dubrovno-Lyady-Krasny-Smolensk. This was under the 57th Corps, with the 29th Motorised Division leading the 17th and 18th Panzer. In the centre the 56th Corps advanced from Mogilev through Mstislavl-Khislavichi-Prudki, with the 10th Panzer Division, SS *Das Reich,* and Guard Battalion *Gross Deutschland*. To the south, up the winding valley of the Oster, came the 24th Corps, with the 10th Motorised Division, the 3rd and 4th Panzer, and the cavalry division.

In fact, the degree of concentration was higher than is suggested by the ratio of seven divisions to a starting line of sixty miles, owing partly to the very high quality of training and equipment of the units concerned, and also to the exceptional firepower and mobility of the individual columns. Guderian's thrust was like a three-pronged fork. Each prong was hard and sharp, but between them was air. Miles and miles of flat steppe, grain and grass, "covered" by the Luftwaffe and occasional patrols of armoured cars and motorcycles. Here lay the seeds of

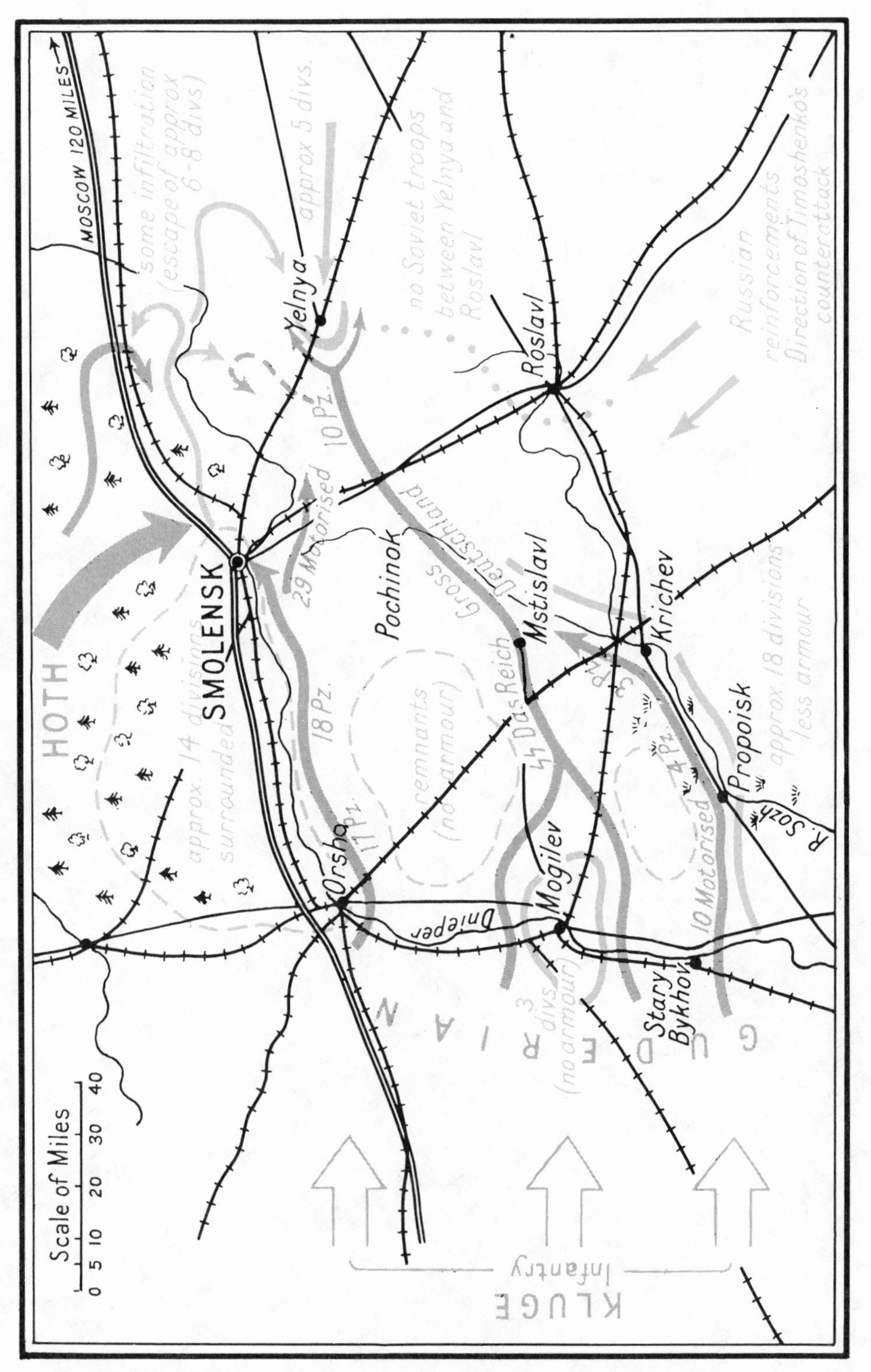

The Dnieper Crossings and the Battles of Smolensk and Roslavl

trouble that might grow with alarming suddenness if the Russian Army were to recover its balance or one of the columns to suffer a check.

The original intention of the Russians had been to establish their defensive line from Vitebsk south to the Dnieper, and then along the left bank of that river as far as Kremenchug. To hold this position they had assigned the fresh troops of the High Command reserve group and placed them under Budënny. But the almost total disintegration of their western front had compelled the *Stavka* to commit these formations piecemeal during the last days of June, and formal recognition of this came on 2nd July with the assignment of the whole of Army Group Reserve to the western front, where it was placed under Timoshenko and the existing command structure dissolved.

Timoshenko struggled desperately to restore some order to the chain of command along his shuddering, concave front. He sent his Chief of Staff[4] to take charge of the right wing, retaining his own headquarters and responsibility on the left, along Guderian's southeastern flank. Far behind them Zhukov was scraping the barrel to form yet another "Soviet reserve front" to protect Moscow along the line Ostashkov-Rzhev-Vyazma. For nearly three weeks the pressure on Timoshenko was too strong for him to recover his balance and concentrate. It was a pressure that emanated as much from the *Stavka* as from the enemy, and that was heightened rather than relaxed by the introduction of dual command[5] and the appearance of Bulganin at his headquarters on 16th July.

Throughout the first two weeks of that month the melancholy tale of squandered lives and equipment had continued. On 6th July the 5th and 7th Mechanised corps had been thrown piecemeal against Hoth, and were chewed to shreds in three days. On the 11th a personal message from Stalin insisted on the "protracted defence" of Smolensk, but four days later Lukin's 16th Army, entrusted with this task on the "direct orders" of GOKO, had been smashed. Around Mogilev the whole of the 13th Army (Lieutenant General Gerasimenko) was trapped and an-

[4] Lieutenant General G. K. Malandin.

[5] "A sign, as always, that the officer corps needed a touch of the Party whip." (Erickson 603) The new instruction on military commissars reinstated the commissar as "the representative of the Party and the Government in the Red Army," bearing with the commander full responsibility for the unit's conduct in battle. "The Commissar was to warn the Supreme Command and the Government against commanders and political workers [who are] unworthy of the rank . . . to wage a relentless struggle with cowards, the creators of panic and deserters."

nihilated. But the Russians were fighting with a crude heroism that drew the admiration even of Halder in his nightly record, and their "savage determination," of which he would frequently complain, was contributing to a gradual erosion of the Wehrmacht's own strength. The wear and tear on men's nerves, as well as on weapons and machinery, was on an altogether different scale from the "manoeuvres with live ammunition" held in the West the previous summer. "We have had our turret hatches closed for ten days," wrote a sergeant in the 6th Panzer Division, "my tank has been hit seven times, and the inside stinks to heaven." Another account describes the fate of two Russian tanks which had succeeded in breaking out of an encircled pocket but had then broken down. One was destroyed, and two of the crew of the other were shot down as they tried to bale out and escape on foot. This tank lay there, hermetically sealed and apparently lifeless, for ten days, while the encirclement battle pursued its course, but the Germans were concerned to find that

> No supplies reached us without first receiving, during their voyage, well-placed salvoes of artillery. We would change the times of delivery, but this served to no purpose. Often our positions themselves were heavily pounded. In the depth of night patrols came through the forest to throw hand-grenades perfectly directed through our gun-slits. We asked ourselves how the devil this was possible. The mystery was only cleared up by chance. The intact tank had been stripped of everything which was usable: tyres [sic] magnetos, pistons, cables, etc. One day an army cook in search of some equipment happened to force a gap through which to peer. Half overcome by the stench which escaped from it, he saw two squatting skeletons. We were able to get them out. Can anyone imagine the guts of those two men, one of whom, a captain, had lost an eye—cloistered up with a corpse among so much filth! Some provisions, which reached them during the night, helped them to endure and, although they were wounded, to send information to their troops by means of their wireless set.

But if the strain on the Germans was novel and significant, on the Russians it was critical. The High Command reserve had simply melted away, leaving a rump of twenty-one divisions—all hurriedly raised, and with only a sprinkling of trained officers and N.C.O.'s. These units had drawn their equipment from depots in the Moscow region at the beginning of July, and concentrated at Vyazma and Bryansk for extended training. They were short of ammunition and had a reduced complement of artillery—all of it horse-drawn (except for the 160-mm. guns, which were drawn by agricultural tractors pressed into service from the collectives). This lack of mobility was accentuated by the shortage of armour. There were only two tank units,[6] the 104th and 105th, and one motorised division, the 204th. Moreover, of these only the 105th, at Vyazma, had a proportion of T 34's.[7] This lack of mobility is also the most probable explanation for the delay in committing these reserves to the battle, and for their deficiency in armour. The original plan, a set-piece counter-offensive across the Dnieper after the Germans had been halted there, was modified after the crossings at Kopys and Mogilev had fallen. Timoshenko then hoped to strike at the root of the German salient, putting in the reserves at Stary Bykhov and Propoisk and converging southward from Orsha with the very strong forces that were grouped along the upper Dnieper between there and Smolensk.

But by the night of 16th July this plan, too, had been nullified by the speed of the German advance. Hoth's break-through across the Duna put the whole Smolensk army in jeopardy as he wheeled down to converge on the northern prong of Guderian's fork, while in the south the fall of Stary Bykhov and the advance of the Panzer screen to the confluence of the Sozh and the Oster brought the 3rd and 4th Panzer to the brink of the concentration area for the Bryansk army—at Roslavl.

For the next week this race for position was to assume an importance greater even than the course of the fighting in the breach itself. The Russian foot, with some help from an egregious railway system, and under regular attack from the Luftwaffe in daylight, could average little more than twenty miles in a day; the German tanks could do twice this distance, even against opposition. To be able to deploy their full weight the twenty-one fresh Russian infantry divisions were dependent on the railway junctions at Yelnya and Roslavl, which allowed them freedom to switch along the southern and eastern side of the German bulge. But on 18th July the leading motorcyclists of the 10th Panzer, Guderian's central "prong," were in sight of Yelnya and had reached the right bank of the Desna, a few miles to the southeast. All those Russians originally in their path had been sucked into the vortex of the Smolensk *Kesselschlacht* to the north or left standing on the Oster, sixty miles to the west.

Now the threat to the cohesion of Timoshenko's army was very real.

[6] Termed tank "brigades," but in size rather larger than a brigade, rather smaller than a division, strength about a hundred tanks.

[7] The divisions of these two reserve armies were the 64th, 53rd, 19th, 120th, 124th, 129th, 29th, 158th, 128th (Vyazma), and 132nd, 6th, 161st, 160th, 55th, 56th, 1st workers', 148th, 145th, 135th, 140th, 232nd, and 46th at Bryansk.

That night and the following day in a shade temperature of 80 degrees the Russian forced march continued. But on the evening of 19th July only two divisions had got through Spas-Demiansk (nearly thirty miles from Yelnya), and the 10th Panzer had entered the town in force, after a twelve-hour battle against workers' battalions and a few decimated regular units that made up the "garrison." The effect of this spectacular advance was twofold. First, both opposing commanders formed conclusions as to the state of the battle which, although reasonable on the evidence before them, we now can see to have been false; second, their action, following on these conclusions, was to have an effect on the OKW appreciation of the campaign which was disconcerting, retardatory, and ultimately fatal.

Guderian, whose resentment at being subject to Kluge's restraints has already been illustrated, was now employing a variety of subterfuges to give himself the excuse, or the opportunity, to side-step the orders of his army commander. But if he was to use the whole of his Panzer army as a spearhead, then he badly needed some of Kluge's infantry divisions, both to weld the fetters his tanks cast around the Russian infantry and to defend the flanks of the salient. There was one way of getting those infantry divisions—an appeal to Bock over Kluge's head; but to back such an appeal Guderian had to show more than a victory, he had to show an opportunity of boundless scope, one in which days and hours were vital. Now, with the second encirclement at Smolensk and the capture of Yelnya, he believed that he had been granted this. He seemed, in fact, to have brought about the very situation for which Bock and Halder,[8] and Brauchitsch himself, had been hoping.

[8] Halder started recording his irritation with Hitler within a few days of his own disastrously inaccurate forecast of 3rd July. He grumbled, "The Supreme Commander places no trust in the commands in the field, or in the education and training of the senior officers!" His views on the centre-thrust controversy were that Directive 33 would lead to "a bogging down of the current stirring operations," (25th July) and that "The operations ordered by the Führer will lead to a scattering of forces and to stagnation in the decisive direction, Moscow. Bock will be so weak that he will not be able to attack." (28th July) Halder seems to have put his views with force to Brauchitsch, but both he and ObdH showed considerable diffidence in protesting to Hitler. Standing at the apex of the *Heeresleitung,* they seem also to have embodied the opinion of the majority. Only Rundstedt, who favoured restricting offensive operations to the northern front, and Kluge (who strongly disliked Guderian) were against them, although it is probable that the commanders in the Army Groups North and South—Leeb, Reichenau, Kleist, and Hoepner—were not averse to the proposed accession of personal strength which Directive 33 offered.

Brauchitsch has since said that he "postponed clarification" of this question—the course of operations following on the shattering of the Russian border deployment—in the hope of "effecting a timely agreement." But no clarification was required. Hitler may have kept his dreams of Super Cannae to himself and his table talk, but he never made any secret of his aversion to a direct march on Moscow, and continued to proclaim it even after the decision had been codified by Directive 33. "At present," wrote Halder on the 23rd, "the Führer is not at all interested in Moscow, only in Leningrad." And two days later a reference to the importance of Moscow was "summarily rejected" by him. The most that Brauchitsch could manage was permission to delay the execution of Directive 33 because ". . . the fast forces of Army Group Centre to which the Führer has assigned objectives urgently need a 10–14-day rest period to restore their combat effectiveness."

But this course of action had two serious drawbacks. First, even if Bock and Guderian managed to bring about a situation in which they were given a "go-ahead," the ensuing offensive would still have a dangerously improvised character. Second, the Panzer groups *at some time* were going to need the "rest period" which they had spent in furthering their own purposes, and the longer this was put off the more serious loomed the threat of a complete breakdown.

Since 1945 the protagonists of the single narrow thrust to Moscow have enjoyed a free run for their views. It is always easier to extol the virtues of a hypothetical alternative than to justify a cautious and disappointing reality. It is also the case that those who were against the centre thrust are all dead. Keitel, Jodl, Kluge, Hitler himself—they had no time to publish justificatory memoirs. Only Blumentritt, Kluge's Chief of Staff, has survived, and under interrogation he was noncommittal.[9] A dispassionate survey of the facts will show how perilous the Germans' position was. They had no more than ten divisions across the Dnieper, and these had now penetrated a further 120 miles beyond the river. The main crossings, at Orsha and Mogilev, were still in Russian hands and held by garrisons themselves larger than the whole German spearhead; while to the north and south of the salient four Russian armies had the force, if not the ability, to converge and crush its roots. Moreover, all the

[9] "Guderian's plan was a very bold one," he told Liddell Hart. "There were big risks in maintaining reinforcements and supplies. But it might have been the lesser of the two risks. By making the armoured forces turn in each time and forge a ring around the enemy forces they had by-passed, a lot of time was lost."

equipment was in need of repair. Every tank had travelled on its own tracks from the Polish boundary, and the wheeled transport, on which the divisions depended for every gallon of fuel and every round of ammunition—besides such vital functions as the transport of maintenance and bridge-building engineers, and the medical services—was being knocked to pieces on the rough going.

The German intelligence service had been fairly accurate about conditions in Russian-occupied Poland, but badly at fault about those beyond the original Russian frontier. Blumentritt has described how, "We were not prepared for what we found because our maps in no way corresponded to reality. The great motor highway leading from the frontier to Moscow was unfinished—the one road a Westerner would call a 'road.' On our maps all supposed main roads were marked in red, and there seemed to be many, but they often proved to be merely sandy tracks. Nearly all the transport consisted of wheeled vehicles which could not move off the roads, nor move on them if the sand turned into mud. An hour or two of rain reduced the Panzer forces to stagnation. It was an extraordinary sight, with groups of them strung out over a hundred-mile stretch, all stuck—until the sun came out and the ground dried."

Moreover, it was proving difficult enough to keep the small-calibre field guns of the divisional artillery supplied with ammunition—much less to bring up the heavier pieces which would be needed for any prolonged battle of position, and for which the Stukas were already proving themselves to be an inadequate substitute. In truth Kluge's "silken thread" was stretched to breaking, but a more apposite analogy would have been that of a cyclist on a high wire. The 2nd Panzer Army had to keep up its momentum—and its balance—or fall over. And now Timoshenko, with his twenty-one fresh divisions, was making ready to throw a log in its path.

That the Russians regarded their own position as extremely serious can be gauged from their having already committed four fresh tank brigades (all newly equipped with the T 34) into the Smolensk *Kesselschlacht* instead of husbanding them for the counteroffensive with the Vyazma and Bryansk armies. There is some evidence that GOKO was already looking ahead to the prospect of a winter counteroffensive. But any further disintegration of the front had to be prevented so that the reserve armies could be built up undisturbed.

It seems probable that the Russians overestimated the strength of the forces opposing them, and that the events of 18th-19th July had disillusioned them of the prospects of an immediate counteroffensive. But for Timoshenko it was now more important than ever that the divisions trapped at Smolensk be relieved and the connection with Orsha, and the front along the northern Dnieper, restored. The four fresh tank brigades which had been sent from the Reserve Army to the Smolensk area a week before had gone already—with the completeness of ice cubes in a cauldron of molten lead—and there was little more behind them. Moreover, the configuration of the railway system made it impossible to switch the Bryansk army around to face due west.

Accordingly, Timoshenko ordered that both relieving armies from Spas-Demiansk and Roslavl go straight into the offensive as they arrived at the scene of battle, and gave instructions to the forces at Orsha and Mogilev, inside the Smolensk *Kesselschlacht,* to attempt a breakout to the south. These attacks against the enemy right and rear were intended to relieve the pressure in the *Kesselschlacht,* and evidence that they were doing so was not long in coming. On 22nd July, Guderian reported that ". . . all units of the 46th Panzer Corps were at that time engaged and, for the time being at least, committed," and from the 47th Panzer Corps that ". . . nothing further could be expected for the time being." To complete the concentric pressure on the German bulge the Russians trapped at Smolensk were counterattacking fiercely to the south. The town was under continuous shellfire, and the road and railway were useless to the Germans. The 17th Panzer, moved here from Orsha, was heavily engaged, its commander, General Ritter von Weber, being mortally wounded.

The first result of these attacks was the escape to the east of a substantial portion of the Russian divisions in the *Kesselschlacht.* At least five slipped out during the night of the 23rd, and the remains of three more during the 24th. Hoth, who had detected the first creakings there as early as the 19th, had sent two Panzer divisions even farther east, in a long "left hook" that was intended to trap any survivors of the original encirclement. But this column had been delayed by the conditions. "It was appallingly difficult country for tank movement—great virgin forests, widespread swamps, terrible roads and bridges, not strong enough to bear the weight of tanks. The resistance also became stiffer, and the Russians began to cover their front with mine fields. It was easier for them to block the way because there were so few roads."[10] This trap, too, would fail to close unless the 2nd Panzer Army could advance from the south to force it shut. Guderian claims that with this in mind he gave the orders for an advance toward Dorogobuzh as early as 21st July. He had spent that day touring the forward positions in his wireless command truck, and while he was doing so repeated messages came in from army group headquarters that the SS *Das Reich* Division should be directed on Dorogobuzh. However, this unit was carrying the brunt of the counterattack at Yelnya—now coming in at an accelerating pace as the new Vyazma army arrived on the scene, and it would have been impossible to disengage it without giving up possession of the town. Guderian's own preference was for sending *Gross Deutschland,* at that time less heavily engaged, and for taking the 18th Panzer out of the line at Kuzino and using this as his mobile reserve. He sent orders for this move at midday from *Gross Deutschland* headquarters at Vaskovo, where he was taking lunch.

However, on his return to group headquarters at Choclovo, Guderian was confronted with a fresh development. "Von Kluge, in his anxiety for the left flank of my Panzer group along the Dnieper, saw fit to intervene personally and ordered the 18th Panzer Division to stay where it was. As at Bialystok he did not inform me of this direct action on his part. As a result of this the force needed for the attack on Dorogobuzh was unfortunately not available."[11]

For the next two days Guderian devoted all his energies to having his army commander overruled and regaining control of the 18th Panzer. And during this time the Russian attacks against Yelnya and northwest from Roslavl mounted in force. The 10th Panzer returns for 24th July show that over one third of its vehicles had been lost. The divisions

[10] Blumentritt, *Interrogation.* In fact, the "stiffer resistance" to which Blumentritt refers came from two small striking forces, drawn from troops in rear positions and each commanded by a general whose name was, in later years, to acquire a particularly ominous ring for the Germans—Koniev and Rokossovski.

[11] Guderian, *Panzer Leader,* Cassell ed., 179. Guderian had to wait a long time before he got his own back on Kluge. But on 21st July, 1944 (the day following the *attentat*), he was personally promoted to Chief of the General Staff by Hitler. During the afternoon "The conversation turned on individuals. My requests . . . [concerning certain other appointments] were approved. In this connection I remarked that the new Commander-in-Chief West [Kluge] *did not have a lucky touch in commanding large armoured formations,* and I therefore proposed he be given another assignment." My italics.

claimed fifty Russian tanks destroyed on that day, and even after making allowance for exaggeration in the heat of combat it is plain that the Russian effort was increasing daily. Eighteen fresh divisions had already been identified between Cherikov and Yelnya, and Vietinghoff reported that the Russians were "attacking from the south, east and north with very heavy artillery support. On account of a shortage of ammunition, which was now making itself felt for the first time, the corps could only defend its most important positions."

At this point, paradoxically, the extempore character of the Russian counteroffensive began to take effect deeper than the surface fluctuations of the fighting suggested, and deeper, too, than would have resulted had the Russians not, with a mistaken sense of urgency, been so precipitate in its launching. For the first of the 4th Army's regular infantry divisions was now drawn into action across the Dnieper. By the evening of 25th July there were three (the 263rd, 292nd, and 137th), and two days later this had become nine (joined by the 7th, 23rd, 78th, 197th, 15th, and 268th). And these units had come not to relieve the 2nd Panzer Army but to reinforce it.

Thus strengthened, the 2nd Panzer Army would be able to win any action that might follow. But the battle it was going to fight was in essence a local one. It had no place in the strategic development of the campaign as originally envisaged by OKW. In this way the desperate Russian attacks "off the march," costly[12] and ill planned though they were, had an importance that was, in the long run, crucial. For by contesting the initiative in those days of late July, in that decisive central front, the Russians had introduced an element of practical uncertainty —uncertainty of purpose and of opportunity—which was to aggravate the divisions between the counsels at OKH and OKW.

On 27th July a conference of army commanders was called at Bock's headquarters at Novy Borisov. Guderian attended with his Chief of Staff, Freiherr von Liebenstein, and was not long in discovering that ". . . the relationship between the commander of the Fourth Army and myself had . . . become strained to an undesirable degree," owing to "divergences of opinion concerning the situation." Kluge was grum-

[12] General Kachalov, commander of the Vyazma group, had been killed in action on 22nd July.

bling on about his long left flank, offered along the length of the Dnieper, and claimed to regard the "threat" to the Smolensk area as "very serious." Guderian's own opinion (almost as much at fault, as we now know) was that ". . . our most dangerous enemy was now to the south of Roslavl and east of Elnya [Yelnya]." But with greater truth he went on to contend that ". . . as a result of units being retained on the Dnieper west of Smolensk, crises and losses had occurred in the Roslavl area which could have been avoided."

Neither Kluge nor Guderian was given time to air his views at any length. Instead the assembled commanders were subjected to a lecture. This took the form of a memorandum prepared by Brauchitsch and read out by Colonel Horch. (Halder did not attend, owing possibly to his reluctance to endure a cross-examination about a course of action in which he had little faith.) The substance of this was that any immediate advance toward Moscow, or even Bryansk, was ruled out. The first task was the final reduction of the Russian 5th Army, which pivoted on Gomel and still bulged out in a great salient to the south, with its westerly tip inviolate in the marshlands of the Pripet. This meant in effect that the 2nd Panzer Army would be swung around through more than 90 degrees, to advance in a southwesterly—i.e., homeward—direction.

These orders were frustrating for Guderian, who must have felt the centre of gravity of the campaign shifting from beneath his feet. He has since claimed that Hitler "preferred a plan by which small enemy forces were to be encircled and destroyed piecemeal and the enemy thus bled to death," and then, having credited the Führer with this most uncharacteristic timidity, goes on to assert, "All the officers who took part in the conference were of the opinion that this was incorrect." In fact, though, it is highly unlikely that professional opinion was as unanimous as Guderian maintains. The elimination of the major enemy salients was an orthodox prerequisite of any further deep advance, and both the evidence which he himself cites[13] and that of the subsequent "Führer conference" on 4th August would seem to indicate that, if anything, the majority were inclined to orthodoxy.

[13] Aside from his personal recollections of the interchanges at the 27th July conference and two conversations, on 29th and 31st July, with Schmundt and Bredow (see pp. 93-94), this amounts to a document "from a reliable Service source" quoted on p. 183 of Guderian's memoirs. This was an OKH appreciation couched in the most general terms, and of such a vague and optimistic nature that anyone could find in it support for his own particular arguments.

However, in spite of the disappointing directive concerning future operations, Guderian could find some compensation in two important administrative changes which occurred at the same time. The Panzer group was renamed *Armeegruppe Guderian,* and together with this formal recognition of its commander's power and personal magnetism came an official severing of Kluge's apron strings: "The Panzer Group is no longer subordinate to Fourth Army."

With this new freedom Guderian immediately set about implementing his own plans—or rather such a version of them as he considered feasible within the loose and indeterminate framework of the OKH directive. "Regardless of what decisions Hitler might now take," he wrote in his diary (and when we read of Hitler's outbursts against his generals in 1944, it is a valuable corrective to remember what they themselves were writing of Hitler in 1941), ". . . the immediate need was to dispose of the most dangerous enemy threat to . . . the right flank" of the Panzer group, a "need" which entailed the mounting of an offensive on an axis at some 90 degrees to that agreed on at the conference of 27th July.

Guderian's plan for the Roslavl attack was immediately accepted by Bock, and in the light of what followed it is reasonable to assume that there was at least a tacit understanding between them that they would continue to try to bring about a situation in which the advance on Moscow would resume its momentum. On 27th and 28th July, Bock transferred to Guderian's command an additional six infantry divisions, of which two were to be put into the Yelnya salient to allow the extrication of the tank forces there. However, the Panzer divisions thus relieved were not sent down to the Roslavl area, but withdrawn to Prudki-Pochinok for rest and maintenance. This, together with the retention of the Yelnya salient under the command of the *Armeegruppe,* must count as additional evidence that Bock still hoped to resume an easterly advance as soon as the Roslavl battle was ended.

While these preparations were in train, Guderian had a visit from Colonel Rudolf Schmundt, Hitler's chief adjutant. Schmundt held an equivocal position in the German hierarchy, but one for which it is not hard to find duplicates. He was an ardent National Socialist, and devoted to the *persona* of Hitler, but enjoyed at the same time the friendship, respect, and in some measure the confidence of the Army. Conflicting loyalties, a certain stupidity, and perhaps an element of snobbery allowed him to take their confidences at face value, without allowing his suspicions to be aroused, and twice he was to be the unwitting tool of

Tresckow in preparing the ground for an assassination attempt on the Führer.[14]

The ostensible reason for Schmundt's visit was to bring Guderian the oak leaves to the Knight's Cross, but he lost no time in raising the subject of the "intentions" of the *Armeegruppe*. According to Guderian, Schmundt told him that Hitler had not in fact made up his mind to give priority to a single objective, but that he had three "in view." These were Leningrad, whose capture was necessary to free the Baltic and ensure the supply route from Sweden and the provisioning of Army Group North; Moscow, "whose industries were important"; and the Ukraine. Guderian thereupon urged Schmundt "with all the force of which I was capable" to advise Hitler in favour of a direct push to capture Moscow, "the heart of Russia." He also took the opportunity of bypassing the normal ordnance staff channels by making a special plea for new tanks and engine spares.

Two days later another visitor to *Armeegruppe* headquarters brought additional evidence of indecision and, with it, some justification for the disregard which both Bock and Guderian were displaying for the "Gomel Plan." This was Major von Bredow, the OKH liaison officer, who reported, "It is now considered that the original objectives for October 1st, the line Lake Onezhskoe-the Volga, cannot be reached by that date. On the other hand, it is believed with certainty that the line Leningrad-Moscow and to the south can be reached. The OKH and the Chief of the General Staff are engaged in a thankless undertaking since the conduct of all operations is being controlled from the very highest level. *Final decisions have not yet been taken concerning the future course of events.*"

On 1st August, Guderian launched his Roslavl offensive. The remnants of the twenty-one fresh Russian divisions which had charged so valiantly into the unprepared attacks of the previous week were now reduced in effective strength to about twelve, and with only one tank formation, the decimated 105th Brigade, to support them. They were completely exhausted, short of ammunition, and disposed in a fragmentary, unbalanced remnant of an offensive pattern that was fatally vulnerable to a co-ordinated *Panzerblitz*. For his attack Guderian disposed of fourteen divisions, of which four were armoured.[15]

[14] See pp. 98, *et seq,* and 306, *et seq.*
[15] Halder shows the following total strengths in Army Group Centre area on 1st August.

On the extreme right of the attack the 4th Panzer broke into open country within a few hours and ripped across the Russian front, some twenty miles behind but more or less parallel to it, south of the Oster. The division advanced nearly thirty miles, and by the evening of 2nd August was astride the Roslavl-Bryansk road southeast of the town and had entered its outskirts. In the meantime the 29th Motorised, the left claw of the pincer, pushed steadily down the valley of the Desna. The Russian centre, under attack by seven fresh German infantry divisions, crumbled away, the exhausted soldiers being blown back like chaff against the steel cutters of the tanks that lay in their rear. Retreating at a pace no faster than a man's run, they found Roslavl barred to them by the guns of the 4th Panzer and 29th Motorised and swung back north to Yermolino, where they ran head on into the 292nd and 263rd Infantry. Here in this swampy and desolate region a "pocket" formed, with the Germans daily bringing up greater artillery concentration and the Russians struggling ever more feebly to break out. Finally, by 8th August, *Armeegruppe* reported that resistance had been "eliminated."

Some measure of the parlous state of Russian equipment can be gauged from the fact that the Germans only claimed two hundred guns captured (they were in the habit of including the trench mortar in the total gun returns), and Russian strength in the pocket had been over seventy thousand.

In fact, the battle of Roslavl may be said to have ended even earlier than the 8th, with the capture of the town itself on the 3rd, for on that day Guderian had ordered a Panzer striking force of three divisions

---

German: 42 Infantry divisions
9 Armoured "
7 Motorised "
1 Cavalry "

Russian: 26½ Infantry divisions
7 Armoured "
(deficient, as concentrated mainly in Smolensk pocket, from which personnel but not equipment was being evacuated).
1 Cavalry division.

This estimate of the Russian strength is based on units identified either in combat or by intelligence from, e.g., air reconnaissance. It does not take account of the *Stavka* "strategic reserve" in the Urals, of which some fourteen infantry divisions were now on their way to the Oka. But in fact, the Ural area held no "strategic reserve" in the generally accepted sense, being little more than a training and transit area for the troops already being transferred from Asiatic Russia, and there is little doubt that at this stage the Germans enjoyed a quantitive as well as a qualitive superiority in the sector.

away from the main battle to probe south toward Rodnia.[16] Thus it was one of the swiftest as well as one of the most complete of the Wehrmacht's victories in the East. Once again there was an empty gap in the Russian front, once again a great haunch of the Red Army had been sliced off and thrown into the mincing machine. But the question of exploiting this victory remained unsettled. The more so because the battle itself had come out of an earlier version of this same indecision.

Guderian himself had no doubts as to the right course, and insofar as his proximity to ObdH allowed it, there is little doubt that Bock supported him. References to "the high road to Moscow" occur constantly in Guderian's own despatches at this time.[17] It would appear that he regarded the Roslavl operation as a preliminary move to clear his right flank, which carried an incidental advantage: namely, the appearance of being connected with the OKW "Gomel Plan." But in so tenaciously pursuing his own schemes Guderian remained saddled with the same basic problem—the inflexible laws of Liddell Hart's "force to space ratio." Moscow was still 150 miles due east—the butt end of his salient was nearly fifty miles across, its flanks over a hundred deep. By his brilliant *coup de main* at Yelnya the German commander had precipitated Timoshenko's counterattack and thereby drawn the infantry of Army Group Centre across the Dnieper and into the battle, but at the same time the Panzers had remained in action, and so the very factors which had kept the centre of gravity in his hands for those three extra weeks now made it inevitable that there would be a pause. Before taking another "leap" the Panzer divisions had to be overhauled, rested, and their stocks of fuel and ammunition brought up to establishment. The rest period of whose urgency Brauchitsch had been persuading Hitler was long overdue. Guderian had borrowed time, and the debt had to be repaid.

Moreover, the strategic redeployment of the Wehrmacht was already taking effect in other sectors. The dutiful Hoth (whose 3rd Panzer Army had not, be it noted, been renamed *Armeegruppe Hoth*) was already swinging around to face the Valdai Hills and take up a new role as Leeb's "right hook" in a renewed assault on Leningrad.[18] It would be

[16] The 3rd Panzer, 10th Motorised, and 78th Division.

[17] Instructions to the 137th Infantry Division, 0230 hours, 2nd August, 1941, "to continue advance during the night so as to reach the high road to Moscow as soon as possible." Guderian, 3rd August, to General Geyr (commanding the 9th Corps), "I pointed out to him the great importance of holding the Moscow Highway." Guderian, 188.

[18] Halder noted Hoth's transfer to the control of Strauss (9th Army) on 27th July.

no longer possible for Guderian to drive on along, leading, as it were by the hand, the rest of Army Group Centre. The final thrust—if such it was to be, must have the full direction and support of OKW. All this Guderian pleaded to Bock, and Bock put, with some qualifications, to Halder, and Halder expressed with vigour and lucidity to Brauchitsch, but with greater discretion (it may be thought) to Hitler.

Then, after further delay, another conference was called at Novy Borisov. For the first time since the opening of the campaign the army commanders were to report to the Führer, who was to attend in person.

five | THE LÖTZEN DECISION

There were others besides Guderian who were waiting with heightening anxiety for the Führer's arrival at Novy Borisov. Already during that first victorious and intoxicating summer of 1941 the headquarters of Army Group Centre had become "the immediate centre of active operational conspiracy—a nest of intrigue and treason." Superimposed on the strictly professional heart-searching and disputation which were part of the centre-thrust controversy and which affected every member of the staff, there was an intense—even grotesque—activity by a group of officers with political and constitutional objectives.

Bock's G.S.O. I, Major General Henning von Tresckow, and his A.D.C., Fabian von Schlabrendorff, had conceived the idea of literally taking Hitler for a ride. Once inside the net of Army Group Centre's security system, the Führer's car was to be diverted and the occupants detained. Hitler would be subjected to an extempore trial, and an order of deposition, or even of execution (although the concept had not yet been mentioned specifically), would be pronounced. How matters were to develop thereafter is not clear. Nor, indeed, is it likely that the conspirators had planned any further than the removal of the Führer's *persona* and the relief, thereby, of the obligations inherent in the loyalty oath. Certainly this particular attempt, and its successors which emanated from that same headquarters, enjoyed none of the carefully planned supporting processes which were triggered off by the *attentat* of 20th July, 1944.

It may well be asked how the idea, much less the execution, of a *Putsch* could enjoy serious consideration at such a time, when German arms seemed everywhere to be invincible. The answer, surely, is that the plotters were the personification of all that was best of their country's qualities, a rational intellectualism allied to a selfless bravery. Their intention was for the creation of a "decent Germany," that national entity whose elusive quality has for so long been the despair of European politicians, and being Germans, they naturally held military strength and constitutional order to be corollaries of "decency." At the front, five hundred miles inside Russia, they were in a better position to appreciate the realities of the campaign. They could see that the irresistible force of the Wehrmacht had at last collided with an immovable object, and ". . . when our chances of victory are obviously gone, or only very slim, there will be nothing more to be done."

The officers privy to this conspiracy were so numerous and occupied positions so close to the army group commander[1] that it is impossible to believe Bock was unaware of what was going on. Yet the notion of a plot to kidnap the head of state while he was visiting his armies on the field of battle, being if not encouraged, then tolerated by the Commander in Chief, is so alien to the practices of a Western democracy that we must find it hard to believe. To comprehend how such a state of affairs could exist one must recall the atmosphere of nightmare and fantasy that pervaded the Third Reich. Private armies, personal hatreds, and rivalries; fresh and vivid memories of denunciation and betrayal; of blackmail and violence; of failure, humiliation, and imprisonment. What we understand by a tradition of moderation was, where public affairs impinged on individual conduct, nothing more than a *laissez-passer* to the concentration camp.

Practically every general in the Army was approached at some time or other by the plotters. Not one of them lifted the telephone to Himmler. Even Brauchitsch[2] had gone no further than to tell General Thomas, ". . . if you persist in seeing me [on this subject] I shall have to place you under arrest."

The gulf between Army and SS made a denunciation unthinkable—besides, there was always the consideration that the messenger with ill

[1] Actively involved were Bock's two personal A.D.C.'s, Graf Hans von Hardenberg and Graf Heinrich von Lehndorff. Others prominent in the conspiracy were Colonel Freiherr von Gersdorff, Colonel Schulze-Brüttger, Lieutenant Colonel Alexander von Voss, Major Ulrich von Oertzen, Captain Eggert and Lieutenant Hans Albrecht von Boddien.

[2] See p. 22.

tidings sometimes loses his own head. Nonetheless, as the generals listened to their juniors and heard the excited pleas of those civilian emissaries who journeyed to their headquarters on passes provided by cells in the Abwehr or the Foreign Office, other thoughts must have been passing through their minds. If there was a plot in the wind, a possible change of regime, was it not their duty to the Reich to remain in office? Here, after so long, was a whiff of Seeckt's time; the possibility of a Reichswehr once again returning to its rightful position as arbiter of the country's political destiny. This concept was to provide the generals with a psychological element of excuse which they were soon to find very welcome. A hint of uncertainty, and the mystique of duty gathered strength. The Army was above politics, certainly, but reserved to itself the right of intervention when events demanded it. Under these circumstances the generals should play for time, even if this meant accepting orders which in the normal course of events they would have countered by resignation. This kind of muddled thinking, aggravated by a nagging fear of being implicated too early, was to plague the peace of mind of the "Eastern marshals" and erode their powers of command and decision throughout the campaign.

Bock himself was one of Seeckt's protégés. He was no innocent in the world of clandestine *raisons d'état,* and twenty years before had been one of the original organisers of the "Black Reichswehr." But now his ambition, compounded by vanity and egotism, caused him to reject the notion of political intrigue. Sheer brilliance at arms would, he believed, grant him the powers which might or might not follow by the less certain route which Tresckow and Schlabrendorff proposed. For he, Bock, was to be the captor of Moscow. Then, verily, he would be pre-eminent among the marshals; the slayer of Bolshevism; the first soldier of the Reich, cast (as he believed himself) in the Hindenburg image.

This is not to deny that had the plotters succeeded in their attempt to kidnap Hitler, Bock would have followed up by arresting all the SS in his area and proclaiming a "military government." But in fact, he rated their chances as "outside probability." [3] Bock's paramount concern at this time was to operate within the existing framework of command, to retain the mass of German striking power under his own hand, and to gain permission to continue his march directly on Moscow.

[3] An example of Bock's excessive "realism" in political matters. In 1943 when formally approached by Thomas for his support, he declared that this would be forthcoming only if Himmler were also a party to the plot.

Without the help of their chief the amateur plotters at his headquarters never got started. Three times the Führer's impending arrival was announced from Rastenburg. Three times it was cancelled. On 3rd August a convoy of SS arrived, bringing their own staff cars. And when Hitler's aircraft finally landed they escorted him over the three miles from the airstrip to headquarters. Until the convoy actually drew up outside Bock's building, it was not known in which car the Führer was travelling. For the duration of his visit none of the young officers at Army Group Centre got close enough to Hitler to be able to point their revolvers with any accuracy—far less to set in motion the elaborate plans for "detention" and "trial" which they had been harbouring for so long.

Nor did events conform any more closely to Bock's own ideas. Instead of being confronted by a resolute and unanimous body of professional opinion, Hitler interviewed his commanders singly and alone, so that none could be sure what the others had said, what they had been offered, or what they had given away. The Führer installed himself in Bock's map room with Schmundt and two SS adjutants. He then sent for Heusinger,[4] Bock, Guderian, and Hoth, in that order, and asked them their "opinion." The result of these tactics was that Hitler at the outset established his customary moral ascendancy. The army group commander and his two Panzer lieutenants were, indeed, united in their recommendation to advance directly on Moscow, but under Hitler's cross-examination certain inconsistencies cropped up. Bock said that he was ready to advance immediately; Hoth said the earliest date by which his Panzer group could start was 20th August; Guderian claimed that he would be ready on the 15th. Bock, in his anxiety to avoid any administrative excuse for a halt, said that the forces of Army Group Centre were adequate for the task. Guderian, partly from a genuine concern and partly no doubt in an effort to cripple the scheme whereby the Panzers were to be diverted on a great southward encirclement sweep, "stressed the fact that our tank engines had become very worn as a result of the appalling dust," and asked for replacements.

After hearing them out Hitler had his commanders called into the map room together and delivered an address. He declared that Leningrad was the primary objective at that time. After this had been achieved the choice would lie between Moscow and the Ukraine, and his inclination was to favour the latter on strategic and economic grounds. A long

[4] Colonel Heusinger, Chief of the Operations Department at OKH, who was representing Halder.

and well-reasoned exposé of these followed. In essence the Führer's attitude was founded on defensive considerations: the capture of Leningrad would shut the Russians off from the Baltic and secure the iron-ore route to Sweden; the capture of the Ukraine would provide the raw materials and the agricultural produce Germany would need for a long war; the occupation of the Crimea would neutralise the threat from the Russian Air Force against the Ploesti oil fields. There was also the consideration that ". . . Army Group South seems to be laying the foundation for a victory in that area," a reason which can hardly have been welcomed by Bock, the less so because of his own (unfulfilled) expectations of lavish praise[5] for the achievement of Army Group Centre.

Hitler would never admit as much to a professional soldier, and unfortunately, we have no fragment surviving of his conversation among his intimates at this time, but it is most likely that he was already seriously taken aback by the strength of the Russian resistance. The ghost of Napoleon stood at his elbow, as it did at some moment for every German officer in the East, and he was determined to resist the temptation of a march on Moscow until he had laid (as he believed) a secure strategic foundation.

The only clue to this attitude—but a significant one—was dropped at this same conference. Guderian was asking for completely new tanks, and not simply replacement parts, to be sent up to the front. Hitler refused on the grounds that the new tanks were being used to equip fresh divisions in Germany, and said, "If I had known that the figures for the Russian tank strength which you gave in your book were the true ones, I would not—I believe—ever have started this war."

Now followed an agonising period, two and a half weeks long, of interregnum. Army Group Centre, its leadership hamstrung, floundered on the brink of a tremendous opportunity, while opposite them, for nineteen perfect campaigning days, the Russians worked unmolested to rebuild their shattered front.

For almost seventy miles along the Desna, between the southern corner of the Yelnya salient and the Bryansk bend, Timoshenko's defence can

[5] This probably put Bock in a very bad temper, but there is no evidence to corroborate Schlabrendorff (*Offiziere gegen Hitler*) in his contention that Hitler upbraided and insulted Bock on this occasion. There is no reason he should have done so; indeed, it would have been inconsistent with the whole tone of the conference (at which Schlabrendorff was not present).

hardly be said to have existed at all. A few units, none above brigade strength, which had slipped out of the Roslavl pocket, were drifting back across to the eastern bank of the river, and at the bridges there were some "workers' battalions," raised locally and without any heavy equipment. There was practically no artillery, and not a single tank in working order between Spas-Demiansk and Bryansk. The whole region, nominally the responsibility of the 43rd Army (which had lost the majority of its headquarters staff at Roslavl), was in anarchy, with local party officials taking the law into their own hands, attempting military as well as civil administration, compensating for their clumsiness in this unfamiliar element with a Draconian severity toward "deserters" and "malcontents." Higher direction there was none, save the perennial bleat from the *Stavka* that the enemy "must be counterattacked wherever encountered." The plight of the Russians was worsened by a total lack of mobility. Even had men and guns been available, there remained no means of moving them—except the forced march. Every vehicle—civilian, agricultural, and military—had already been spent in the death rides of July, when the Red Army had loaded up and driven head on into the advancing enemy.

This, indeed, was the moment for Super Cannae. A wedge of Panzers, driven hard into this gap, might yet have levered the whole creaking gate off its hinge. But the state of the German armour, worn down in the battles of July, made this a dangerous concept; and now Hitler's directive made it virtually impossible from an administrative point of view. In spite of this OKH and the staff, instead of formulating a new policy and throwing themselves vigorously behind the plan for the capture of Leningrad, lingered on with the cherished notion of a direct march on Moscow. They used their waning powers to thwart the expressed "general intention" of the Chief of OKW (Hitler) and to divert and confuse the issue on a tactical level. Brauchitsch managed to extract from Hitler the concession that a defensive posture by Army Group Centre was "only temporary" and the important (because vague) permission to "make attacks . . . with limited objectives which *might improve its positions for subsequent operations.*"

After the conference at which Hitler agreed to this, Halder wrote:

> In themselves these decisions represent a cheering progress, but they still fall short of the clear-cut operational objectives essential to a sound basis for future developments. With these tactical reasonings as a start-

> ing point, the Führer was deftly steered towards our view-point on operational objectives. For the moment this is a relief. *A radical improvement is not to be hoped for unless operations become so fluid that his tactical thinking cannot keep pace with developments.*

There can have been few more concise expressions of contempt for Hitler's military ability, or more clearly indicated intention to force his hand by a delicate sabotage of the orders of the Supreme Command. What in fact Halder was "hoping for" was exactly the situation which Guderian had, half deliberately, been trying to precipitate by his deep advance to Yelnya and his operations at Roslavl. If the Supreme Command would not, or could not, give the orders, then it must be drawn into committing itself by the headlong rush of "tactical developments."

The generals' attitude to their Führer at this time is the sounding board for Hitler's own ruthless contempt, which was to echo about their ears, magnified tenfold, after the winter debacle. "While flying back [from the conference] *I decided in any case to make the necessary preparations for an attack towards Moscow.*" Guderian feels sure enough of his case to have committed this calculated insubordination to paper ten years later, and from what followed there is no doubt whatever that his army group commander was in complete agreement with him. Halder's diary, his guarded references to conversations with "ObdH"—as he would refer to Brauchitsch—and everything that has been written by those other commanders and staff officers like Blumentritt who survived the war point to a general conspiracy to thwart Hitler's intention, if not by direct disobedience, then by circumvention of unwelcome instructions.

This conspiracy, fumbling and barely articulate though it may have been, was nonetheless obstructive enough to result in a completely disastrous effect on the German campaign. For in dealing with the various hypotheses which offer themselves, we can now see that the Germans did the one thing which was fatal—namely, nothing. The probable outcome of a direct thrust on Moscow has already been discussed. It remains to be said that had the generals accepted Hitler's orders and put themselves wholeheartedly into the preparation of an immediate drive on Leningrad, that city would probably have fallen by the end of August. This would have left time for an autumn campaign against Budënny and consolidation on the Donetz before winter. It is hard to avoid the view that the Russian capital, isolated on both flanks, would then have fallen to the Germans in the first rush of their spring campaign. But instead, as will be seen, Army Group Centre procrastinated. The Panzers were not taken out of the line, yet they did not move forward; some divisions were sent north to Leeb, others were released with great reluctance by Bock for a southward move; but neither was done in the strength required. And while these hesitations and delays dragged on, the days—the priceless summer meridian of dry going and soft temperatures—slipped past.

The Russians were well aware of their vulnerability in the Roslavl gap, but petrified by their lack of mobility, they had neither the men with which to plug it nor the machines to move them. In the first days of August a substantial body of the force encircled at Smolensk managed to cut its way through the German ring at Yermolino, and these divisions were all packed into the front around the Yelnya salient. Both the 66th and 67th Panzer corps were still tied down here, and although three fresh infantry divisions had been put into the line, Lemelsen had managed to extricate only two armoured formations (the 29th Motorised and 18th Panzer) for "rest." In this way, by continually reinforcing their position at Yelnya and keeping up a succession of local counterattacks, the Russians managed to hold firm at the northern end of the breach. To the south the 5th Army with its satellites was urgently piling men up along the Sozh, oblivious to the yawning cavity on its right wing and keeping up pressure on the newly arrived infantry of the German 2nd Army.

The effect of the Russians' keeping their nerve (and once again it is impossible to say whether this arose out of calculated generalship or simply in conformity with the general orders that no more ground was to be yielded) was that the width of the Roslavl gap remained constant—at about fifty miles. Its exploitation required as a preliminary the breaking down of one or both of the buttresses, at Yelnya and along the Sozh, which constricted it.

For an operation on this scale Bock and Guderian no longer disposed of the force—much less the authority—required. Yet even after spending two days in the Yelnya area, where he saw at first hand how his men were being compelled to give ground by the accumulating Russian strength, Guderian

> told my staff to prepare for an advance on Moscow as follows: the Panzer Corps were to be committed on the right, along the Moscow highway

> [i.e., directly into the Roslavl gap] while the infantry corps were to be brought forward in the centre and on the left wing. By attacking the relatively weak Russian front on either side of the Moscow highway, and then rolling up that front from Spas-Demiansk to Viasma, I hoped to facilitate Hoth's advance and bring our forward movement into the open.

In the meantime a clumsy attempt by the German 34th Infantry Division to force the Sozh below Krichev had been repulsed. The strength of the Russian reaction had caused considerable alarm at 2nd Army headquarters, and on 6th August, Guderian received a "request" from OKH that at least two Panzer divisions be detached from his command and placed under the 2nd Army for an attack on Rogachev. After a telephone conversation with Bock ("Both staffs regarded the resumption of the advance on Moscow as the primary objective"), Guderian refused the "request" on the grounds that the long approach and return march, of 250 miles, would be too great a strain on tanks which had had so little maintenance.

For the next few days Bock, in spite of his private support of Guderian's scheme, continued to pass on messages from OKH that the Panzer group should "at least send a few tanks off to Propoisk" (in the 34th Division area). Guderian himself admits that

> before the attack on Moscow could be launched or any other major operations undertaken, one condition must be fulfilled: Our deep right flank in the Krichev area must be secured.

But he was loth to part with any units from his own diminishing strength to allow another commander to attempt this. Finally, under continuing pressure from OKH, he decided to clear the flank himself, and sent the 24th Panzer Corps due south toward Krichev and the left flank of the 2nd Army.

Not unnaturally, OKH was becoming increasingly restless at the continuing disobedience of its forward commanders. On 11th August, Army Group Centre was formally notified that the plan of Colonel General Guderian (for an advance along the Moscow highway) was rejected as being "completely unsatisfactory."

Bock thought it prudent not to demur, and "agreed to the cancellation" (of the plan). Guderian, however, was furious, and retorted with the threat to abandon the Yelnya salient, "which now has no purpose and is a continual source of casualties." This was unacceptable to OKH, and even Bock had the temerity to assure Guderian, "It is far more disadvantageous to the enemy than it is to us."

Matters remained in a state of deadlock for some days, during which OKH, Guderian claims, "deluged us with a positive stream of varying instructions—which made it quite impossible for subordinate headquarters to work out any consistent plan at all." During this time the concentration of the Panzer army was being hourly diluted as the mass of Freiherr von Geyr's 24th Panzer Corps moved south. Soon there were barely as many German troops in the Roslavl gap as there were Russians defending it. The 29th Motorised was ordered from its short-lived "rest" to this area, and both *Gross Deutschland* and *Das Reich* were moved straight there from the north of the Yelnya salient as soon as they were relieved by regular infantry. Already "tactical developments" were relegating the Roslavl gap from being the point of opportunity to the status of a "quiet area" where tired formations could recuperate.

While these manoeuvres and discussions proceeded, the matter was being pursued at the highest level by Halder, who for the last week had been compelled to listen thrice daily to Bock's grumblings over the telephone. The Chief of the General Staff had managed to persuade ObdH, but Brauchitsch himself was too nervous about Hitler to put the matter to him directly until he could be sure of support from someone else at OKW. Together the two men approached Jodl and put their case. After "a lengthy discussion" the Chief of the OKW staff declared himself to be strongly impressed and promised to use his influence with Hitler. Accordingly, Brauchitsch drafted a long memorandum on the desirability of an immediate offensive toward Moscow and formally "submitted" it to OKW through Jodl.

Ten days of inactivity had now passed since the conference at Novy Borisov. The route march of the 24th Panzer Corps had ended in victory at Krichev, where the three Russian divisions guarding the Sozh were broken up, and sixteen thousand prisoners were taken. But the result was that Guderian fell victim to the same sort of situation he had exploited to his own ends less than a month earlier. A local victory, achieved in a vacuum of indecision, threatened to draw along behind it a train of exploitation which this time would be fatally distracting to the Panzer group.

On 15th August, Army Group Centre asked Guderian to release one Panzer division to the 2nd Army, "to stiffen the drive on Gomel." Guderian telephoned Bock, and all the old arguments were run over once again—the more fluently, we may think, as both men were basically in agreement. The 24th Panzer Corps had not passed a single day out of action since the start of the campaign, it was urgently in need of a rest period for tank maintenance, and so on. But, Guderian ended, "as a single division cannot operate through the enemy's lines," then if the operation *had* to be undertaken it should, once again, be done with the whole Panzer corps. It seems likely that Guderian calculated that this argument would deter Bock, for any continued southerly move by the 24th Panzer Corps would still further upset the balance of the tank spearhead and must delay—perhaps by as much as ten days—its concentration for a renewed offensive.

Sure enough, the army group commander assented and gave up his plan, but hardly had Guderian put the telephone down when another order came through, this time directly from OKH, to the effect that one Panzer division was to be sent to Gomel forthwith. With a disregard for OKH instructions which by now must have seemed habitual, Guderian ordered Freiherr von Geyr to move off with the whole corps, thus removing three of his best divisions (the 3rd and 4th Panzer and 10th Motorised) and virtually dividing the Panzer group into two halves.

"I shall pass over," Guderian has written, ". . . the fluctuations of opinion at Army Group Centre as expressed during the telephone conversations of the next few days," and the historian is left without details of this stage in the dispute. But the actual movement orders of the various formations in Army Group Centre do exist, and they reveal the extraordinary degree of atrophy and indecision which had taken hold of the German Army during that critical period. Although the 24th Panzer Corps made good progress on the left flank of its attack front (i.e., in the area where the Russian line along the Sozh ended *in vacuo* at the southern end of the Roslavl gap), it was held up by fierce resistance on its left. But here the 2nd Army, instead of attacking in support of the Panzers, was actually in the process of extricating itself altogether from its positions. By 18th August the Panzer columns moving south were suffering from a disturbance of their rear communications by the slow-moving infantry of the 2nd Army, which was marching diagonally across in a northeasterly direction. When Guderian insisted that the army group should reverse these orders and compel the 2nd Army to join in the attack, the staff at Novy Borisov agreed to do so. But the following day, with the 24th Panzer Corps' communications still suffering, Guderian contacted 2nd Army headquarters directly and was told that nothing could be done as ". . . it was the Army Group itself which had ordered the move of formations to the north-east."

On 20th August, Bock surfaced again, and told Guderian that ". . . attempts to press on southwards [with the 24th Panzer Corps] were to be discontinued. He wanted the whole of the Panzer Group withdrawn to rest in the Roslavl area, so that he would have fresh troops at his disposal when the advance on Moscow for which he was hoping was resumed." The Alice in Wonderland atmosphere was strengthened by Bock's contention that ". . . he had no idea why Second Army had not made better time in its advance; *he had been continually urging it to hurry*."

While the commanders in Army Group Centre had been performing their ponderous minuet and variations, two developments had occurred which finally extinguished any hope of an immediate march on Moscow. First, the offensive against Leningrad, which had gradually got under way in the days following the first Novy Borisov conference, began to run into trouble. On 15th August the Russians had put in a series of counterattacks against Leeb's right flank at Staraya Russa and the Germans had been compelled to give ground.[6] The immediate result was that Hoth had to send another Panzer corps[7] north to reinforce Leeb, and this reduced the strength of Army Group Centre by a further three divisions. Hoth himself, less volatile and more obedient than Guderian, had his Panzer army in a better state of concentration and readiness. His presence and his tanks were absolutely essential to any large-scale operation by Army Group Centre, yet already they had been almost halved by OKW movement orders directing them northward. Bock's strength was running out with the dispersal of his Panzer groups, and although he could still talk of the "resumption" of the advance on Moscow, the reality behind this idea was weakening daily. (Within ten days he was to be complaining to Halder that ". . . he could foresee the end of his Army Group's ability to hold out.")

But although the physical means with which to achieve its purpose were slipping from its grasp, OKH continued to press its plan. On 18th August, Brauchitsch finally plucked up courage to submit his "appreciation" to Hitler. Jodl, as usual, had ratted and withdrawn his support, and Hitler rejected the memorandum in its entirety. In his own hand the Führer wrote out a long reply, which contained both tactical criticism

[6] It was that evening that Halder made his celebrated diary entry: "We reckoned with 200 Russian divisions. Now we have already counted 360. Our front on this broad expanse is too thin, it has no depth. In consequence the enemy attacks often meet with success."

[7] The 39th, consisting of the 12th Panzer and 18th and 20th Motorised Divisions.

and strategic direction. The armoured columns of the centre, Hitler alleged, had not ever succeeded in surrounding the enemy sufficiently. They had been allowed to push too far ahead of the infantry, and *been permitted to operate with too great a degree of independence.* The plans for the future which Hitler set out and codified as Directive No. 34[8] showed that the preparations for an assault on Leningrad had been allowed to pass into limbo and that the maximum effort was to be confined to the south.

With the publication of this directive the centre-thrust plan was dead, officially. But for another week the officers of Army Group Centre, abetted by Halder, continued to nurture the scheme and to obstruct its alternatives. On 22nd August, Guderian was once again asked to "move armoured units capable of fighting" to the Klintsy-Pochep area, on the left flank of the 2nd Army. And for the first time the concept of cooperation with Army Group South was mentioned. Once again Guderian retorted that ". . . the employment of the Panzer Group in this direction is a basically false idea," and that splitting it up would be "criminal folly."

The following day Halder travelled to Bock's headquarters, and he, Bock, and Guderian discussed at length "what could be done to alter Hitler's 'unalterable resolve.' " Halder felt that one of them should go to Hitler and, "speaking as a general from the front, lay the relevant facts immediately before him and thus support a last attempt on the part of OKH to make him agree to their plan." After "a great deal of chopping and changing," during which, presumably, he was weighing the chance of changing Hitler's mind against that of falling out of favour by opposing him, Bock suggested that Guderian should go. Halder and the Panzer group commander left at once, and took off in a Ju 52 for Lötzen that same afternoon.

They arrived at the airfield just as it was getting dark and reported to Brauchitsch. ObdH, as his subsequent actions show, was in a nervous condition, and this seems to have worsened during the absence of his punctilious *éminence grise.* Guderian has related that Brauchitsch's first words as they entered the room were, "I forbid you to mention the question of Moscow to the Führer. The operation to the south has been ordered. The problem now is simply how it is to be carried out. Discussion is pointless." Guderian then said that in that case he would fly

[8] Reproduced as Appendix.

back to the Panzer group immediately since any conversation that he might have with Hitler would simply be a waste of time. No, no, Brauchitsch replied; he must see Hitler, and he must report on the state of the Panzer group—*"but without mentioning Moscow!"*

The interview took place before a large audience. Neither Brauchitsch nor Halder was present, although there were several officers of OKW, including Keitel and Jodl. Hitler listened in silence to Guderian's account of the state of the Panzer group, and then asked him:

"In view of their past performance, do you consider that your troops are capable of making another great effort?"

Guderian replied:

"If the troops are given a major objective, the importance of which is apparent to every soldier, yes."

"You mean, of course, Moscow?"

"Yes. Since you have broached the subject, let me give you the reasons for my opinions."

Guderian then delivered his argument, to which Hitler again listened in silence—an unexpectedly meek attitude, which may have encouraged the Colonel General to indulge in certain exaggerations (as, for example, his assertion that ". . . the troops of Army Group Centre are poised for an advance on Moscow"). When Guderian had finished, Hitler replied, and went into great detail over the economic background to his decision. "My generals know nothing about the economic aspects of War," he said, and it is certainly true that they seldom gave any sign of it.

It appears that Hitler had already held forth to his audience many times on this very subject. Guderian noted, ". . . I saw here for the first time a spectacle with which I was later to become very familiar: All those present nodded in agreement with every sentence that Hitler uttered, while I was left alone with my point of view."

All the same, one must ask, did not Hitler finish by persuading even Guderian? Guderian himself claims, ". . . I avoided all further argument [as] I did not then think it would be right to make an angry scene with the head of the German State when he was surrounded by his advisers." This may well have been entirely true, but it is followed by the admission (or excuse), "Since the decision to attack the Ukraine had now been confirmed, I did my best at least to ensure that it be carried out as well as possible. I therefore begged Hitler not to split my Panzer

Group, as was intended, but to commit the whole group to the operation."

An order confirming this was immediately promulgated, and reached Army Group Centre the next day. To what extent this decision—to make the whole Panzer group march south instead of husbanding some of the divisions at rest in the centre—was responsible for the failure of the attack on Moscow when finally it was launched is hard to determine. There is no doubt that it played its part, supplementing the fatal delays which preceded it. Halder's view was that it was a bribe by Hitler to induce Guderian to acquiesce in the plan and when the Panzer group commander told him the news ". . . he suffered a complete nervous collapse, which led him to make accusations and imputations which were utterly unjustified." The two officers "parted without having reached agreement," as Guderian's understatement puts it, and their relations were never the same for the remainder of Halder's term as Chief of the General Staff.

After Guderian had returned to his headquarters, Halder telephoned Bock and told him that Guderian had let them all down, then got hold of Brauchitsch and urged that since they could not take responsibility for the course of operations as prescribed by Hitler they should both resign.

The unfortunate ObdH, fresh from an interview at which he had been charged with "allowing the Army Groups too much latitude in advancing their particular interests," was against this. He attempted to calm his Chief of Staff, telling him, "Since a relief of office would not in fact eventuate, the situation would remain unchanged." For two days Halder hesitated, then Hitler made it up with Brauchitsch, declaring, "He did not mean it that way," and it was too late for the Chief of Staff to resign alone.

By now, too, the whole pattern of the front was changing. The squeal and clatter of tank tracks; the twenty-mile dust clouds; the song of men on the march; as the new offensive gathered momentum, these things erased the memory of the preceding weeks. OKH was busy with tactical planning. Guderian was driving south in his armoured radio truck, directing a whole new series of encirclement battles. Hoth was away with Leeb. Only Bock was left, with his infantry, to brood on what might have been. As for the personal side of the affair, it seemed nothing more than a passing tiff, a petty squabble between the generals, jointly and severally, and Hitler.

But in fact it had been nothing of the kind. This was a catastrophic dispute whose consequences, both upon the course of the war and upon Hitler's own relations with his commanders, can hardly be measured.

six | LENINGRAD: HYPOTHESIS AND REALITY

As the weight of the German tank forces wheeled south, drawn by the gravitational pull of Budënny's mass, the course of events on the northern wing ran smoothly along the lines laid down in Hitler's directive.

Leeb's armies had succeeded in smashing the two Russian "fronts" opposite them, and had punished the Soviet commanders so severely that the *Stavka* had, in effect, been compelled to constitute the Leningrad theatre as a separate command, operating in independence, if not isolation, from the rest of the battle front. After the 41st Panzer Corps had penetrated the extempore screen thrown up along the course of the Stalin Line, there remained only one position between the old Estonian frontier and the outskirts of Leningrad itself where a stand was possible. This was the Luga River, running southeast from Narva, toward Novgorod, at the tip of Lake Ilmen.

The Luga position was divided into three sectors,[1] but each was little more than a corps in strength—in fresh troops, at least—plus what could be made of the stragglers and beaten-up formations that had been car-

[1] Excluding irregular and militia formations, the strengths were

Kingisepp sector:	Major General V. V. Semashko, three rifle divisions and coast defence artillery
Luga sector:	Major General A. N. Astanin, three rifle divisions
Eastern sector:	Major General F. I. Starikov, one militia "division," one mountain brigade with artillery.

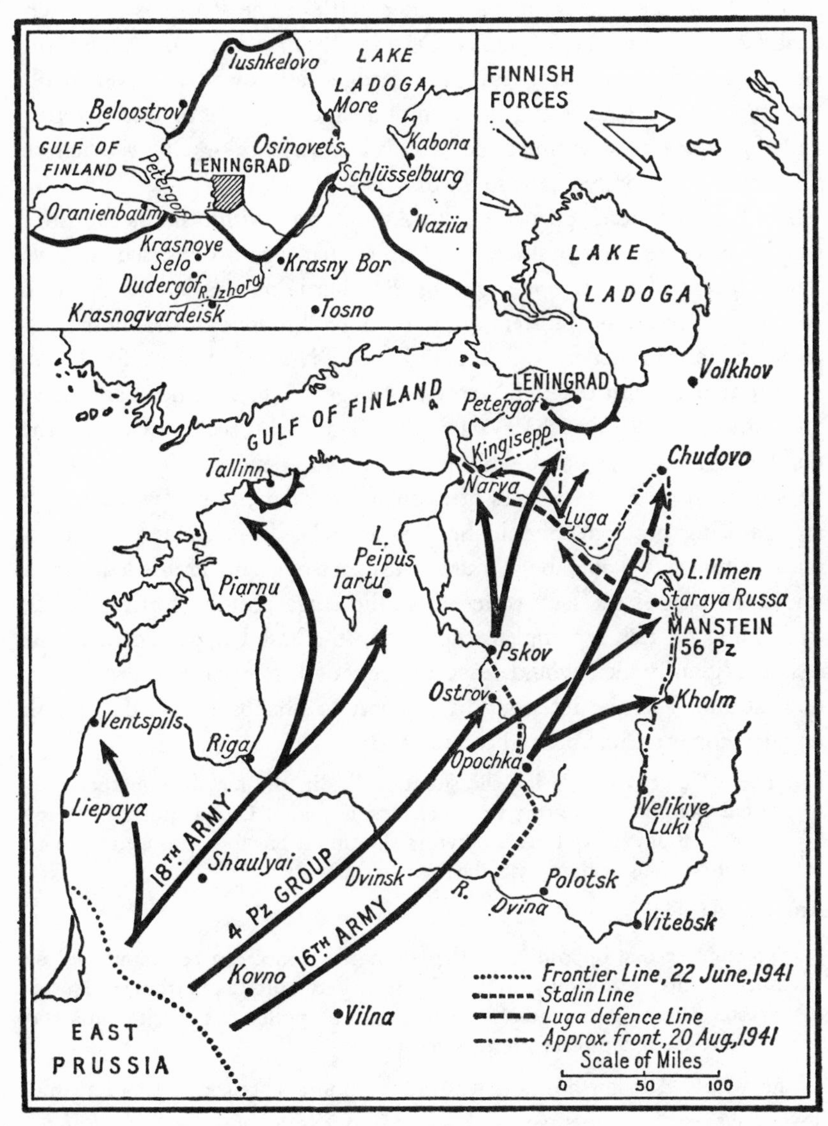
Iushkelovo
LAKE LADOGA
More
Beloostrov
GULF OF FINLAND
Osinovets
Kabona
LENINGRAD
Petergof
Schlüsselburg
Oranienbaum
Naziia
Krasnoye Selo
Krasny Bor
Dudergof
R. Izhora
Krasnogvardeisk
Tosno
FINNISH FORCES
LAKE LADOGA
LENINGRAD
Volkhov
GULF OF FINLAND
Petergof
Kingisepp
Chudovo
Tallinn
Narva
Luga
L. Peipus
Tartu
L. Ilmen
Piarnu
Staraya Russa
MANSTEIN
56 Pz
Pskov
Ostrov
Kholm
Ventspils
Riga
Opochka
Liepaya
18TH ARMY
Velikiye Luki
Shaulyai
4 Pz GROUP
Dvinsk
R.
Polotsk
16TH ARMY
Dvina
Vitebsk
Kovno
Vilna
EAST PRUSSIA
Frontier Line, 22 June, 1941
Stalin Line
Luga defence Line
Approx. front, 20 Aug, 1941
Scale of Miles
0
50
100

Leningrad

ried back on the tide from the frontier battles. The Russians had practically no artillery, and no tanks. During the first week of August the Germans had been filling up their bridgeheads over the river while Popov, short of equipment, ammunition, and—still more disconcerting and, for a Russian commander, a novel experience—*men,* watched in impotence and reported daily to the *Stavka.*

The German attack opened on 8th August, and within hours the Luga position was creaking ominously. Hoepner had again dispersed the two Panzer corps of his *Gruppe,* putting Reinhardt on the left of the 18th Army at Narva, and Manstein (now reinforced by one of the most sinister formations in modern war, the SS Police Division) at Luga. SS *Totenkopf* remained with the 16th Army at Novgorod, to spearhead and drive on Chudovo. Within three days the Kingisepp sector was on the point of giving way, and with his reserves expended and nothing in Leningrad itself except a mass of popular irregulars, Popov was faced by an agonising problem. Should he "withdraw"—if such a term was applicable to what would be a murderous, footsore, and embattled retreat under the lash of the Luftwaffe—from the Luga position forthwith? Or by holding on risk a complete rupture by the 41st Panzer Corps in the coastal region, which would leave the rest of his precious cadre formations stranded in the interior? In a report to Shaposhnikov, dated 11th August, Popov's Chief of Staff complained:

> The difficulty of restoring the situation lies in the fact that neither divisional Commanders, army Commanders, nor front Commanders have any reserves at all. Every breach down to the tiniest has to be stopped up with scratch sections or units assembled any old how.

And two days later:

> To suppose that opposition to the German advance can be maintained by militia units just forming up or badly organised, units taken from the North-Western Front command after they have been pulled out of Lithuania and Latvia . . . is completely unjustified.

The whole Russian position was on the point of cracking at this time. Two days earlier Hoepner had started to disengage the 56th Panzer Corps and move it north to support Reinhardt, whose own tanks had at last managed to keep the Kingisepp breach open. But on 12th August the Russian 48th Army, in response to the urgings of the *Stavka,* had moved around the south shore of Lake Ilmen and attacked the right flank of the 16th Army. The Russian force was mainly cavalry, and short of heavy equipment, but it was fortunate in its timing and direction. For the only German unit in its path was the 10th (infantry) Corps, which was the extreme right marker of the 16th Army, itself the flank army of the whole of Leeb's group. The area between the 10th Corps and the northernmost units of Bock's army group was a desolate region of swamp and forest, roadless and practically unmapped.

The 10th Corps was soon under severe pressure, and Leeb responded, perhaps too generously, to its calls for help. Manstein was ordered to turn about and place himself under the 16th Army. The result of this was that the 56th Panzer Corps passed the vital days from 14th to 18th August, while the entire Luga front was on the point of disintegration, in marching and countermarching, and finished up in a position on the flank, 150 miles from the centre of gravity of the battle.

The counterattack of the Russian 48th Army had saved the Luga front from annihilation. But reality was bad enough as the whole Russian line gradually crumbled away. In the second week of August, Narva, Kingisepp, and Novgorod all fell and SS *Totenkopf* broke through on the southern flank and captured Chudovo, on the main Leningrad-to-Moscow railway.

In Leningrad itself a million civilians worked twenty-four hours in a day on a wide defence perimeter which was being thrown up around the city. The Party mobilised every one of its enormous human pool into labour or paramilitary organisations.[2] Exhortatory proclamations filled the air and plastered the walls:

[2] There are three generic types. The largest—the *Opolchenye,* or "People's Army"—a more or less enthusiastic rabble, indifferently armed, with only a memory of weekend training in the *Osoaviakhim* and virtually without signals or communication equipment. (Sirota says that in addition to "some" rifles and machine guns ". . . the workers were armed mainly with Molotov cocktails and some hand grenades; they also had some 10,000 shotguns, and about 12,000 small-calibre and training rifles donated by the City's population.") The cream of the *Opolchenye* were organised on a divisional basis, with improved equipment and some cadre officers and N.C.O.'s. These formations were termed "Guards" divisions by Voroshilov.

(At almost the same time a general reorganisation of the Red Army was selecting those units which had particularly distinguished themselves in battle and using them as cores around which to build up elite formations to which it was applying the same designation—"Guards." There is thus some confusion over the origin of the term, although in the later part of the war it came to refer only to regular units.)

The third group comprised the so-called "destruction battalions" of Party, Komsomol, and NKVD personnel. These units had been formed to fight German parachutists, and were well equipped. From the political element in their constitution it seems that they were conceived for general "internal security" duties, as well as for combat with the enemy.

> A threat hangs over Leningrad. The insolent Fascist Army pushes towards our glorious city—the cradle of the proletarian revolution—our holy duty is to bar the road to the enemy at the gates of Leningrad with our breasts. [20th August]

> Comrade Leningraders! Dear friends! Our dearly beloved city is in imminent danger of attack by German Fascist troops. The enemy is striving to penetrate into Leningrad . . . The Red Army is valiantly defending the approaches to the city . . . and repelling his attacks. But the enemy has not yet been crushed, his resources are not yet exhausted . . . and he has not yet abandoned his despicable [sic], predatory plan to capture Leningrad. [21st August]

> The enemy is at the gates of Leningrad! Grave danger hangs over the city. The success of the Red Army depends on the heroic, valiant, and firm stand of each soldier, commander, and political worker, and also on how active and energetic the assistance given to the Red Army by us Leningraders is. [2nd September]

To anyone experienced in the duality of meaning in all Communist texts, the message was clear and urgent. The threat of total defeat hung over Leningrad.

There now occurs an episode of the greatest interest, whose background remains to this day shrouded in obscurity.

We know, and these pages will attempt to show, how the decline in their military fortune fostered plotting and treachery within the ranks of the Nazis. The totalitarian constitution is no less vulnerable to personal intrigue for succession to power than the looser fabric of a democracy, and it is tempting for the historian to discern evidence of such discord in the Kremlin. But in spite of the wave of "de-Stalinisation" and the denunciations of the 20th Party Congress, very little has come to light of the personal conduct of the Soviet leaders when *in extremis*—with the exception of this one episode during the battle for Leningrad.

On 20th August, Voroshilov and Zhdanov set up a "Military Soviet for the Defence of Leningrad." Within twenty-four hours Stalin was on the telephone, "expressing dissatisfaction" that the Soviet had been set up without his having been consulted. Voroshilov protested that it "corresponded to the actual requirements of the situation," but Stalin was in no mood for listening to Party jargon of this kind and suggested "an immediate review of the personnel"—i.e., the resignation of Voroshilov and Zhdanov. This was effected immediately, but the incident was not yet over. At the end of August two members of GOKO, Molotov and Malenkov, arrived, charged with "organising the defence." Some days passed, during which the military situation worsened rapidly, and then Voroshilov was relieved of his "responsibility" for military operations and kicked upstairs to GOKO in Moscow. The General who succeeded him in command was the Chief of Staff of the Red Army, the "fireman" who in his day was to visit and stabilise practically every dangerous sector of the Eastern front, Georgi Zhukov.

Some authorities attribute this shuffling around to a dispute between Voroshilov and Zhdanov, and even go so far as to mention a rumour that Voroshilov favoured the surrender of the town after the Germans had captured Schlüsselburg and completed its encirclement. It is said that Zhdanov then appealed over Voroshilov's head to Stalin. But from what we know of the character of those concerned, and in the light of the telephone conversation of 21st August, it seems more probable that Stalin had suspected that the Leningrad Military Soviet—whatever its advantages from the aspect of military administration—might then or at some later date form the nucleus of a breakaway administration which could menace his own internal authority or attempt negotiation with the enemy. There is a tradition of nonconformity at Leningrad—and the Communist doctrine teaches that the internal enemy is always the most important.

In fact, it is impossible to find evidence of anything more than local grumbling and discontent at any time during the siege of the town. And during the autumn of 1941, as the Germans daily moved closer, the entire population seems to have been united at every level. They were told:

> The Germans are out to wreck our homes, seize our factories and mills, plunder our public property, drench the streets and public squares with the blood of innocent victims, torment the civilian population, and enslave the free sons of our country . . .

They believed it, and they were right.

The city, now on the point, it seemed, of falling into their hands like an overripe melon, posed a delicious problem for the Germans—one which promised the opportunity of slaking even their insatiable thirst for blood.

The "problem," of course, was that of the civilian population. Hitler's first "firm decision" was to "level the town, make it uninhabitable and relieve us of the necessity of having to feed the population through the winter." Once the town had been razed, the site could be turned over to the Finns. Unfortunately, though, the Finns were very reluctant to have anything to do with the scheme, or even to annex any Russian territory, whether "razed" or otherwise. Then, also, there was the question of world opinion. A massacre on this scale would need a little explaining—even to those who looked on Hitler as the hammer of Bolshevism. Accordingly, Goebbels was instructed to manufacture a Russian plan—which was, as soon "discovered"—for the Soviets to destroy the city themselves.

The German military were against getting "involved" with the civilian population at all. Warlimont went into the question at length, and prepared a memorandum. "Normal" occupation was rejected. But it might be acceptable to evacuate the children and the old people (presumably to "shower baths") "and let the remainder starve." This, though, could lead to "new problems." Perhaps the best solution would be to seal off the whole town, surrounding it with an electrically charged wire fence, guarded by machine guns. But there remained the "danger of epidemics" (curious how often the German plans for mass extermination dwell on the "danger" of epidemics) "spreading to our front." In case this solution should be adopted, corps commanders were alerted to the need for using artillery against civilians trying to break out of the city, as Warlimont thought it "doubtful whether the infantry will shoot at women and children trying to break out." In any case, "disposal of the population cannot be left to the Finns."

There was also the possibility of making propaganda capital out of the affair—an offer to

> the philanthropist Roosevelt to send either food supplies to the inhabitants not going into captivity, or to send neutral ships under the supervision of the Red Cross, or to ship them off to his continent . . .

Naturally, though, any response to this which threatened to assume real shape could not be accepted. The proper solution was to

> Seal off Leningrad hermetically, then weaken it by terror [i.e., air raids and artillery bombardment] and growing starvation. In the spring we shall occupy the town . . . remove the survivors into captivity in the interior of Russia, and level Leningrad to the ground with high explosives.

Jodl, Warlimont's immediate superior, gave his approval to the memorandum, commenting that it was "morally justified," as the enemy would mine the town on leaving (an interesting example of the German mind at work; Jodl seems to have dug up this justification quite independently of the Propaganda Ministry scheme described on the previous page) and also because of—again—the "serious danger of epidemics." Jodl did, however, dwell briefly on a somewhat quaint alternative—that the population should somehow be allowed to flee in mass panic into the interior of Russia, arguing (with no great lucidity) that this would "increase chaos, and to that extent facilitate our administration and exploitation of the occupied areas."

As was not uncommon throughout the course of the Russian campaign, the German direction of military operations suffered from overlap and contradiction with personal and political factors.

The first stumbling block was the attitude of Mannerheim and the Finns. After the Luga front had disintegrated, Keitel wrote to Mannerheim, asking that the Finnish Army begin to exert "real pressure" along the Karelian Isthmus, and also that it cross the Svir River to the northeast of Lake Ladoga.

On 28th August, Mannerheim had rejected the plan, only to have it pressed on him immediately. He again rejected it (on 31st August), and remained adamant even when subjected to a personal visit by Keitel, who came over to plead with him on 4th September.

From a military aspect this obstinacy on the part of their allies was exceedingly troublesome to the Germans. The Wehrmacht was fully deployed, and had no strategic reserve at all. Only a form of grand tactical reserve was available by a process (of which there had already been examples) of stripping armour and mobility from one army group to reinforce another. Consequently OKW had literally no means of exerting pressure on the Russians' northern flank except by persuading the Finns to do it for them. They had no forces to put into Finland and no sanctions with which to enforce their request. Thus by the beginning of September there was already some hard factual backing for the decision to "hermetically seal," rather than to assault, the town.

Hitler, who had been watching with impatience the flank battles develop, was already beginning to look forward to the capture of Moscow. His mind, running several weeks ahead of the actual development of operations, nevertheless forecast their course with remarkable accuracy. On 6th September he issued Directive No. 35, which provided for the return of the *Panzergruppen* to Army Group Centre and the preparation for a final attack on the Russian capital. Owing to the depleted condition of many of the Panzer divisions, it was possible to deploy only a weight equivalent to that of the first attack by attaching the whole of Hoepner's *Gruppe* to Army Group Centre, in addition to those of Hoth and Guderian. The directive also ordered the 8th Air Corps to vacate its bases in Estonia and fly south to strengthen Bock—a move which would reduce the air strength at Leeb's disposal to less than three hundred machines, of which the majority would be short-range fighters or noncombatant transport and liaison aircraft.

Hitler's intention was that Leningrad be reduced in status to "a subsidiary theatre of operations," and that the siege perimeter be entrusted to six or seven infantry divisions.

A force of this size, while strong enough to keep three million starving civilians behind an electric wire fence, was far from adequate to cope with Voroshilov's army—ill equipped and exhausted though it was by that time. And even after the fall of Schlüsselburg (which did not in fact occur until three days after the issue of Directive No. 35) the siege perimeter—largely due to the intransigence of the Finns—remained loose enough to allow the garrison a dangerous amount of movement.

Anticipating this, and forewarned by rumours from OKH of the impending directive, Ritter von Leeb had been preparing a plan for a direct assault on the town. He had hoped to launch his attack on 5th September, the day before the issue of the directive, but the 41st Panzer Corps was so worn out by its exertions that it required a three-day period for refit while infantry cleared the east bank of the Neva and fought to hold the desperate counterattacks Voroshilov was ordering from the 54th Army—now in position to the south of Lake Ladoga.

On 9th September, Reinhardt was ready and the attack went in: the 1st Panzer following the left bank of the Neva, the 6th Panzer on either side of the main railway to Leningrad from the south. Both divisions were soon enmeshed in a net of anti-tank ditches and straggling earthworks which had been thrown up by the construction battalions and *Opolchenye* during the previous weeks. These defences were often poorly sited and crudely finished, but they were extensive.[3] The Russians were

[3] Karasev, 73, and other sources (see Goure, *The Siege of Leningrad*, 313, n. 39) give details. "Earth walls" over 620 miles in length; over 400 miles of anti-tank ditches;

seriously deficient in artillery, and indeed, in all arms not produced on the spot at Leningrad and its environs. But they had a large number of medium and heavy mortars whose weight of fire, at the ranges of that first day's battle, was nearly as effective as regular field artillery. On the coastal sector, between the sea and Krasnoye Selo, the twelve-inch guns of the Black Sea fleet pounded away at the German rear. Over the battlefield KV tanks roamed singly and in pairs, manned often by testers and mechanics from the Kirov factory, where, at that time, they were still being produced at the rate of about four per day. This was the kind of close infighting in which Russian qualities—courage, obstinacy, cunning in camouflage and ambush—more than counterbalanced the deficiencies in command and technique which had crippled them in the open battlefields on the frontier and on the Luga.

The Panzers, in contrast, were suffering as armoured troops always do when they encounter close defences after weeks of mobile fighting. Like the British 8th Army when it hit the Tunisian mountains after months in Libya, or the "Desert Rats" pinned down in the Normandy *bocage,* the tank commanders took fearful punishment as they sought to adapt their tactics in an unfamiliar element. On the first day of the assault four successive commanders of the 6th Panzer were casualties.

Leeb's committal of over half his tank strength[4] to an attack on closely fortified lines *after* he had been ordered to release it for operations on another front is a further example of the casual insubordination of the "Eastern marshals." By taking advantage of a proviso in the directive that the redeployment be subject to "first achieving a close encirclement," he was trying to finish the job on his own, before Army Group North was reduced in status and power to the level of a couple of army corps. Leeb's action is still less excusable when we recall that the purpose of the redeployment, namely the use of the two Panzer corps in a fresh series of mobile battles in the centre of Russia, had been made clear to him, and he must have realised that their committal to a frontal attack on a major enemy fortress could only result—whatever the outcome of the opera-

---

185 miles of wood abatis; 5,000 earth, timber and concrete pillboxes; over 370 miles of wire entanglements. But Goure also quotes a private informant as complaining, ". . . the fortifications were dug in the wrong places, with pitifully poor equipment, and for very little purpose."

[4] The 41st. Panzer Corps held over 60 percent of Hoepner's tanks. The 56th Panzer Corps, which had not been affected by the OKW directive, was understrength and located west of Lake Ilmen, where it had remained after checking the diversionary attack of the Russian 48th Army.

tion—in the tank force being handed over to Army Group Centre in a thoroughly worn-out and depleted condition.

In fairness to Leeb, it must be remembered that every senior German officer regarded the war against Russia as won. It was no longer "whether," but "how?" And, still more important, "who?" Who is going to get the credit for this prodigious feat of arms? Even Guderian, who was one of the first to feel the cutting edge of the Russian military revival, had believed at the time of his own "independent" leadership no more than that it was his policy which would bring victory that year—it can never have occurred to him that the alternative was total defeat and the sacking of Berlin. It was this conviction that made the intrigues at OKH, the disregard and "mislaying" of unwelcome instructions, such a personal affair in the first summer's campaign. After the fall of France, Manstein had complained:

> Hitler's appointment of a dozen Field-Marshals simultaneously was bound to detract from the prestige of a rank which had previously been considered the most distinguished in Germany. Hitherto . . . one needed to have led a campaign in person, to have won a battle *or taken a fortress* to qualify for this dignity.[5]

It was certainly human—to put it no higher—that Leeb, who had been an army commander of no particular distinction during the battle of France, should wish to win his promotion twice over by taking the most important "fortress" of the Eastern campaign. And indeed, just as in Bock's command the previous month, a disregard of Hitler's orders, and those from OKW, seemed at first to be justified. By the evening of the second day (10th September) the Germans had penetrated as far as the last line of Russian defences, which ran along the crest of some shallow eminences—known as the Dudergof heights—about six miles southeast of Leningrad. During the night many of the tanks of the leading division, the 1st Panzer, lay out on the battlefield, forward of the main German positions, and fought throughout the hours of darkness to beat off the succession of counterattacks which the Russians always put in during the night. By the lurid glare of blazing gasoline bottles and sodium flares they broke up one Russian formation after another as these

[5] Manstein was decorated Field Marshal the following year, after both "leading a campaign in person" (the Crimea) and "taking a fortress" (Sevastopol). Neither task was of outstanding difficulty or importance. The real services he was to render the Reich were to come in the spring of 1943. (See Ch. 14.)

assembled to charge the Germans established in the positions captured during the day. At first light the Stukas returned to the battlefield, and the 41st Panzer Corps braced itself (and how many times hereafter is this phrase to recur!) for "one last heave." The 1st Panzer had lost so many tanks that there was only one battalion left with over 50 percent effective strength, yet it gradually inched its way forward during the day, and by 4 P.M. had scaled "Height 167," a hill of about 450 feet and the highest point in the Dudergof ridge.

> In front of the victorious troops stood the City of Leningrad in the sunlight, only twelve kilometres away, with its golden cupolas and towers and its port with warships that tried with their heaviest guns to deny us possession of the heights.

On the left flank of the Panzer corps the infantry was slowly edging its way forward, and once the Russian guns and observers had been cleared off Height 167 it was able to make better progress, entering the suburban districts of Slutsk and Pushkin, and on the evening of 11th September, Krasnoye Selo.

By 12th September, the fourth day of the assault, it was not so much apparent as embarrassingly obvious to OKH that a full-blooded engagement was raging in a theatre from which it was trying to draw reinforcement. Halder teleprinted Leeb that the city "was not to be taken, but merely encircled. The attack should not go beyond the Petergof-Pushkin road" (a line which had already been passed). However, the following day Hitler issued a new directive. Whether this was at the prompting of Keitel, who was himself a proponent of the Leningrad school and a friend of Leeb's, or whether because the idea of a spectacular victory there had caught his fancy anew, the Führer now declared:

> In order not to weaken the attack . . . [the air and Panzer forces] must not be withdrawn before a close envelopment is achieved. Therefore the date set by Directive 35 for the withdrawal can be set back by a few days.

As the definition of "a close envelopment" in this latest directive was "within artillery range," and as no field piece in the German Army could shoot from one end of Leningrad to the other, this amounted, in effect, to an order to break into the city itself. In the next four days the German grip slowly tightened. Pulkovo was captured in the centre; Uritsk, close to the gulf coast, connected to the centre of the town by a four-mile-long "promenade"; and Alexandrovka, where the tram line from the Nevsky prospect reached its terminus. But in the ebb and flow of close combat that delicate point had been reached, which so often eludes perception by either side, when the attacker is expending force in a diminishing ratio to the successes he gains. An attack from three sides on the Russian positions at Kolpino, by the 6th Panzer and two infantry divisions, was repulsed, and the same day OKW seems to have tired of the whole affair, for it ordered the withdrawal "forthwith" of the 41st Panzer Corps and the 8th Air Corps. During the night of 17th September the 1st Panzer began to load its surviving tanks onto rail cars south of Krasnogvardeisk and the 36th Motorised took to the road for Pskov. Only the stricken 6th Panzer was left for a few days to extricate itself and lick its wounds. Gloomily Halder noted that evening:

> The ring around Leningrad has not yet been drawn as tightly as might be desired, and further progress after the departure of 1st Panzer and 36th Motorised from that front is doubtful. Considering the drain on our forces on the Leningrad front, where the enemy has concentrated large forces and great quantities of material, the situation will remain critical *until such time as hunger takes effect as our ally.*

The attack by the 41st Panzer Corps is another example of the way in which the army commanders played fast and loose with Hitler's directives when it suited them and they thought that they could get away with it. Leeb nearly did get away with it—in the sense that he forced the hand of the Supreme Command and then led it around (however briefly) to support him.

But the effect on the campaign as a whole was unfavourable. There was a delay of nearly ten days in moving Hoepner's *Gruppe* south—and this at a time when days were already beginning to assume a vital importance. And when the Panzers finally left Leningrad they were in no state to fight another battle. They needed refit, replacements, rest. They needed, in essence, still more time.

This attack was the first and the only occasion on which the Germans attempted to take the city by storm. The leading Western authority on the siege takes the view that ". . . by withdrawing the armoured divisions just at the moment when the capture of the City seemed certain, Hitler had saved Leningrad." But is this really a valid assertion? Only hindsight reveals that the city was "saved" by factors operating in the autumn of 1941. At that time every reasoned approach must have led to the conclusion that a long siege would be successful in the end. And, in fact, the plight of Leningrad did worsen steadily until it was relieved in 1943—at a time when the whole strategic initiative had passed to the Russians. Furthermore, to argue that Leningrad was "saved" in 1941 begs the question that its capture "seemed certain" by the 41st Panzer Corps. This, too, is highly doubtful. Although the Germans were gradually constricting the fieldworks on its outskirts and had penetrated them with a few tanks in places, there still lay before them the prospect of a long and savage street battle in a town of immensely strong stone buildings, intersected by a maze of canals and small waterways. On ground of this kind a mass of irregulars armed with gasoline bottles and sticks of dynamite can, as the siege of Madrid had already demonstrated, swallow up corps of professionals.[6]

> If the insolent enemy tries to break through to our City he will find his grave here. We Leningraders, men and women, all the patriots of the City, acting as one—from the smallest to the biggest—will throw ourselves into the deadly fight with the Fascist robbers. We will fearlessly and unselfishly defend each street, each house, each stone of our great City.

No army which rests its quality on training, technology, and firepower—as the German Army did—should ever allow itself to be drawn into terrain where this quality is at a discount. Whether from an appreciation of this fundamental law of tactics (which he was to ignore, with such disastrous consequences, the following autumn) or whether from less rational motives, Hitler's basic inclination was against a direct assault on Leningrad.

The general simplification that Hitler's decisions were always at fault, and its converse, that the alternative courses of action would always have succeeded, is as fallacious as its companion generality—that all differences of opinion found Hitler on one side and the whole body of the generals unanimous on the other.

From the German point of view, the real irony of the Russian campaign is that the time when Hitler's grasp of military affairs was at its

[6] Although the main effort of the labour battalions had been put into the outer fieldworks, work had been in train on fortifying the city itself since the middle of August. Fadeev quotes one of the workers at the Kirov factory: "We decided on a circular defence. We fortified the whole area so that if necessary we could defend ourselves. In addition to the *Opolchenye* we formed other volunteers' units. The others could do as they pleased but we, the Kirov workers, were not going to give up our plant."

surest and his powers of judgment were at their most rational was the period when he had the greatest difficulty in getting full obedience from his commanders. It was only after he had disciplined them and was a master unto himself down even to battalion level that the Führer's military intuition became really unbalanced—and then, to justify his iron grip, he could always point to that moment in the campaign when his own plans were constantly thwarted by the personal pride and ambition of his "Eastern marshals."

seven | SLAUGHTER IN THE UKRAINE

In contrast to their efforts to break the Russian northern flank and reduce Leningrad, the Germans' operations in the south were dazzlingly successful. All the objectives Hitler had outlined in Directive 33 and enlarged on at various conferences with the commanders were achieved. The Pripet Marshes were cleared; the Dnieper bend was occupied; the river itself was crossed by tank spearheads which drove deep into the Donetz basin; and the industrial complex of the Ukraine was denied to the enemy, either through dispersal or seizure. Above all, the mass of the Red Army in the south was battered to pieces in a colossal "annihilation battle" which cost the Russians nearly a million casualties.

Yet the operation remains essentially, because strategically, a failure. It did not win the war for the Germans, and today we can see that it was not necessary, even as a prelude to such a victory. Indeed, the Ukrainian campaign went far to lose Hitler the war, in that its conception and pursuit denied him all chance of subduing the Russians before the winter—and thereby for another six months, until the high summer of 1942.

The reasons for this are varied and disputable. But the most compelling of them all is to be found in the mundane field of service and logistics. The very scope of the Germans' success, the depth and force of their tank drives, and the tireless manner in which they forced the pace had thrown a severe and cumulative strain on the machines they were using. The task of converting the Russian railway system to the European gauge was barely started—at the beginning of August, Halder complained that there were only six thousand kilometres converted in the whole of the occupied territory—and the "tank transporter" concept (the practice of carrying the tanks on truck-drawn trailers to save wear on their tracks) was in its infancy. The result was that the Panzers were moving everywhere under their own power, and after two months of fighting they were in urgent need of attention. Engines, driving gear, track sprockets and grousers, even secondary moving parts like the turret traversers, were all in need of attention, choked by the dust and dangerously overstressed on the endless bad going. Guderian's first reaction on being told the new plans for his *Gruppe* (a round march of over six hundred miles) had been grimly prophetic. "I doubt if the machines will stand it," he said, "even if we are unopposed."

The Russian front in the south had begun to give way at the beginning of July, when Kirponos, with his tank strength exhausted, had been levered off the Rovno-Dubno-Ternopol axis and forced back at a pace, which daily gathered momentum, into the wide-open spaces of the Ukraine proper. At the same time the Balkan satellite armies which had been ranged along the Prut under Schobert were prodded into action, and began to move slowly into Bessarabia and the area of the "Odessa Military District." The effect of these developments was that the active front in the south more than trebled its length at the very moment that the frontier armies were beginning to give way.

On 10th July the *Stavka* had fused the two commands[1] and placed them under a new hand—that of Marshal of the Soviet Union S. M. Budënny, who brought as his Commissar, charged specially with the "organisation" (i.e., the evacuation) of industry, Lieutenant General N. S. Khrushchev.

Khrushchev we know. Since that day in 1941 much has been written about him, and at this stage there is nothing that need be added save to point to the link between his appointment and the position of an old friend, Anastas Mikoyan, on the supreme "Evacuation Soviet," which had nationwide authority in all matters pertaining to the evacuation of industry. Khrushchev's own tireless energy and resource must be reckoned the inspiration behind one of the most substantial (though least publicised) of the Soviet achievements in 1941—the wholesale dismantling and removal to the east of nearly one quarter of the country's industrial capacity. But Khrushchev's talents were ill complemented by those of his military colleague.

[1] The southern and southwestern (Odessa M.D.) "group."

It is hard to see any positive quality which rendered Budënny eligible for this vital command—unless it is that certain *paysan rusé* cunning which, allied to his jovial mediocrity, had kept him out of trouble during the purges (with Voroshilov he had been the only Marshal to survive)—and only too easy to detect in his career that fortuitous combination of luck and circumstance which brought him to preside over one of the most disastrous land battles in history.

Budënny had been a sergeant major in a cavalry division (not the Cossacks, although later he did not object to a myth asserting this). In 1917 he had joined the "Regimental and Divisional Revolutionary Committee," and after a number of vicissitudes found his way to Tsaritsyn (Stalingrad), where he had met up with Voroshilov. With Stalin and Yegorov (who was to go under in the 1937 purge) he formed the 1st Red Cavalry Army. Two years later, in the war against Poland, he had shown singular clumsiness in his handling of even divisional-size units, failing to co-ordinate with Tukhachevski's attack on Warsaw and being forced into a humiliating retreat. With the exception of a punitive cavalry mission against the Georgians in 1920, Budënny did not hold an active command until the outbreak of war although, benefiting from Stalin's precept that "reliability" is more desirable than aptitude, his ascent, through a succession of staff appointments, had been rapid.

With his love of the cavalry charge, his handle-bar moustaches, and mahogany-butt revolvers hanging at his waist, Budënny resembles nothing more than a Slavic synthesis of Foch and Patton—with the talents of neither and a taste for self-indulgence which, if either of those two great commanders possessed it, has been mercifully kept from posterity.[2] Here, indeed, was an unhappy choice to oppose Rundstedt, one of the coolest brains on the German General Staff, and Kleist, one of its most vigorous Panzer commanders. But Budënny still had something which both his colleagues at Leningrad and in the centre had already dissipated

[2] As First Deputy Commissar for Defence in 1940, Budënny had visited the lately annexed province of Bessarabia, where a party was given in his honour at the distilleries at Kishinev. Toward the close of festivities a canvas screen was ripped off the largest vat, which had been filled with red wine to a depth of one and a half metres. Inside were a number of naked girls desporting themselves in the warm red liquid. Without further ado Budënny and his aides threw off their clothes and joined the nymphs in the pool. The bacchanalia gathered momentum until another guest, disgruntled at being unable to climb into the vat, fired a long burst at it with a tommy gun; three of the occupants were injured, and the wine ran out of the holes and onto the floor of the distillery. The orgy then moved to more comfortable quarters at the back of the building.

—numerical superiority. Stalin's decision that Kiev was to be held at all costs had given top priority in the allocation of men and equipment to the southern sector, and the more highly developed railway system in the Ukraine speeded their concentration (it was at Kiev station that Halder had first noted the practice of dumping the rolling stock and sending the engines back alone).

The configuration of the railways had resulted in two major concentration areas, the first at Kiev itself and the second, drawing reinforcement up the southern route Nikopol-Krivoi Rog, from the Crimea and Azov districts, at Uman. Between these two Budënny disposed of a million and a half soldiers, or more than half the active strength of the Red Army at this time. But even had this force been directed by a commander worthy of its latent strength and unencumbered by the rigidity of *Stavka* interference, the Russians would have been hard put to match Rundstedt's speed once he had broken up Kirponos' armour in the frontier battles. Feeling the weight of the Russian infantry deploying in front of Kiev and daily appraised by the Luftwaffe of its rate of increase, Rundstedt decided to force his armour through to the south and take the option of wheeling either to the Dnieper or the Black Sea. The German commander knew that neither of the concentrations, at Uman or Kiev, had sufficient tank strength to menace his flanks, and that once Kleist had broken through, the Panzers would be able to roam at will. What was urgent was to breach the Russian line before the new masses of infantry and artillery could be defensively deployed.

On 12th July, Kleist had completed his concentration of all three Panzer corps at Zhitomir, and within the next three days he had driven the Russians off the vital twenty-mile stretch of the Berdichev-Kazatin railway, entering both towns on the night of 15th-16th July. This effectively severed the north-to-south communications of Budënny's front and enforced a *de facto* reversion to the division (between "southern" and "southwestern" theatres) which his appointment, still less than a week old, had been intended to eliminate. The only solution for the Russians was a strategic withdrawal, deep into the Dnieper bend, with their flanks hung on Kiev and Odessa. Lengthening the front would have presented no problems to Budënny, who each day was putting more of his freshly mobilised forces into readiness, but it would have faced Rundstedt with a difficult choice: whether to press into the empty but menacing salient between Odessa and the Dnieper or to revert to a frontal assault on Kiev, an operation which promised to be costly and protracted.

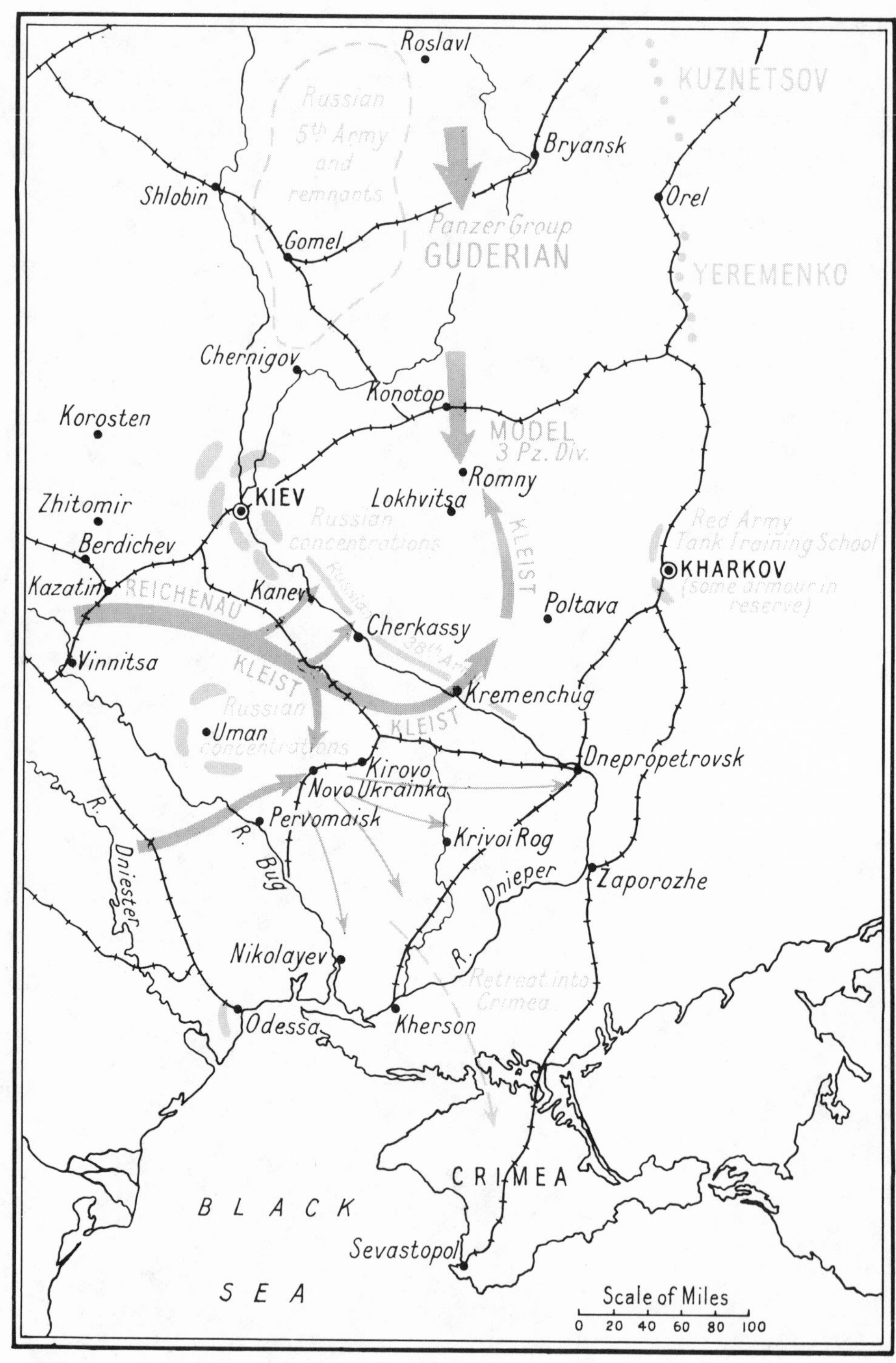

Budënny in the Ukraine (Uman and Kiev)

Indeed, for a few days it seemed as if the Russians might have opted for this course, because on 16th July, after hearing of the fall of Berdichev, Budënny ordered Tyulenev to evacuate Kishinev and bring all his forces back behind the Dniester. But such action was already long overdue, and the Marshal's failure to appreciate the real gravity of the situation is corroborated by his order (on 17th July) that Tyulenev was "to assemble his reserves and concentrate them in the Uman area." Another week passed while a strange paralysis seemed to have gripped the Russian command. Men and guns poured into Uman, the region itself, like the vortex of a tornado, enjoying a sinister calm. But on the fringes of the concentration area the storm warnings were sounding with terrible urgency. Reichenau was driving due east from Kazatin with his motorised corps, including the SS Adolf Hitler and Viking; Stülpnagel had widened the breach by capturing Vinnitsa and Shmerinka, and thereby deprived the armies at Uman of any chance of success in a move northwestward; and most serious of all, all three Panzer corps of Kleist's *Gruppe* were racing eastward, Manteuffel leading, toward Belaya Tserkov and the Krivoi Rog railway.

Kleist entered Belaya Tserkov on the evening of 18th July, and the expected Russian counterattack, staged by six infantry and two cavalry divisions of the 26th Army, had heralded itself with frantic radio conversations *en clair* between the formations concerned all day on 19th July. When the Russian attack started on the 20th, it followed the traditional pattern, a three-minute artillery barrage followed by waves of infantry, as much as twelve in succession. Here and there a few tanks[3] accompanied trucks, crammed with soldiers, which simply drove flat out at the German positions until stopped by a direct hit. With all chance of surprise gone, the Russian attack was broken up in a matter of hours. The progress of the 1st Panzer Army was barely affected.[4] Within five days Kleist had reached Novo-Ukrainka, and on 30th July, Manteuffel's leading tanks burst into Kirovgrad; over a hundred miles *southeast* of Budënny's mass. With German tanks straddling the railways in his rear,

[3] Some measure of the Russian tank losses in the southern theatre so far can be derived from the fact that up to the start of the battle of Uman the Germans claimed to have destroyed 2,287. Yet in the final "count" of prisoners and material in the pocket they only claimed an additional 317. For the 26th Army's counterattack it is doubtful if more than about 50, of all types, were operational.

[4] Nonetheless, Soviet sources claim that these counterattacks held up the 1st Panzer Army for several days and allowed the 6th, 12th and 18th Russian armies to escape encirclement.

Budënny could no longer save the armies at Uman by a retreat to the Ingul. His only chance was a withdrawal down the Bug to Nikolayev. Yet another five days passed, with no move by the Russian commander.

Meanwhile, thirty miles south of Uman, Schobert's 11th Army, with two Hungarian divisions, had succeeded in forcing its way across the Bug at Gaivoron. Frenzied Russian counterattacks and extravagant artillery barrages prevented the mixed assault group from making progress in a northerly direction, and the Russians claimed a victory, but in fact Schobert's spearhead had another target. With two motorised divisions and the Hungarian cavalry he side-stepped the weight of Russian fire before Gaivoron and moved rapidly down the left bank of the river to Pervomaisk. Here, on 3rd August, Schobert made contact with the outriders of Wietersheim's 14th Panzer Corps, which Kleist had turned due south, at right angles to his main axis of advance. A noose had been tied around the whole of the Uman concentration.

For five days the infantry divisions of the 11th and 17th armies marched in concentric arcs across the steppe, beating out thirty or forty miles a day, with their guns and equipment trailing behind in patient horse-drawn convoys, until by 8th August the noose had become a ring of steel, filled out with artillery and broad enough to trap all but the smallest groups of stragglers. And already the centre of gravity had moved hundreds of miles due east, carried on the clattering treads of Manteuffel's Panzers, to the banks of the Ingul, where Budënny should have been making his stand. This period was the high peak of *enjoyment* for the German soldiery. The war was exciting and victorious. Malaparte's description of an evening halt conveys the romance of that drive across the Ukraine:

> During the night-time all fighting ceases. Men, animals, weapons rest. Not a rifle shot breaks the damp nocturnal silence. Even the voice of the cannon is hushed. As soon as the sun has set, and the first shadows of evening creep across the corn-field, the German columns prepare for their night's halt.
>
> Night falls, cold and heavy, on the men curled up in the ditches, in the small slit-trenches which they have hastily dug amid the corn, alongside the light and medium assault batteries, the anti-tank cannon, the heavy anti-aircraft machine guns, the mortars . . . Then the wind rises—a moist cold wind that fills one's bones with an immense numbing weariness. The wind that sweeps this Ukrainian plateau is laden with the scent of a thousand herbs and plants. From the darkness of the fields

comes a ceaseless crackle as the moisture of the night causes the sunflowers to droop on their long, wrinkled stalks. All about us the corn makes a soft rustling sound, like the rustle of a silk gown. A great murmur rises through the dark countryside which is filled with the sound of slow breathing, of deep sighs. Shielded from sudden attack by the sentries and patrols the men abandon themselves to sleep. (There in front of us, concealed amid the corn and within the solid dark mass of the woods—over there, beyond the deep, smooth bleak fold of the valley, the enemy sleeps. We can hear his hoarse breathing, we can discern his smell—a smell of oil, petrol, and sweat.)

Between 15th and 17th August, Manteuffel and Wietersheim (with the 14th Panzer Corps) spread out along the Ingul, capturing Krivoi Rog and reaching south as far as Nikolayev. Russian resistance on the river was slight and, once again, suffered from the absence of central direction. It was put up almost entirely by reservists and troops from training areas east of the Don, who had been ordered to Uman and now found their path blocked by the German tanks. Except where the initiative of local commanders dictated otherwise, the battles took the usual course of piecemeal and repetitive attacks by the Russians which exhausted their own strength in a matter of hours. The two Panzer corps rapidly forced a number of bridgeheads and drove on into the bend of the Dnieper.

Meanwhile the imprisonment of the old "southwestern front" mass at Uman had opened wide the Black Sea flank to the satellite divisions in Schobert's army. The few remaining Russian units which had escaped the trap were rapidly withdrawn into the Odessa perimeter, and the Rumanians, with Hungarian cavalry, were able to advance toward the Dnieper estuary almost unresisted. On 21st August these forces had been stiffened by two German divisions from the 11th Army and had succeeded in forcing the Dnieper above Kherson. Immediately, with his southern flank secure, Rundstedt ordered the Panzer forces to regroup in a northerly direction, pulling Wietersheim back to the Cherkassy-Kremenchug region and directing Manteuffel on to the northeastern corner of the Dnieper bend, between Dnepropetrovsk and Zaporozhe. Within twenty-four hours the 3rd Panzer Corps had crossed the Dnieper above Dnepropetrovsk, and moving southward along both sides of the river, entered the town on 25th August.

Now there was nothing between the German tanks and the whole basin of the Donetz. They had achieved a break-through as complete as that of Hoth and Guderian the previous month at Bialystok. All they required, it seemed, was gasoline and the whole of southern European—and indeed, Asiatic—Russia lay at their mercy. On the day that Manteuffel captured Dnepropetrovsk the Russians blew the Zaporozhe dam, one of the great "engineering masterpieces of the Proletarian Revolution." This desperate gesture took out the source of power for the industries in the Dnieper bend—the majority of which had already been sabotaged by Khrushchev's wrecker gangs—but had little more practical effect than to lower the level of the river upstream and make its crossing easier for Manteuffel's engineers. Nevertheless, at a symbolic level the destruction of the dam underlined two things: the almost suicidal sincerity of the "scorched-earth" policy and the fact that the Russians had little expectation of reoccupying the Dnieper basin for a very long time.

Unlike Bock, Rundstedt had no single objective whose capture would, or ought to, finish the war. Was he to halt on the Donetz? On the Volga? At the Caspian? There was no political or geographic goal. The German commander had to confine himself to the terms of the original *Barbarossa* directive, which had laid down the primary task of "preventing the retreat of intact bodies of troops toward the vast hinterland of Russia." To this end the commander of Army Group South was now moving his armour not westward but northeast. He knew that Guderian, too, had reversed direction, and had fixed his eyes on the largest single prize of the Eastern front, the huge, almost static garrison of Kiev, nearly three quarters of a million strong, under orders from Stalin to "hold at all costs."

It took Kleist less than a week to re-form the 1st Panzer Army on the south bank of the Dnieper, and during that period his reconnaissance detachments were forcing a sequence of small bridgeheads down the whole length of the river from Cherkassy to the bend. This area was the responsibility of one weak Soviet Rifle Army, the 38th, which had some divisional artillery, but no armour at all. Budënny was keeping his strength bunched up east of Kiev, and troop trains from Kharkov were still proceeding there, in conformity with month-old directions which had never been superseded.

The front of the Soviet 48th Army was over 120 miles long, and it was facing, in the last week of August, the Panzer corps of Wietersheim and Manteuffel, two divisions from Kempf's 48th Panzer Corps, and SS Viking. Some of the infantry from Reichenau's 6th Army had already made their way around the Kiev bulge and appeared to the west of Cherkassy, and with the collapse of resistance in the Uman pocket, the whole of Stülpnagel's 17th Army was freed. On 22nd and 23rd August an additional seven infantry divisions began to move northeast toward Kremenchug and the embryo bridgeheads which had already been won for them by Kleist's armour.

The Russians at Kiev, in contrast, were barely moving at all. Such tanks as they had left in working order were immobilised by lack of fuel. Artillery was plentiful, but many calibres were short of ammunition[5] and only horses were available to move them around. The Germans, whose own speed of reaction was multiplied tenfold by the mobility of their armour, roamed at will around the loose perimeter of the Russian concentration, herding the bewildered Soviet infantry like sheep dogs. Over his doomed command, still filling up with men, though now shrinking in area, presided the imbecile Budënny, a general in the worst tradition of 1914-18 whose malevolent fatuity has no rival—even among those competitive effigies of the Great War, Haig, Joffre, and Nivelle. There is some evidence that at this late stage the *Stavka* considered the possibility of withdrawing the southwestern front to Kharkov, or even as far as the Donetz. Shaposhnikov favoured the latter course. But Stalin preferred to stand and fight it out in front of Kiev, and Budënny, who numbered among his limited gifts the ability to divine Stalin's wishes, reported that he had "formidable defensive positions at his disposal."

Although Stalin must carry the bulk of the responsibility for the disastrous handling of the Red Army in the first months of the war, it is no more accurate to saddle him with the whole blame than it is to do so to Hitler, with the German Army, in the last. He had given Budënny nearly a million men; it was not unreasonable to expect this force, even if it could not hold the line of the Dnieper, to give such an account of itself that the German attack might be blunted—for that summer, at least. To draw back such an enormous army in face of the complete air supremacy which Lohr and Kesselring had established would have been perilous in the extreme. Stalin also seems, as always in his career, to have considered the political factor first. Habitually suspicious of army morale and "loyalty," he was not yet convinced of the tenacity and devotion the Army was showing in combat. It is always easier to preserve morale in a

[5] The 6th Army records the total inventory of captured material at the close of the Kiev battle as tanks, 884; canon (sic), 3,714.

static defensive battle than in a long retreat, and there was the additional consideration that to yield up still more territory would encourage the creation of "separatist" movements in the occupied territories—Stalin's mind was too Machiavellian to conceive the insensate brutality of the German administration, too uneasily aware of the agrarian discontent which could have blossomed into rebellion had the invaders given it but a hint of encouragement. Hence the decision to stand and fight it out at Kiev.

This decision might still have paid off had the huge army been properly led. Timoshenko (who was eventually brought down to take charge of the shattered remnants at the end of September) and Zhukov (for whom the future held still more important tasks) could have changed the course, if not the outcome, of the battle. So indeed might several of Budënny's subordinates like Kirponos or Ryabyshev, had they been given the authority in time. But instead the days slipped past, as they had at Uman in August, with the Red Army in a state of fatal immobility.

Some idea of what might have been achieved under aggressive leadership can be gauged from the exploits of the 2nd Cavalry Division, which had found a weakly held sector of Reichenau's flank and started out on a foray southwest of Kiev. Fortune smiled on these bold horsemen, and they managed to avoid the attentions of the Luftwaffe for over a week. In that period the 2nd Cavalry roamed the marshy length of the Teberev, one of the tributaries of the Pripet, falling upon isolated units of German infantry, who were marching along in close order toward a "front" they believed to be forty miles to the east. On the last day of August the Russian cavalry had another stroke of luck, surprising the map depot of the 6th Army while it was in evening bivouac at a village just off the main highway from Korosten to Kiev. One of the Germans who survived has described the scene:

> We had no proper sentries . . just a few men strolling about with their rifles slung over their shoulders, as the whole of 16th Motorised was meant to be between us and the Russians. There was quite a lot of fraternisation with the villagers; I remember that some of them had never seen a lemon before. Then the inhabitants began to withdraw to their houses . . . we thought this seemed peculiar, and soon the village was empty of Russians.
>
> A short time afterwards there was the sound of horses, and . . . a dust cloud to the south. Some people said that it was a supply column for one of the Hungarian divisions. Then they were upon us . . . like an American film of the Wild West . . . sturdy little horses riding at a gallop

> through our camp. Some of the Russians were using sub-machine guns, others were swinging sabres. I saw two men killed by the sword less than ten metres from me . . . think of that, eighty years after Sadowa! They had towed up a number of those heavy, two-wheeled machine guns; after a few minutes whistles began to blow and the horsemen faded away; the machine gunners started blasting us at very close ranges with enfilade fire . . . soon tents and lorries were ablaze and through it the screams of wounded men caught in the flames . . .

But neither local victories such as this one nor the stubborn valour of the Russian soldier in close combat could halt the strategic development of Rundstedt's offensive. While Kleist was building up his strength in the south Dnieper bridgeheads, Guderian was driving two Panzer corps, the 24th and the 47th, at maximum speed toward the Desna.

It is now clear that the wheeling of Guderian's *Panzergruppe* through 90 degrees, to drive due south into Budënny's rear, took the Russians completely by surprise. The gap between the two masses—Timoshenko's, exhausted by the fighting at Smolensk and Roslavl, and Budënny's, crouching inert before Kiev—was over 120 miles. Some degree of protection was still granted by the remnants of the Russian 5th Army and the troops clustered around Gomel—but only against attack from the west. Guderian's tanks were striking down a longitude eighty miles in the rear of the 5th Army, carving their way through a desolate region of deep forests, log roads, and marshy scrub. Guderian was leading with two Panzer divisions, following roughly parallel routes, but some thirty miles apart. Each was commanded by a general who was later to achieve particular distinction: the 3rd, under Model, "the Führer's Fireman" of 1944; and the 17th under the unfortunate Ritter von Thoma, who was to take command of the Afrika Korps when Rommel fell ill at Alamein. As early as the third day of the march, after covering sixty miles, Model captured the 750-yard bridge over the Desna at Novgorod-Severski, and forced the last major natural barrier between the *Panzergruppe* and Kleist.

Russian histories are still reluctant to allow Budënny (who is alive) to carry his full share of the blame for the Kiev disaster. Instead they attribute the prime responsibility to Kuznetsov and Yeremenko (now dead), who commanded the troops along Guderian's flank. But what did these forces amount to? Guderian's own situation map for 31st August shows only nine Soviet infantry and one cavalry division along the whole stretch between Roslavl and the Novgorod bridge, and it can be reckoned unlikely that any of these were much over brigade strength. Furthermore, the whole German force was mechanised; the Russians, who had first to concentrate before they could attempt to sever the German column, could move no faster than the pace of a walking infantry man—no faster, that is to say, than the armies of the Grand Duke at Tannenberg or, come to that, than those of Darius or Hannibal.

It is thus all the more curious to see the record of Guderian's perpetual complaints about the inadequacy of his own strength and the dangers to his flanks. As early as 26th August he had been explaining to Paulus (at that time Oberquartiermeister I at OKH),[6] who was on a tour of the forward area:

> . . . the enemy along our deep left flank is too strong to be ignored . . . he must be defeated there before we can pursue our southward advance.

However, when Paulus repeated these remarks to Halder the latter did nothing—a consequence which Guderian attributes to

> the general animosity towards myself that reigned in those quarters [i.e., the personal ill feeling between himself and Halder].

However, as Halder's attitude was based on more comprehensive intelligence data than that of the Panzer commander, and as it was borne out by the sequence of events, it is possible that it was less subjective than Guderian claimed at the time.

In spite of this rebuff Guderian continued to make a nuisance of himself to OKH. First he tried to get the cavalry division transferred from the 2nd Army (under whose command it had been operating since 27th July), drawing the rebuke: ". . . the movements of Second Army are to be regarded as simply tactical"; next, a request that Vietinghoff's Panzer corps, which had been left behind at Yelnya, should be sent down to stiffen the *Gruppe;* Halder refused it. Two days later, though, the cavalry division was transferred to Guderian, and he was told that Vietinghoff could spare *Gross Deutschland* and SS *Das Reich*. It appears that this did not satisfy the Colonel General, for the next day he was on the telephone, not to Halder but to Bock, insisting that the whole of Vietinghoff's corps be sent south, together with the 7th and 11th Panzer, and 14th Motorised, as he knew that these divisions "were not committed

[6] Deputy Chief of Staff to Halder.

at the time." This was hardly a tactful approach, and it was aggravated by the fact that an intercept station of OKH had listened in to the conversation. "A positive uproar" resulted. Lieutenant Colonel Nagel, the OKH liaison officer with Guderian's headquarters, was recalled to the army group at Borisov, where as ill luck would have it, Brauchitsch had arrived to oversee the last stages of the encirclement. Here he was dismissed, on the grounds that he was a "loudspeaker and propagandist." The following morning Guderian received a message to the effect that OKH was "dissatisfied" with the handling of the *Gruppe,* and ordering him to contract his front of operations by bringing the 47th Panzer Corps back across the Desna. Guderian records:

> . . . these orders were cast in an uncouth language, which offended me.

It seems as if both Halder and Brauchitsch realised that the Russian strength on Guderian's left was much feebler than he himself professed to believe. They may have thought, too, that his demands for reinforcement presaged some new independent action intended to force their hand, like the origins of the Roslavl operation a month earlier, which would also fatally handicap the next stage of the offensive—which everyone at OKH now regarded as inevitable. At all events, on 7th September, three days after the emission of the "uncouth" orders, Brauchitsch saw Guderian at 2nd Army headquarters at Gomel and told him that it was necessary to contract the front of the Panzer advance because of the need to keep as many divisions as possible fresh for the attack on Moscow, scheduled for early October.

On 12th September, Kleist finally broke through the exhausted 38th Army and debouched from his bridgeheads at Cherkassy and Kremenchug. (This day, the same as that on which Reinhardt cracked the Leningrad perimeter [pp. 124-25] can be reckoned the low point in the fortunes of the Red Army for the whole war.) Model, whose 3rd Panzer Division had been leading Guderian's thrust, was racing south, having emerged from the forest and swampland of the Seim, and running on the dry corn fields of the steppe. The Panzers made contact at Lokhvitsa on 16th September, and the outer ring was closed in the largest encirclement achieved by either side in the entire campaign.

In the cauldron the Soviet confusion was total. Budënny himself had been relieved by Stalin's order on the 13th and flown out to a sinecure on the "reserve front." Without even the semblance of a central command, the mass of surrounded men reacted to the independent (and often contradictory) orders of their separate corps and army commanders. Some accounts have it that Budënny had ordered a withdrawal on 8th September, then rescinded it the next day. Vlasov claims that he tried repeatedly to get Stalin to authorise a withdrawal, but that this permission was only granted two days after Model and Kleist had closed the ring, on 18th September.[7]

In fact, the Russians had neither the ammunition, the fuel, nor the co-ordination to attempt a breakout. With a stubborn pride they fought until what little they had was exhausted. In those last chaotic days whole battalions would attempt mass counterattacks, advancing with their last five rounds against the German artillery that blasted them down over open sights. When the Germans came to them, the Russians fought to the last, even with their teeth, as Krylov had forecast. Stalin presided

[7] The Soviet Official History (Vol. 2, p. 107) quotes the text of an appeal from the Military Council of the SW Front to Stalin, dated the 11th September, 1941:

> The Military Council of the South-Western Front considers that in the present situation it is essential to authorize general withdrawal of the front to a rear line.
>
> The Chief of the General Staff, Marshal comrade Shaposhnikov, on behalf of the Stavka of the Supreme High Command, in response to this proposal, gave an order to withdraw two rifle divisions from the 26th. Army and use them to liquidate the enemy who have broken through from the Bakhmach-Konotop area. At the same time, comrade Shaposhnikov indicated that the Stavka of the Supreme High Command considers withdrawal of the forces of the SW Front to be premature at present.
>
> For my (sic) part, I presume that by now the enemy's intention to envelope and surround the SW Front from the Novgorod-Severski and Kremenchug directions has become fully apparent. To oppose this idea a strong group of forces must be created. The SW Front is not in a position to do this.
>
> If, in its turn, the Stavka of the Supreme High Command is not able at this time to concentrate such a strong group, then it is high time the SW Front withdrew. The measure which the military council of the front is to carry out, viz, the moving out of two divisions from the 26th. Army, may be only a way of covering this. Besides, the 26th. Army is very much weakened. On 150 km. of front only three rifle divisions are left.
>
> Delay in withdrawal of the SW Front may lead to losses of troops and great quantities of materiel . . .
>
> Budënny. Khrushchev. Pokrovski.

This evidence could be cited by those who wish to exonerate Khrushchev from responsibility for the Kiev disaster, but even the 11th September (less than 24 hours before the collapse of the 48th Army at Kremenchug) was perilously late to start withdrawing, especially under the conditions of total air superiority which the Germans enjoyed.

over their death, for specially trained electricians had rigged up apparatus which played recordings of his speeches to the defenders of key positions. Malaparte has described how

> During the fighting the words of Stalin, magnified to gigantic proportions by the loudspeakers, rain down upon the men kneeling in holes behind the tripods of their machine-guns, din in the ears of the soldiers lying amid the shrubs, of the wounded writhing in agony upon the ground. The loudspeaker imbues that voice with a harsh, brutal, metallic quality. There is something diabolical, and at the same time terribly naïve, about these soldiers who fight to the death, spurred on by Stalin's speech on the Soviet Constitution. By the slow deliberate recital of the moral, social, political and military precepts of the "agitators" [8]; about these soldiers who never surrender; about these dead, scattered all around me; about the final gestures, the stubborn, violent gestures of these men who died so terribly lonely a death on this battlefield, amid the deafening roar of the cannon and the ceaseless braying of the loudspeaker.

After five days of slaughter the first surrenders began. By the time that the area was pacified, over six hundred thousand men had been sent into captivity.[9] Nearly one third of the Soviet Army, as it had been at the outbreak of war, was eliminated. As they counted the spoils, the Germans took pains in the categorising and recording of every item. Photographers and artists crowded onto the battlefield, and have left us an immense mass of documentation; great jumbled groups of charred and gutted trucks, tanks blackened by fire, their armour rent and twisted from 88-mm. shells. Enormous piles of small arms, rifles stacked thirty or forty feet high, rows and rows of field guns, each with the breech dutifully blown open with its last shot. In profusion, too, are the pictures of the dead. In lines, in heaps; stretched out and tranquil, huddled in agony; contorted, mangled, and burned. Sometimes it is plain that they have died in combat. At others, as the captions primly assert, as a result of "punitive" measures. In sorting through this quantity of "record" material so lovingly assembled, the sense of Teutonic sadism, of the Ger-

[8] The "agitators" were junior commissars specially charged with the dissemination of propaganda among the soldiers.

[9] The Soviets deny the German claim that over 600,000 were captured at Kiev. They say that the strength of SW Front before the start of the Kiev battle was 677,085, and that 150,541 were brought out. Thus they admit to losses of 527,000, including casualties in almost one month of fighting, but deny that these taken prisoner at Kiev could have come to much more than one-third of the German figure of 665,000 which may well have been inflated by inclusion of Opolchenye and civilian males.

man exultation in violence and brutality, is overpowering. Pictures are specially selected for their horrific character. Great as the victory was, the Germans were at pains to make it seem even more savage and merciless than its reality.

Of all the subjects none are more pathetic than the prisoners. Those long, patient columns winding their way in hopeless dejection across the cratered soil. In the eyes of the Russians there is that dumb, oxlike resignation of men who have fought for their homeland and lost. Can they guess what lies ahead of them? Deliberate starvation, camps ravaged by typhus, a twenty-hour day of forced labour at Krupp's under the stock whip of the SS? "Medical experiments," torture, four years of contrived brutality of the most horrifying and inexcusable kind. Did they with some intuitive shudder realise that out of a given thousand less than thirty would ever see their homes again?

But as we ask these rhetoric questions, let us pose one more. Did the Germans, as they watched those black caterpillars wending their exhausted path across the steppe, realise that they were sowing the wind? The first reaping, more terrible than anything they had ever experienced, was less than twelve months away.

eight | THE START OF THE MOSCOW OFFENSIVE

At the end of September 1941, as the last shots died away in the Kiev pocket and the hatches were battened down on the cattle trucks carrying Russian prisoners westward, the Germans were vexed by a continuing problem. The Bear was dead, but he would not lie down. Russian losses will never be known with complete accuracy, but the OKW estimate of two and a half million men, 22,000 guns, 18,000 tanks, and 14,000 aircraft was not a propaganda figure, but based on an intelligence collation of all unit reports. It corresponded, almost exactly, to the figure of Russian strength which these same intelligence experts had prepared at the start of the campaign. What, then, was holding the Red Army up?

The strategic objectives, such as they were, with which the Wehrmacht had begun were largely fulfilled; for Leningrad had been isolated and neutralised; the Ukraine had been opened to the German economy as far as the Donetz—and denied to that of the Russians. In the Bendlerstrasse work had already begun on a draft "Occupation Planning Staff Study," which forecast the withdrawal to Germany of about eighty divisions (of which half were to be demobilised); the military government would keep at its disposal

> strong mobile forces in the principal industrial and communication centres; each group, besides its normal occupation duties, would be able to send out fast battle groups into the unoccupied hinterland to destroy any attempts at resistance before these could become dangerous.

Yet in the front line the prospects looked very different. The German soldiers felt themselves deep in an alien land. The sameness of the territory was broken only by the rivers. The Dnieper, the Don, the Mius, the Sal, the Donetz, the Oskol, the Terek, the Sozh, the Oster, the Desna, the Seim. Over all these the patient engineers of the Wehrmacht had laid their bridges, by the banks of each their comrades had been buried. Beyond each lay the enemy, always in retreat, but always shooting. So often the day's fighting would end with the Russians once again on the horizon, the T 34's with their sinister hooded turrets just perceptible through Zeiss lenses, luring them even deeper, it seemed, into the east. From their allies, Rumanians and Hungarians who did not regard themselves as supermen, disquieting sentiments began to infect the Germans fighting alongside. That you had always to kill a Russian twice over; that the Russians had never been beaten; that no man who drew blood there ever left Russia alive. And every German, whatever part of the front he fought in, noticed with an uneasy mixture of horror and admiration the conduct of the Russian wounded.

> They do not cry out, they do not groan, they do not curse. Undoubtedly there is something mysterious, something inscrutable, about their stern, stubborn silence.

As if from a spiteful determination to *compel* their enemies to show weakness, the Germans rendered no medical services to the prisoners and kept their food rations to a minimum. Dwinger has described how

> Several of them burnt by flamethrowers, had no longer the semblance of a human face. They were blistered shapeless bundles of flesh. A bullet had taken away the lower jaw of one man. The scrap of flesh which sealed the wound did not hide the view of the trachea through which the breath escaped in bubbles accompanied by a kind of snoring. Five machine-gun bullets had threshed into pulp the shoulder and arm of another man, who was also without any dressings. His blood seemed to be running out through several pipes . . . I have five campaigns to my credit, but I have never seen anything to equal this. Not a cry, not a moan escaped the lips of these wounded, who were almost all seated on the grass . . . Hardly had the distribution of supplies begun than the Russians, even the dying, rose and flung themselves forward . . . The man without a jaw could scarcely stand upright. The one-armed man clung with his arm to a tree trunk, the shapeless burnt bundles advanced as quickly as possible. Some half a dozen of them who were lying down also rose, holding in their entrails with one hand and stretching out the other with a gesture of supplication . . . Each of them left behind a flow of blood which spread in an ever-increasing stream.

The uneasy feeling that they were fighting something of almost supernatural strength and resilience was widespread among the German soldiers, particularly the infantry, and can be traced in their letters and diaries—alternating with periods of triumph and exultation. But this mood took a little longer to penetrate OKH. It was not until the end of August that the possibility of a winter campaign—as distinct from the police duties of an occupation force—was accepted there. On 30th August, Halder ordered:

> In view of recent developments which are likely to necessitate operations against limited objectives even during the winter, the Operations Department will draft a report on the winter clothing that will be required for this purpose. After approval by the Chief of the Army General Staff, this report will be passed to the Organisation Department for the necessary action.

The victory at Kiev had encouraged many of the General Staff to believe that one more *Kesselschlacht* would finish the Russians off, and that they could winter in Moscow. Only Rundstedt was wholeheartedly against this course, recommending that the Army stand on the Dnieper until the spring of 1942. As this would have entailed withdrawals on the central front, it was *ex hypothesi* unacceptable to Hitler, and the fact that the most of the senior generals, Bock, Kluge, Hoth, Guderian, all held responsibilities (and nurtured ambitions) in the central sector ensured that Rundstedt remained in a minority among his professional colleagues. Under interrogation by Liddell Hart after the war, Rundstedt claimed that Brauchitsch agreed with him, but as his interrogator dryly observes,

> from contemporary records it does not appear that Brauchitsch really shared this cautious view when in more eager company.

Certainly the Commander in Chief did not give any indication of his misgivings when he spoke to Freiherr von Liebenstein and the other chiefs of staff at Bock's headquarters on 15th September. The purpose of the new operations, he told the assembled commanders, was "the destruction of the final remnants of Army Group Timoshenko." Three quarters of the German Army on the Eastern front was to be committed, including all the Panzer divisions (except those in Kleist's *Gruppe,* which were to continue clearing the Ukraine). Hoepner had been brought down from the north and would be put in the centre; on either side of him would be the 9th and the 4th, and the 2nd Army; and on the extreme flanks, once again, would be the Panzer armies of Hoth and Guderian.

The attack frontage was unusually wide. Between Hoth's starting line, north of Smolensk, and Guderian, on the left bank of the Desna, were over 150 miles. The German plan was that Hoepner's thrust would break the Russian front in two, and that the shattered fragments would polarise around the communications centre of Vyazma (assigned to Hoth) and Bryansk (the objective of Guderian). Once these twin *Kesselschlachten* had been subdued, there would be no obstacles to a direct advance on the Soviet capital.

For the rank and file the Führer had prepared an order of the day which informed them:

> After three and a half months' fighting, you have created the necessary conditions to strike the last vigorous blows which should break the enemy on the threshold of winter.

Coming after earlier and more strident exhortations, perhaps, it was an exhortation from which few can have derived much comfort.

As ill luck would have it, the Russian front opposite Bock's concentration was passing through a command vacuum during the very days when the Germans were completing their final dispositions. For Timoshenko had been ordered south, to build up some sort of a screen out of the shattered fragments of Budënny's old group. Koniev had been appointed to the "western front," while Yeremenko remained in command of the "Bryansk front." Co-ordination between these two was far from perfect, and the many gaps which weakened their joint front were being plugged directly from Moscow by troops of the "reserve front," responsible to Zhukov. While the "reserve front" was mainly concentrated around the arc of an inner defensive zone, Yeltsi-Dorogobuzh, with two armies astride the Yukhnov approach, Yeremenko was planning an independent counterattack at Glukhov—the very point which Guderian had selected for his own thrust.

The total Russian strength, including the active forces of Koniev and Yeremenko and the "reserve front," was fifteen infantry armies.[1] In

[1] In front of Moscow these were (north to south) Koniev's western front (22nd, 29th, 30th, 19th, 16th, and 20th Armies), 24th and 43rd from the reserve front, and the Bryansk front (50th, 3rd, 13th and "Operational Group Yermakov"). Behind them was the "Vyazma Line," occupied by the 31st, 49th, 32nd and 33rd armies of the reserve front.

numbers this came to over half a million, but their equipment varied enormously. Almost all were short of artillery, though mortars and smaller arms were plentiful. Their mobility factor was very low—even horses were beginning to be scarce, and were being drafted into mixed horse-drawn and mechanised "corps" (actually of about divisional strength) which were useful in reconnaissance or rear-guard actions but lacked the weight and firepower needed for prolonged fighting. Still more serious was the decline in the quality of the human material; for although the higher tactical leadership was improving as it emerged from the terrible crucible of the early battles, the Red Army man was left with little save his own personal courage and physical toughness to help him against the most experienced and highly trained soldiers in the world.

From the point of view of equipment and training, the armies with which the *Stavka* found itself fighting the opening stages of the battle of Moscow were the weakest the Soviets ever put into the field. Nearly all were reservists; what little they could remember of their basic training would have been along principles very different from those which had become *de rigueur* against the Panzers. A "refresher" course—with the majority so illiterate that they could not read a blackboard; a few days in overcrowded barracks or being driven like cattle across firing ranges with live ammunition singing over their heads; then the journey westward, the endless waiting in railway sidings, the shock and violence of air attack. This amorphous mass was soon to be subjected to an assault technique—the *Panzerblitz*—whose practitioners had brought it to a state of perfection, and to which no army in the world had yet found the answer.

But there was one reserve pool still left to the Russians, and it contained some of the finest units in the whole Red Army; these were the twenty-five infantry divisions, and the nine armoured brigades of General Apanasenko's "Far Eastern front." Apanasenko's command had been fully mobilised on 22nd June, and as the western frontiers began to cave in a Japanese attack was expected hourly. Then as days lengthened into weeks and the Siberian campaigning season shortened, the tension slowly relaxed and the *Stavka* began to contemplate the heady possibility of using these troops, highly trained and inured to cold as they were, at some moment of crisis in the West.

Stalin's experience of Japanese conduct along his Far Eastern frontiers over a long period in the 1930's, and of the force and suddenness with which they provoked and exploited "incidents," made him extremely reluctant to weaken his defences there. Following the safe course of attributing to others the same malignant and cynical process of thought which he himself applied to a given problem, the Russian dictator had for long resisted Shaposhnikov's advice that men should be brought west along the Trans-Siberian. That Stalin finally relented was due to the assurances which the *Stavka* was receiving from the Sorge network in Tokyo, whose reliability—unlike the Soviet-Japanese Neutrality Pact—he had solid grounds for accepting.

The Soviet Union, by reason of the appeal and catholicism of the Communist faith, has always possessed an advantage in matters of espionage and subversion over nations which have to rely on the baser (as Communists claim) motives of patriotism or avarice. In its confrontation with Germany, Russia received help, literally immeasurable, from three separate organisations.

The first of these was the *Rote Kapelle,* a spy cell operating deep within the German Air Ministry and including Schulze-Boysen, a senior intelligence officer of the Luftwaffe. Two other Soviet agents, Dolf von Schelia of the Foreign Office and Arvid Harnack of the Economics Ministry, fed Schulze-Boysen information from their own departments, which he communicated to Moscow by a secret radio service. *Rote Kapelle* was particularly valuable in supplying information concerning the disposition of the Luftwaffe, strengths and objectives for particular operations, and even details of individual air raids. It is also credited with passing the first news of the German decision not to push Kleist on into the Caucasus after the fall of Kiev, and of Hitler's preference for investing rather than assaulting Leningrad.

Second and, at this early stage of the war, undoubtedly the most important, was the Sorge group in Tokyo. Sorge was on the staff of the German Ambassador. He had access to, and reported on, every secret document which passed through the Embassy bag. Sorge also had a direct entrée to the working and decisions of the Japanese Cabinet through an associate, Hozumi Ozaki, who was an aide to Prince Konoye. As early as 25th June, Sorge reported on the Japanese decision to move into French Indochina, and during the summer all evidence from this source pointed to a Japanese preference for the soft pickings of the Dutch East Indies over the barren territory of Mongolia.

The third source from which the *Stavka* derived information concerning its enemy's plans was its Swiss agent "Lucy." Lucy's identity has never been established, but his—or her—importance was crucial.

> Information of such an accurate and incredibly well-informed nature streamed to Moscow that suspicions were aroused that this was merely an agent of the *Abwehr* engaged on an elaborate process of "disinformation" aimed at luring the Soviet command into a giant trap. In what remains an astonishing performance, and one finally accepted by Moscow as genuine, "Lucy" supplied up-to-date data on the German order of battle, with day-to-day changes, as well as being able to answer enquiries about high-level matters dealing with the German Army. Such was "Lucy's" role that one highly valued Soviet agent considered that ". . . in the end Moscow very largely fought the war on Lucy's messages."

The Germans, in contrast, had very little idea of what was going on in Moscow. The gross, and now painfully apparent, errors in their earlier estimate of Soviet strength had discouraged OKH from making any assumptions other than those justified by actual field data—interrogation of prisoners, unit identification, and so forth. Captured Russians were usually too frightened to talk, and generally so ignorant of matters outside their own platoon that, even had they the inclination, their information would have been of negligible value. The Luftwaffe did its best, but once again the primitive character of the Red armies, their absence of a heavy "tail," their propensity toward moving by night, made it difficult for an intelligence forecast to be built up outside the sphere of local tactical dispositions. While the Germans retained the initiative, and a complete ascendancy over the battlefield, this handicap was of no very great importance. But once their advance slowed down and their forces were overextended, ignorance of their opponent's real strength and intentions was to bring them close to disaster.

The Russians' field intelligence was inferior to the Germans' during the first months of the war. They took fewer prisoners, and in the chaos of the seven-hundred-mile withdrawal there was neither the time nor the apparatus for sifting and analysis of reports. But by the autumn of 1941 the Russians were beginning to enjoy the benefit of reports, which were to grow in scope and accuracy, from the Partisan bands that were operating behind the German lines.

"The Partisan movement was the 'fourth arm' in the Great Patriotic War." This is the standard claim by Soviet military historians, and until the 20th Party Congress the credit for its inauguration was given to Stalin's speech of 7th July. But the facts are that no comprehensive plan for guerilla warfare in occupied territory existed. The Soviet dictator's reluctance to encourage independent paramilitary organisations was partly responsible, together with the customary aversion of dictatorships to admit that territory might ever be ceded—lest this should lead to undesirable political speculation by the local population. Even the Party-dominated *Osoaviakhim* was geared to the requirements of the Red Army in "orthodox" warfare and the maintenance of security behind the lines.

Consequently the origins of the movement were (dialectically speaking) unheroic and slightly disreputable. Bands of soldiers, cut off from their units, men often who had slipped out of some Panzer encirclement at night and found themselves impossibly far from their own lines, usually with a good leavening of deserters and local militia who had simply gone to ground in country they knew well, these were the stuff of the marauding groups which roamed White Russia in the summer of 1941, pillaging and looting certainly, but exchanging fire with the Germans only when necessary to save their own skins. These men were cut off in every sense. They had no desire to rejoin the main body of the Red Army, where, regardless of their individual histories, they were likely to be shot as deserters or relegated to punishment battalions.

Their fate at German hands lacked even this element of uncertainty. When the first rumours of Partisan "warfare" reached Hitler, he welcomed them. "It has its advantages: it gives us a chance to exterminate whoever turns against us." The SS was nominally responsible for "order" in the capital territories, and the Regular Army was brought in on the act by an OKW order of 16th July, 1941.

> The leading principle in all actions and for all measures that must be resorted to is the unconditional security of the German soldier. . . . The necessary rapid pacification of the country can be attained only if every threat on the part of the hostile civil population is ruthlessly taken care of. All pity and softness are evidence of weakness and constitute a danger.

As in the case of the "Commissar order," the fiction of legality could be dispensed with, in favour of

> The spreading of that measure of terror which alone is suited to deprive the population of the will to resist.

Instead of leading to "rapid pacification" of the country the repressive measures on which the Germans embarked with such relish fanned a hot wind in which the Partisan movement grew apace. Farms and villages no longer bolted their doors and hid their food; the inhabitants who had first received the invader with curiosity and, almost, relief turned against him with universal hatred. In particular the "national" character of the struggle, which Stalin was now exalting over ideological and Party doctrines, acquired a particular significance. The wandering brigands observed the brutality with which their innocent kinsfolk were treated, and the news spread. They began to strike at the Germans not simply for food and ammunition, but for revenge. A new dimension of cruelty began to throw its shadow over the war in the East.

At the same time the *Stavka* came to appreciate the military significance of the mass of men who were left behind the German advance (the lowest estimates indicate that there were never fewer than 250,000 armed Russians at loose in the occupied territories) and to take progressive measures to organise and encourage them. Trained "agitators" were dropped by parachute, regional commands established, discipline was reasserted, wireless and explosives were supplied. Within a few months the bands were thinking of themselves not as repudiated stragglers but as national heroes.

The German reaction was predictable. Terror (described in this document as "unusual hardness") must be intensified, and it must be universal. To ease soldierly consciences the High Command decreed:

> In every instance of active opposition against the German occupation authorities, *regardless of the specific circumstances,* Communist origin must be assumed.

Once having got things onto an ideological footing, it was then easier to order:

> For the life of one German soldier, a death sentence of from fifty to one hundred Communists [i.e., Russians] must be generally deemed commensurate. *The means of execution must increase the deterrent effect still further.*

To this end it was ordered that firing squads should aim at or below the waist—a practice which resulted in the majority of the victims being buried alive and in agony from stomach wounds. So much for increasing "the deterrent effect." There were also practical grounds to "justify" the order, for

> In cases where children are included among the hostages [aiming at the normal height] such persons may escape execution altogether . . .

> and would have to be despatched by hand of officer in charge of the burial party.

Gradually the motives behind the terror campaign altered. At first, confident in approaching victory, the Germans took a sadistic pleasure in the repression. How agreeable to combine duty and sport; to bask in the glow of the crusader while enjoying the particular physical pleasure which so many Germans derive from the infliction of pain. In the long summer evenings "man hunts" used to be organised on the slightest pretext, villages surrounded, set alight, and the inhabitants "beaten" like rising game birds and cut down in the streets. Then it was rewarding to loot the dwellings for "souvenirs" and to send these, and photographs of the scene, to friends back in Germany.

> . . . here is a lock of hair from a Russian guerilla girl. They fight like wild-cats and are quite sub-human [*Untermensch*].[2]

But gradually it dawned on the Germans that the war would not be over so soon; that they were few, the Russians many, the territory vast. Fear and guilt compounded exultation as the activity of the Partisans and the sullen hatred of the civilian population became daily more obtrusive. The Russian underground repaid its oppressors in the same barbaric coin. A hospital train was derailed at night, and the wounded burned with paraffin; a barracks had its water supply poisoned.

> While we were in the Mogilev region a rumour reached von dem Bach[3] that there was a quantity of gold at the Polyakovo state farm. We went out there and tore the place to pieces looking for it. The head of the settlement begged us to wait as he could get the gold in 24 hours and if the buildings were all destroyed the peasants would have nowhere to spend the winter. At dusk we left with orders that the gold was to be produced the following day or the entire population of the farm would be placed under arrest. Fischer . . . and a detachment of four men stayed behind to keep an eye on things. The next day—no Fischer. We could not get him on the short wave radio as they were with motor-cycle outfits and the farm was nearly 100 kilometres from Mogilev. So the following day we went back there with 6 armoured cars. . . . The place was burned to the ground. One building, the office stood, and in it . . . a leather box,

[2] The term *Untermensch* was soon to embrace a whole philosophy toward the Russians, and in particular toward prisoners of war and forced labour. This subject is dealt with at some length in Ch. 11.

[3] SS Obergruppenführer Erich von dem Bach-Zelewski. His name will be encountered again on these pages.

> very heavy, *Gelb* scrawled on it in white paint. We opened it. Inside were the heads of Fischer, Hahn, Neudeck and Grose.

Thus it can be seen that in two respects, long-range espionage and security in their rear, the Russians already had an advantage over their enemies—and one whose dividends would accumulate as the war lengthened. But on a purely military assessment it must have seemed at the start of the Vyazma-Bryansk battles that the end of the war was still likely in the autumn of 1941. By the end of September, Bock's army group was set up for a tank drive greater even than the first days of *Barbarossa.* Kempf's 48th Panzer Corps had been brought up from Army Group South and put, with the 9th Panzer and two motorised divisions, under Guderian. And the whole of Hoepner's *Gruppe* had been taken from Army Group North. Long as the attack frontage was by normal army group standards, it was but a fifth of the line on 22nd June, yet the number of Panzer divisions taking part was only three fewer.[4]

Facing them was the last of the Russian mass armies (for until 1944 the Red Army was to fight, as much out of scarcities as owing to inexperience, in smaller units and along different tactical principles). Thrown together in urgency and confusion, under a divided command, it was dismally unprepared for its ordeal.

On the third day of the offensive Guderian noted, ". . . a complete break-through has been effected."

Twenty-four hours later the tanks of the 4th Panzer burst into Orel, to find the electricity still connected and the trams running. Precious machine tools were stacked at the railway sidings, awaiting shipment to the Urals. In the centre Hoepner, whose *Gruppe* had been reinforced by SS *Das Reich* and *Gross Deutschland,* had broken the Russian front irreparably into two, forcing the severed mass of Koniev's "western front" around against the upper Dnieper and into the path of the advancing armies of Kluge and Strauss. Still farther north Hoth had wheeled his tanks down onto the Vyazma-Gzhatsk highway, placing a "long-stop" behind the Russian infantry. In these two pockets over five hundred thousand men were pinned down for liquidation. It was the shortest and, it must have seemed, the most surgical of all the amputations performed on the Red Army that summer. Now the way to Moscow was truly

[4] Including satellites, total German strength was 207 divisions—41 more than on 22nd June.

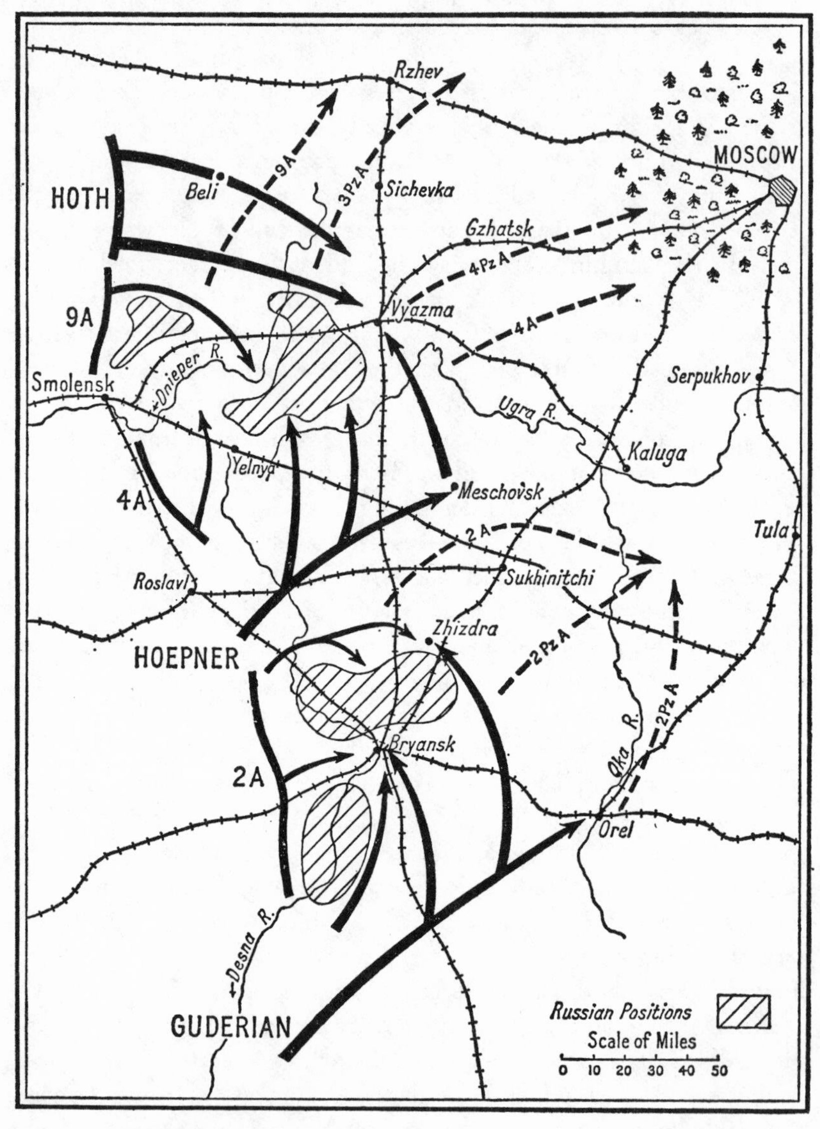

Moscow. The Battles of Vyazma-Bryansk

wide open, and it was no time to heed either the unfavourable weather forecasts or the disquietingly high figures for vehicle breakdowns. An assembly of foreign correspondents in Berlin was told by Goebbels, "The annihilation of Timoshenko's [sic] army group has definitely brought the war to a close."

Certainly there were very few in the German Army who now thought of stopping before the Russian capital was reached. And some minor portents went unnoticed in the general euphoria of victory. On 7th October the first snow fell. It melted quickly. That day Guderian sent army group an inquiry for winter clothing. He was told that he would receive it in due course (actually he was never to get any) and "not to make further unnecessary requests of this type."

And over the broken-up remnants of the Soviet "western front" a new commander was appointed. His name, which went unnoticed by German intelligence, Georgi Zhukov.

nine | THE BATTLE OF MOSCOW

In the second week of October the enormity of their defeat in the Vyazma-Bryansk battles had dawned on the Russians, and with it the immediacy of their peril. On 12th October, Hoepner crossed the Ugra River, now sluggish with forming pack ice, and forced a new and agonising dilemma on the *Stavka*. Would the Germans wheel right, to Kaluga, to bring off one more ruinous encirclement of the bedraggled armies that faced Guderian? Or would they drive direct on Moscow, through Maloyaroslavets? Or would they turn north, joining Hoth's 3rd Panzer Army to crush the right flank of the Moscow armies and expose the whole northeastern screen of the capital's defences?

Directly in Hoepner's path there were three skeletal Russian infantry divisions. They had left their artillery on the west bank of the river, had no tank strength whatever, and barely a brigade equivalent in numbers of cavalry remnants—for the most part exhausted and demoralised from units that had been smashed in the previous week's fighting. Some eighty miles to the north another small force, under Lieutenant General D. D. Lelyushenko, was falling back from Gzhatsk, straining every muscle to hold off the weight of Hoth's armour.

Upon these two groups—"armies" is too grandiose a term, although Lelyushenko in particular carried the staffs of many broken armies fighting with his forward elements—hung the cohesion of the whole Russian front before Moscow. What strength the *Stavka* could still command in this sector was stranded out on the wings, before Guderian at Orel or in the far north around the headwaters of the Volga. Yet even if these divisions are included it is plain that the Germans enjoyed a crushing superiority of numbers to leaven their material ascendancy. Only 824 tanks remained to the entire "western front," and of these it is unlikely that more than half were fully operational, and still fewer were T 34 or KV—the only types capable of taking on the Panzers.

As at Leningrad in September, the local Party Secretariat worked twenty-four hours a day raising and organising "workers' battalions," but once again the grinding shortage of equipment limited their effectiveness. Five thousand rifles and 210 machine guns were released by the Moscow garrison for these new formations, but after this a new order decreed that no further grants of equipment should be made by the regular forces. A typical battalion of 675 men was ordered to the front with only 295 rifles, 120 (captured) hand grenades, 9 machine guns, 145 revolvers and pistols, and 2,000 "Molotov cocktails," the crude gasoline-filled bottles which had to be rolled down a tank's exhausts before they were effective.

Throughout this critical period only one fresh, trained division reached the "western front," the 310th Motorised, which came (without its vehicles) from Siberia. One watcher has described how the leading battalions, fit and strong in their quilted uniforms, rolled into Zvietkovo railway station and "greeted in passing, with hand and caps, their comrades, who had scarcely the strength to answer with a wave of their hands." Indeed, when Zhukov assumed personal command of the battle on 14th October, it must have seemed to him that the twin barrels from which Mother Russia had for so long been accustomed to draw for her security—space and men—were truly being scraped to the bottom.

For a few more days Zhukov could cherish the belief that his situation, though highly dangerous, was controllable. While the flanks held, at Kalinin and Mtsensk, a weak centre was the lesser evil, for the dreaded pincers of Hoth and Guderian would be kept apart, and only Hoepner's tanks remained to back the direct approach of the German 4th and 9th armies. Even if Bock succeeded in covering the whole distance to the capital, there was still the possibility that, like Napoleon's, his tenure of the city could be precarious and short-lived. If the Russian divisions on the flanks could hold their position, for Bock's army to strain forward into Moscow by the direct route would be "to poke his snout into a trap —which we will spring when the blizzards start."

It was apparently this thought which was uppermost in Zhukov's mind, as it must have been in that of every man in the Red Army. There was but six weeks to go before the onset of winter—their last ally, which might yet succour them in their extremity. Each day that passed brought closer the time when the ice wind, now gathering strength over the Aral Sea, would sweep down over Siberia, across the steppe, through Moscow, and onto the battlefield.

The Germans, too, had an inkling of this. Bayerlein has described how in the mornings, as the Panzers started up their engines,

> The already flat rays of the sun, low in the horizon over the plains, misled us. But each evening . . . ominous black clouds would build up far in the distance, towering high above the steppe. These dark masses carried in the stratosphere the rain, the ice and snow of the coming winter. But each morning they were gone or seemed to only to reappear more mountainous than ever in the evening twilight.

But on 14th October the station was transformed when the northern hinge of the Russian front cracked. Hoth's tanks broke into Kalinin, and the 3rd *Panzergruppe,* with the 9th Army on its heels, rolled down the headwaters of the Volga toward the "Moscow Sea," the huge artificial lake from whose eastern end a canal led seventy miles due south to the capital. Within weeks the lake would have been frozen over its entire fifty-mile length, and useless as a defensive barrier. But those days were themselves vital, and now, with the five infantry divisions and the few shot-up packets of armour and cavalry which were stranded on the north shore of the lake, they were gone.

Zhukov realised that above all, he had to keep his command in being. There could be no more standing fast regardless of the consequences, no more trading of lives for time while the reserves accumulated. Because there were no reserves, lives and time were *pari passu* in the scales.

In Moscow itself a sullen fear spread downward from the ranks of high Party officials, who knew the reality behind the communiqués and the exhortatory proclamations which glared down from the drab walls. The cream of manpower was skimmed off into the "workers' battalions," while over half a million of the city's less martial inhabitants were drafted into the outskirts to work day and freezing night on defence positions and anti-tank trenches. From these people, from wounded Red Army men passing back through the railway stations, and along that broad mysterious "grapevine" that flourishes in a repressive society, ugly and disturbing rumours sprouted. The recurrent nightmare of the Tsarist armies, crippling shortage of ammunition, was throwing its long shadow. Horrific stories of the enemies' behaviour to prisoners and civilians alike matched tales of mass execution of "deserters" and "malcontents" by the NKVD. For three days after the fall of Kalinin something akin to panic gripped Moscow. The news that government offices were being transferred to Kuibyshev prompted a mass flight by all those who were capable of movement. At first affecting lesser Party officials and bureaucrats, under the pretext that they had already received orders, the movement spread rapidly to their families, to officials of key organisations such as the ration and postal bureaus, and even to the police and part of the militia on internal duties.

With distribution broken down and the streets empty, looting and plundering began. The sound of small-arms fire by day and the glare of German incendiaries by night completed the picture of a capital on the verge of disaster.

Stalin himself remained. It would be understandable if he ruminated on the peace of Brest-Litovsk, which Lenin had signed in 1918 to save the Bolshevik regime from destruction by the German Army. There is no doubt that Stalin's nerve gave way on occasion, just as his clumsy interference in military operations had upset Shaposhnikov's conduct of the retreat. He is on record with some uncharacteristically wild statements at moments of stress. "American troops under American command will be welcome on any part of the Russian front"—to Harry Hopkins on 30th July; "A British Expeditionary Force might operate from Persia and join in the defence of the Ukraine"—to Stafford Cripps after Uman. And, even more melodramatic, following the fall of Kiev: "All that Lenin created—we have lost for ever." But no record can be found of any diplomatic approach, of peace overtures, however tentative or indirect, by the Soviet Union at this time. The dictator of all the Russians had too many enemies to risk altering the *status quo*.

On 19th October, Moscow was declared to be under a state of siege, and special reinforcements of NKVD security troops were brought in to restore order. From that time the momentary flickers of panic died away. The atmosphere in the city, blacked out, with snow lying in the streets, the shops closed, air-raid sirens wailing, was of a kind of militant despair.

This despair was grounded in the hopelessness of personal salvation from any quarter. It was the antithesis of the apathy and resignation that lay behind the French collapse in 1940. Then a people sacrificed their country and institutions for their own personal safety. The pleasures of wine, adultery, and civilised conversation could, it seemed, be preserved simply by refusing to fight. But the Russians of 1941 knew these things only dimly. Privation and sacrifice were, and for centuries had been, their habitual condition; and now in the German invader they had a focus for all their misery and resentment. There was also a deeper inspiration to their resistance.

> Even those of us who knew that our government was wicked, that there was little to choose between the SS and the NKVD except their language, and who despised the hypocrisy of Communist politics—we felt that we must fight. Because every Russian who had lived through the Revolution and the thirties had felt a breeze of hope, for the first time in the history of our people. We were like the bud at the tip of a root which has wound its way for centuries under rocky soil. We felt ourselves to be within inches of the open sky.
>
> We knew that we would die, of course. But our children would inherit two things: A land free of the invader; and Time, in which the progressive ideals of Communism might emerge.

If, as Hitler claimed to believe, the Will was all-important, the Germans had already lost the war. For what could they put against this? Greed for territory and *"Sklaven,"* a contrived doctrine of racial "superiority"; some muddled prejudices about "Bolshevism." These things were valueless against the deep patriotism, the submissive faith in the dialectic, of the Russians. The Wehrmacht was living by the sword. If, and when, the sword should blunt . . .

The problem before Zhukov and the *Stavka* was one of finesse. They had to keep some sort of a front in being until winter arrived, and their resistance had to be flexible enough to evade encirclement yet sufficiently vigorous to delay the enemy whenever his condition, or a break in the weather, offered the opportunity. Of the two classic pillars of Red Army doctrine (and of the Imperial Army before it) one, mass, had already been broken beyond repair; the second, the fighting retreat which lured an invader ever deeper into the east, was limited by the urgent necessity of keeping him at arm's length from Moscow. To this end the Russians were fighting with small *ad hoc* groups of mixed arms, strong in cavalry but rarely greater than a brigade in strength, which were conducting mobile operations between a net of defended localities manned by local militia and "workers' battalions" themselves having orders to fight to the end.

In the north and centre the country was so heavily wooded that the Panzers rarely got the opportunity to fan out. Accustomed to the puny *Bauern* of Western Europe, the Germans were bewildered by the vast forests in which they struggled day after day. Now it was dark for fourteen hours out of twenty-four. While the German columns were halted, Soviet cavalry threaded its way along trails behind the "front," laying mines and mortaring supply convoys. Even Hoth's *Gruppe,* which had broken through, it seemed, at Kalinin, was reduced to a walking pace by the end of October.

Thus, although the Germans were closest to Moscow in the north and centre—at Mozhaisk they could see the anti-aircraft fire over Moscow on a clear night—the real danger for the Red Army was farther south, where the country was more open and where, almost without tanks, Zhukov was faced by the whole of Guderian's 2nd Panzer Army. At this stage in the battle Zhukov had only one independent tank force left, the 4th Armoured Brigade of Colonel Katukov. This had been newly equipped with T 34's in September, and was staffed by pupils and instructors from this tank training school at Kharkov. It had already enjoyed two narrow escapes from encirclement, having first been ordered to Budënny at Kiev, but arriving on the scene two days late, after Model had closed the trap; then entrained at Lgov for the "western front," it had passed through Orel on the very day that Guderian broke through. With the exception of a sharp brush with *Gross Deutschland* at Bielopolie on 20th September, the brigade had barely fired its guns, and after the Bryansk encirclement had opened the Russian southern flank it was the only force with any striking power left between the Oka and Mtsensk—a void nearly seventy miles wide. Ordered to turn around at Mtsensk and hold up Guderian's advance on Tula, Katukov had delivered a sharp attack against the 4th Panzer on 6th October, causing that division "to go through some bad hours, and suffer grievous casualties." Instead of pressing his initial advantage Katukov then withdrew, prudently taking the view that to keep his force in being was more important than a glorious death ride against the whole enemy *Gruppe*. Guderian recorded, "This was the first occasion on which the vast superiority of the T 34 to our own tanks became plainly apparent . . . The rapid advance on Tula which we had planned had to be abandoned . . ."

After licking its wounds for a couple of days the 4th Panzer had resumed its advance on Mtsensk, and entered the outskirts of the town on 11th October. Geyr (the corps commander) had wished to relieve the

4th Panzer with the 3rd, and a part of 10th Motorised, but the bad condition of the roads made this impossible without a halt of at least two days and pulling the division back to Orel, so the 4th Panzer had struggled on, through periodic snow showers, which melted as they touched the ground. The mud was now so thick that movement off the road was impossible, and along it vehicles were averaging six miles in an hour. On the evening of 11th October, as the vanguard of the 4th Panzer probed cautiously into the burning suburbs of Mtsensk, the division was strung out over fifteen miles of single-track road, and with its supporting artillery and infantry almost outside the range of radio contact. This was the moment for Katukov to strike again. The T 34's, moving rapidly across the earth, which was already freezing again in the lowering temperature of dusk, their wide tread carrying them in places where the PzKw IV's floundered on their armoured bellies, attacked the German column with dash and violence, slicing it up into sections which they systematically reduced. The gunners of the 4th Panzer, whose morale had already suffered in their first encounter with Katukov five days earlier, once again saw their shot bouncing off the sloping armour plate of the Russian tanks.

> There is nothing more frightening than a tank battle against superior force. Numbers—they don't mean so much, we were used to it. But better machines, that's terrible . . . You race the motor, but she responds too slowly. The Russian tanks are so agile, at close ranges they will climb a slope or cross a piece of swamp faster than you can traverse the turret. And through the noise and the vibration you keep hearing the clangour of shot against armour. When they hit one of our panzers there is so often a deep long explosion, a roar as the fuel burns, a roar too loud, thank God, to let us hear the cries of the crew.

The 4th Panzer was virtually destroyed, and the defence of Tula granted another precious respite. But aside from its tactical significance Guderian drew an ominous conclusion. "Up to this point we had enjoyed tank superiority. But from now on the situation was reversed."[1]

[1] Guderian records, "[I] made a report on this situation, which for us was a new one, and sent it to Army Group. I described in plain terms the marked superiority of the T 34 to our PzKw IV and drew the relevant conclusions as they must affect our future tank production. I concluded by urging that a commission be sent immediately to my sector of the front, and that it consist of representatives of the Army Ordnance Office, the Armaments Ministry, the tank designers, and the firms which

One of the many paradoxes of the Eastern campaign is that this moment, when the Russians were at their weakest, should have seen the first real doubts rising in the German Army. While Guderian was busying himself with practical complaints, other officers took advantage of the enforced halt in literary reflection. Blumentritt has described how, although there was little opposition, ". . . the advance was slow, for the mud was awful, and the troops were tired."

> Most of the commanders were now asking: "When are we going to stop?" They remembered what had happened to Napoleon's army. Most of them began to re-read Caulaincourt's grim account of 1812. That book had a weighty influence at this critical time in 1941. I can still see Von Kluge trudging through the mud from his sleeping quarters to his office, and standing there before the map with Caulaincourt's book in his hand. That went on day after day.

During the last three weeks of October weather conditions—heavy rain, snow showers, damp and penetrating mists—made movement almost impossible on two days out of three. Conditions varied along the arc of the German attack and added to the difficulties of co-ordinating the armoured flanks with Kluge's infantry mass in the centre. At the northern end of the front, from Kalinin to Mozhaisk, freezing temperatures would sometimes persist all day. Then the Germans could press hard against Zhukov's screen as it covered the withdrawal from the Rzhev-Gzhatsk bulge. But even here the barometer was erratic, and a twelve-hour thaw with rain would throw the advancing columns into disorder. On such days a single battery, a hastily laid mine field in a defile between swampy woodlands, could hold up a whole Panzer corps.

General Bayerlein[2] (who commanded a mixed combat group in the 39th Panzer Corps) has given one of the best descriptions of this stage of the advance from the German side. His group consisted of about twenty-

---

built the tanks. . . . it could examine the destroyed tanks on the battlefield . . . and be advised by the men who had to use them what should be included in the designs for our new tanks. I also requested the rapid production of a heavy anti-tank gun with sufficient penetrating power to knock out the T 34."

[AUTHOR'S NOTE: This commission was in fact constituted remarkably quickly, and visited Guderian's headquarters on the 20th November. The part it and other influences played on German tank design is dealt with in Ch. 15.]

[2] Lieutenant General Fritz Bayerlein, soon to be engaged in more agreeable climes. He went to the Afrika Korps in February 1942. He commanded the unsuccessful attack at Alam Halfa, and later the redoubtable *Panzerlehr* in Normandy and the Ardennes.

five PzKw III's and IV's, with a few Czech tanks, which were used to stiffen the motorcyclists in reconnaissance; a *Panzerjäger* company with twelve 37-mm. anti-tank guns; an artillery battery with four 105-mm. guns; and two *Panzergrenadier* companies of infantry, some in half tracks and some in trucks.

> By the beginning of November [he writes] we had reached an extensively wooded area east of Ruza and north of the main Smolensk-Moscow highway. . . . After unceasing rainfall the ground became soggy and afterwards intermittently, lightly frozen. According to the map good roads should have been available. This turned out to be an illusion. The Ruza-Voronzovo highway went over a bad forest road and was only usable in the beginning. . . . The tanks could only labour forward step-by-step in the sticky morass. The movement of wheeled vehicles was impossible. The attack, however, had to be pressed forward under all circumstances.
>
> After only about 10 km, near Panovo, even the tanks were stuck . . . The Pioneers had to build a corduroy road of saplings over a 15 km stretch from Voronzovo to Panovo, but even on this, because of its unevenness, travel was only possible for full or half track vehicles. . . . it took several days to bring up the infantry and secure Modenovo against counter-attack.

The Russians were now withdrawing fast. Zhukov had fixed the bounds which were to be defended until the last, and until these were reached he had no intention of risking any more disastrous encirclements. But their rear guards never left a position until they were forced off it; and the moment the Panzers halted, whether because of fatigue, shortages, or climate, the Russians turned and harried them without respite. Bayerlein goes on:

> The Russians would attack during the night . . . and it was necessary that the task group be constantly on the alert defensively. In order to do this it was necessary to maintain the tanks' engines at the right temperature. Every four hours the motors were run for 10-15 minutes until they had reached a temperature of 140° Fahrenheit. These periods commenced for all tanks at exactly the same second in order to minimise interference with the forward listening posts, whose task was aggravated by the dense ground mist that rose from the marsh, particularly at night-time. We discovered that the transmissions must also be operated while the engine is idling, otherwise by a sudden start the metal parts of the power train from the engine to the drive sprockets would be damaged [because of the low viscosity of the oil at those temperatures].
>
> After a few days' halt the Russians would have full knowledge of the positioning of our entire defences. . . . They employed the whole civilian

> population; women, children and cripples, who at first did not seem at all suspicious . . .

Bayerlein also complains of the effect on morale of the *Katyusha* rocket mortars, which were now beginning to be used in numbers for the first time; and also of a revival of activity by the Red Air Force. "They would attack with single aircraft of any type, even under the most unfavourable weather conditions, when we ourselves were getting no cover from the Luftwaffe."

> For the tanks' crews it was the beginning of a bad time. The constant confinement to the inside of the tank was not possible without a reduction of the combat efficiency of the soldiers, in as much as it was narrow and cold. Earthen bunkers could not be constructed in the ever-soggy and muddy ground. The wooden houses which were made available as billets were in the course of time systematically destroyed, one after another, by enemy fire, and burned to the ground.
>
> The supply difficulties took on formerly unheard-of proportions. Through the constant warming of the engines there was a higher rate of gasoline consumption. Constant defensive combat caused a profuse ammunition expenditure. . . . For days on end there were no hot meals for the combat troops—intestinal diseases and disorders were the result.

The Germans' difficulties were multiplied by their order of battle, which was becoming unbearably congested as the actual battle front shrank in size. There were only three major approach routes to the capital, and the few secondary ones were narrow, vulnerable, and weather-bound for days at a time. The result was that Hoth, Strauss (9th Army), and Kluge were all competing for use of two main roads, the Smolensk-Moscow "highway" and the Moscow-Klin road. Hoepner and Weichs (2nd Army) had to share the Moscow-Kaluga road, and Guderian, until he had captured Tula, would have no hard-surface communication routes whatever. After three weeks of struggling through mud and mine fields with their supply lines entangled, their vehicles breaking down, and casualty returns mounting, Bock saw that they would have to regroup before the final march on Moscow. On 27th October, Goebbels told a somewhat startled press conference (who had heard him proclaim only a fortnight before that the war was over), "Weather conditions have entailed a temporary halt in the advance."

Against this background of disillusion and failing momentum a Chief of Staff conference was called at Orsha, the headquarters of Army Group

Centre, for 12th November. This was to be one of the decisive moments in the history of the German Army. The question before its senior officers was a simple one: were they to follow the prudent dictates of their military conscience, take up "winter quarters," rest and refurbish their forces, and plan the next stage of the campaign at their leisure? Or were they to gamble the remains of their own strength against an imponderable—the strength remaining to the Red Army and a certainty—the violence of the Russian winter? Of course it is true that there are occasions in every campaign when the state of the enemy may present a tactical opportunity which demands vigorous, indeed reckless, exploitation. But was this really one of them? The evidence that the Russians were on the point of collapse was very tenuous, and based in large part on calculations which had already been proved false. As for the winter—here the evidence for the last 150 years was unanimous, yet no provision at all had been made to fight a war of movement through those months, other than the first tentative enquiries about clothing "requirements." [3]

The Orsha conference was an OKH affair, called by Halder and attended by the Chiefs of Staff (not the commanders) of the subordinate armies in Bock's army group. Although Orsha was the site of Bock's headquarters, the conference did not take place there but in Halder's special train, which was standing on a siding at the railway station. And although a formal "discussion" was invited following the address with which Halder opened proceedings, it is clear that this was intended to be concerned with details rather than principle; for the Chief of the General Staff had brought the "Orders for the Autumn [sic] Offensive, 1941," with him, and these were distributed, without amendment, when the meeting broke up.

The decision which Halder announced—to resume the advance on Moscow—has often been held up as one of the many instances of Hitler's forcing the generals into actions of which they disapproved. But like so many of the other "examples" of the Führer's destructive interference, a moment's objective scrutiny will show another side to the affair, that it could equally well be cited as illustrating the characteristic inflexibility of the German General Staff.

At the end of October there was a good deal to be said for making one last attempt to reach the Soviet capital. Both Halder and Brauchitsch had been trying (in their own personal and circumspect fashion) to

[3] See p. 157.

persuade Hitler to concentrate on Moscow since the start of the campaign. In reports, in conversation, in memoranda, this course had been urged to the exclusion of others. After the battles of Vyazma-Bryansk the last obstacle (on Bock's figures, at all events) had been removed. There was also the consideration that if the attempt were not to be made, then consolidation of a "winter line" would entail withdrawals; minor ones, it was true, but in the course of straightening out the front ground bought with German blood would have to be ceded. How could anyone at OKH explain this to the Führer so soon after the greatest victory of the whole campaign?

All this is understandable, and Halder and Brauchitsch must have come to their decision to stage a new offensive at some time between 26th and 30th October, as the orders for redeployment of the army group were sent out at that time. The two infantry armies of Strauss and Weichs were moved to the flanks; Reinhardt (who had taken over Hoth's *Gruppe*) and Hoepner were placed side by side on Kluge's left, and Guderian was moved closer in, to take station, as he had on 22nd June, on Kluge's right.

But while these movements were taking place, the assumptions upon which the OKH plan was based were daily shedding their validity. The impact of the weather, in terms of its effect on morale and on the efficient working of equipment, was already greater than expected; far from waning, Russian resistance had intensified. It must have been clear to Halder, several days before he journeyed to Orsha, that to reach Moscow before Christmas was going to be a major operation, and a very difficult one. If Halder had private doubts, there would still be an opportunity to change his recommendation should the army group Chiefs of Staff be unanimous or a majority of them reject the plan. This likelihood was increased by certain changes Hitler had made in the original scheme when Brauchitsch discussed it with him. The Führer had become uneasy at the manner in which the Panzer divisions were bogging down in the Istra forests, and favoured a wide sweep behind Moscow rather than a direct march on the capital. "The town will fall," he had told Mussolini, "without the loss of a single man." This plan looked good on the wall map at Rastenburg, but completely ignored the condition of the troops and the state of the terrain—and thereby made it easier for the generals to present a united opinion and throw out the idea of an attack on Moscow, in any form.

There are only two records of what went on at the Orsha conference. One is to be found in Halder's diary. It is cursory, and insofar as his own attitude to the operation emerges, it is seen to be inconsistent. The second account was given by Kluge's Chief of Staff, Blumentritt, under interrogation in 1946. This makes it plain that there was an ample body of professional dissent if Halder had needed support.

> The Chief of Staff of Army Group South, von Sodenstern, expressed a most emphatic opinion against any further advance. So did the Chief of Staff of Army Group North. At Army Group Centre von Greifenberg took a more indefinite line, pointing out the risks but not expressing opposition to an advance. He was in a difficult position. Field-Marshal von Bock [whose Chief of Staff he was] was a very capable soldier, but ambitious, and his eyes were focussed on Moscow . . .

Some of the subordinate staff officers were more outspoken. On being assigned as objectives the railway junctions at Gorki (250 miles *behind,* i.e., east of Moscow) Liebenstein protested, "This is not the month of May, and we are not fighting in France!" Halder listened impassively to the objections, then closed the discussion by declaring that the offensive was "the Führer's wish," and that it was necessary to capture the railway junctions "as OKH has reports that large Russian reserves, amounting in strength to a fresh army, are on their way from Siberia."

On this highly disquieting note the conference broke up, and the staff officers returned to their armies to prepare for the final battle.

The transfer of troops from the Far East had begun in earnest in the first days of November, and by the time that the German offensive got under way again Zhukov had more than doubled his strength[4] as compared with the initial period at the middle of October, when he had assumed active command. Yet the total Russian strength remained inferior to that of the Wehrmacht, in numbers as well as in weight of equipment, and to match the German strength before Moscow the *Stavka* was coolly running the risk of thinning out other sectors of the front, taking divisions from regions where it could still "deploy" space until winter

[4] The total brought from the Far East in the winter of 1941 included seventeen hundred tanks and fifteen hundred aircraft, and was made up as follows:

Transbaikalia:	seven rifle, two cavalry divisions, two tank brigades
Outer Mongolia:	one rifle division, two tank brigades
Amur:	two rifle divisions, one tank brigade
Ussuri:	five rifle divisions, one cavalry, three tank brigades

set in. From the depleted southern fronts Timoshenko was being ordered to send tanks and artillery to Moscow and was compelled, also, to retain the majority of his divisions in the Belgorod-Yelets area, where they could give indirect support to Zhukov's right flank. In the far north all the local (i.e., other than Far Eastern) reserves were concentrated into two armies, the 4th and the 52nd, which were directly subordinated to the *Stavka* and charged with the double task of reopening the Leningrad-Tikhvin-Moscow railway and staging an offensive of sufficient vigour to prevent reinforcement of Army Group Centre out of Leeb's command.

We now know that it was not until 30th November that Stalin gave final approval to Zhukov's plans for a winter counteroffensive, but the project must have been in the planning stage for several weeks before this. The *Stavka* calculation was simple, with every factor subject to rational prediction, as befitted Köstring's "nation of chess players." By the end of October the two armies had fought each other to a standstill; but the Russians would soon receive help from their classic ally—the ferocious winter, whose severity no European could ever calculate, for which their own soldiers were trained and clothed, and to which they were accustomed since childhood. Yet by itself the impact of the winter would not be enough for an exhausted and outnumbered Red Army to turn the tables on its adversary; the chosen instrument for this task was the agglomeration of hard, long-service divisions from the Siberian Command. In order that the impact of the Siberian troops take maximum effect, it was vital that they be held back until the last moment; and it was at this stage (if the chess analogy may be pursued) that different games might develop. Would the Germans attempt another forward movement? And if they did, would it add to their exhaustion, and thereby to their vulnerability; or would it be so dangerous as to force the Siberian piece into play before the board was ready?

Zhukov and Shaposhnikov expected that the Germans would make one more effort, and they had also correctly anticipated its form—a reversion to the orthodox Cannae-like plan, with the armour concentrated on the flanks. They, too, placed their strength on the flanks; the 1st Shock Army at Zagorsk to the north of Moscow, the 10th Army, the very strong 1st Guards Cavalry Corps, to the south, at Ryazan and Kashira. The 26th Army was held back at Yegoryevsk, east of the capital, and the 24th and 60th Armies of the reserve front were at Onekhovo-Zuyevo.

But the bulk of these forces and all the Siberian units which had been drafted into them were held back from the front itself. They were not to exert themselves blocking the German armour, but to allow Hoth and Hoepner in the north, Guderian in the south, to turn inside them and wheel toward Moscow, breaking their strength against the Russian infantry which manned the inner ring of defences. It was an operation as delicate and as critical as the *manoletina* of a matador who lets the bull brush his side as he withdraws the cape.

On 15th and 16th November, Bock's army group started off on its final heave, or lunge, toward the Russian capital. The ground was white, lightly sprinkled with snow, and hard as rock. The sun, barely perceptible even at its meridian, shone from a sky "neither blue, nor grey but strangely crystalline and luminous utterly without warmth or poetry." The air was still, for the blizzards were not due until December, and the sound of firing, the orange flash of a 75-mm. had an astringent clarity. Across the hard going it seemed for a few days as if the Panzers had recovered their freedom of action. On the northern flank, in particular, where forest trails had frozen hard and ice covered the swamps, the density of their Panzer concentration which had so hampered the Germans in October began to yield results. On 23rd November, Hoth entered Klin. The 7th, 15th and 11th Panzer followed close on one another's tracks in the breach, an armoured jemmy of tremendous strength that soon threatened to break open the whole Russian position in the northwest. Two days later Rokossovski was compelled to abandon Istra, and on 28th November the tanks of the 7th Panzer, trailing plumes of light powder snow as they clattered across frozen reservoirs, reached the Moscow-Volga canal. The division was one of Reinhardt's, made up of the same men who had breached the defences of Leningrad in September and seen the sunlight reflecting from the minarets of the Winter Palace. This time there must be no hesitation.

After twenty-four hours of constant movement and under continuous attack from Russian aircraft, the 7th Panzer surprised a sapper detachment at Dmitrov and forced its way across the bridge there before the structure was demolished. By evening four hundred men had been established on the eastern bank, with some thirty tanks and two batteries of 37-mm. anti-tank guns. All unknowing, they had trespassed into the territory of the Siberian divisions.

Meanwhile Guderian, in the south, was forcing his way up to the Oka

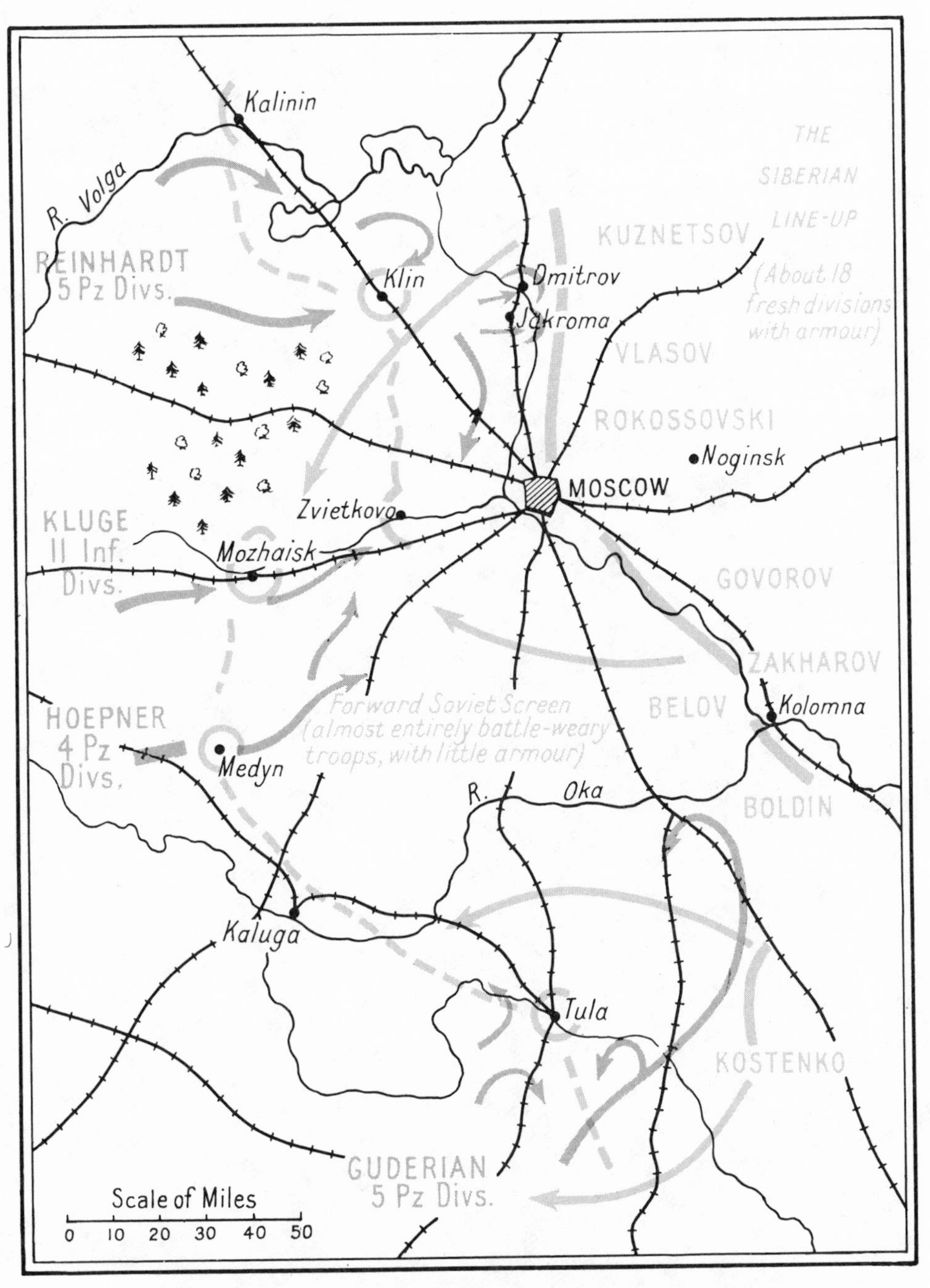

Moscow. Disposition of the Siberians and Zhukov's Counteroffensive

crossings; and here, too, the bull's horn was to graze perilously close to the matador's flesh. The three-week respite which Katukov's exploits had earned for the Tula garrison had been well spent, and the Soviet infantry of the 50th Army, reinforced by some four thousand men in workers' battalions, had transformed the town into a fortress which would have taken an army corps to reduce it. Guderian's tanks had neither the time nor the heavy firepower to attempt such a task. Instead he diverted them east, then north, looping around Tula in an effort to reach the Serpukhov railroad—a turning movement through an arc of 120 degrees. To protect his flank Guderian had directed the 4th Panzer on Venev, and was dropping off his infantry divisions in a protective screen along the upper Don.

For the ordinary German line divisions conditions were already verging on the impossible. Many of the men were without any clothing to supplement their uniforms except denim combat overalls. These they used to pull on over their uniforms, the larger sizes being especially favoured as the soldiers would fill the loose folds with screwed-up paper. "Newsprint was the best, but hard to get hold of. More plentiful were leaflets addressed to the Russian Army. I remember trying for a week to keep warm on a proclamation that 'Surrender is the only sane and sensible course as the Issue has been finally decided.'" Doubtless the Russians enjoyed this irony when they took these men prisoners—though that they maintained their geniality toward soldiers "dressed" in pamphlets threatening the scale of reprisal activity against Partisan bands seems less probable. The impact of the cold was intensified by the complete absence of shelter; the ground was impossibly hard to dig, and most of the buildings had been destroyed in the fighting or burned by the retreating Russians. A doctor with the 276th Division contrasted

> The Russian [who was] . . . completely at home in the wilds. Give him an axe and a knife and in a few hours he will do anything, run up a sledge, a stretcher, a little igloo . . . make a stove out of a couple of old oil cans. Our men just stand about miserably burning the precious petrol to keep warm. At night they gather in the few wooden houses which are still standing. Several times we found the sentries had fallen asleep . . . literally frozen to death. During the night the enemy artillery would bombard the villages, causing very heavy casualties, but the men dared not disperse far, for fear of being picked up by marauding horsemen.

The 112th Division, one of the infantry units guarding the right flank of the 4th Panzer's drive on Venev, had suffered over 50 percent frostbite casualties in each of its regiments by 17th November. On 18th November it was attacked by a Siberian division from the Russian 10th Army and an armoured brigade newly arrived from the Far East with a full complement of T 34's. The Germans found that their automatic weapons were so badly frozen that they would fire only single shots; the 37-mm. anti-tank ammunition—useless against the T 34 at other than point-blank range—had to be scraped with a knife before it would fit into the breech as the packing grease had frozen solid. To these shivering and practically defenceless men the sight of the Siberians in their white quilted uniforms, lavishly equipped with tommy guns and grenades, riding along at thirty miles an hour on the dreaded T 34's, was too much, and the division broke up. "The panic," the army report gloomily noted, ". . . reached as far back as Bogoroditsk. This was the first time that such a thing had occurred during the Russian campaign, and it was a warning that the combat ability of our infantry was at an end, and that they should no longer be expected to perform difficult tasks."

It is evident that the *Stavka* was determined to confine Guderian's offensive, even if to do so entailed the expenditure of some of its cherished reserves of fresh troops. For in the week following the route of the 112th Division, Guderian's intelligence identified three more units which had come from the Far East, the 108th Tank Brigade, the 31st Cavalry, and the 299th Rifle Division. In each case the Russians stayed in action only a short time before withdrawing into the frozen wastes south of the Oka, but their intervention was enough to cramp the development of the 2nd Panzer Army's thrust, which no longer had the appearance of a spear on the map, but more of a shallow boil swelling defensively in every direction around the persistent thorn of the Tula garrison. Something of Guderian's frustration can be gauged from his letters to his wife, in Germany.

> The icy cold, the lack of shelter, the shortage of clothing, the heavy losses of men and equipment, the wretched state of our fuel supplies, all this makes the duties of a commander a misery, and the longer it goes on the more I am crushed by the enormous responsibility which I have to bear. . . .

And again:

> We are only nearing our final objective step by step in this icy cold and with all the troops suffering from the appalling supply situation. The difficulties of supplying us by rail are constantly increasing—that is the

> main cause of all our shortages since without fuel the trucks can't move. If it had not been for this we should by now be much nearer our objective.

By 28th November, however, Guderian was forced to recognise that other factors besides fuel shortage were exerting their influence:

> Only he who saw the endless expanse of Russian snow during this winter of our misery, and felt the icy wind that blew across it, burying in snow every object in its path; who drove for hour after hour through that no man's land only at last to find too thin shelter, with insufficiently clothed half-starved men; and who also saw by contrast the well-fed, warmly clad and fresh Siberians, fully equipped for winter fighting; only a man who knew all that can truly judge the events which now occurred.

On 24th November, Guderian travelled to Bock's headquarters to explain the delays in getting his *Gruppe*'s offensive under way. At Orsha the Panzer commander delivered a long and bitter tirade about the conditions under which his men were having to fight, and ended by claiming, ". . . the orders which I had received had to be changed since I could see no way of carrying them out." (A curious attitude for a member of the *Heeresleitung* or, indeed, for any professional soldier. A more orthodox reaction on receiving orders which one can "see no way of carrying out" is, surely, to resign?) Bock, who was ill, replied that he had "informed OKH verbally of the contents of [Guderian's] earlier reports, and that the OKH were thoroughly aware of the true nature of conditions at the front." Guderian persisted in his "demand" that the orders be changed, and eventually Bock consented to ring up Brauchitsch (who was also ill; he had suffered a serious heart attack on 10th November), and handed the reluctant Panzer commander an earpiece so that he could listen in to the conversation.

Fatigue and ill health had sapped Bock's ambition, but they do not seem to have affected ObdH's caution in his personal relations. In the face of Bock's flat request to cancel the attack and go over to the defensive in suitable winter positions, Brauchitsch was "plainly not allowed to make a decision." Or rather, as an unprejudiced observer might record, he made a decision—but not the one for which Guderian, and to a lesser extent, Bock, had been asking—namely, that the attack was to continue. All that Brauchitsch would agree to was that the 2nd Panzer Army's larger objectives should be placed temporarily in abeyance and that

Guderian could confine his efforts to reaching the line Zaraisk-Mikhailov and cutting the Ryazan railway.

This change was a tacit admission that the southern arm of the pincer attack on Moscow was no longer operational in the full sense. The only hope of reaching the Russian capital lay, therefore, with the Panzers of the 3rd and 4th *Gruppen* in their approach from the northwest, and with Kluge's 4th Army, edging its way forward astride the Smolensk-Moscow highway. Under the terms of the original plan Kluge was not supposed to start his attack in earnest until the two Panzer columns had met east of the capital, but this was plainly impossible if Guderian had orders to go no farther than the line of the Ryazan railroad. Meanwhile Reinhardt, by his audacious thrust across the Moscow-Volga canal, had drawn down on the 3rd *Panzergruppe* the full weight of the Siberian reserve in a succession of savage counterattacks against his flank. In five days the German order of battle in the north had so altered that there were only two Panzer and one motorised division moving forward, and the rest of the armour was involved in desperate protective battles along their northeast flank. Under these conditions it was essential that Kluge attack with all his strength against the Russian centre, or the whole offensive was in danger of withering away. Blumentritt has given a highly subjective account of those critical days at 4th Army headquarters:

> These unpromising conditions [the difficulties of the Panzer divisions in the north] raised the question whether 4th Army should join in the offensive or not. Night after night Hoepner came through on the telephone, to urge this course; night after night von Kluge and I sat up late discussing whether it would be wise or not to go to his assistance. Von Kluge decided that he would gain the opinion of the front-line troops themselves—he was a very energetic and active commander who liked to be up among the fighting troops—so he visited the forward posts and consulted the junior officers and N.C.O.'s. The troop leaders believed that they could reach Moscow and were eager to have a try. So after five or six days of discussion and investigation von Kluge decided to make a final effort with the 4th Army.

This version of events, with its picture of the commanders managing their separate armies independent of any supreme authority, like crusading barons, has no relation to the facts. It may be that Blumentritt feels that the best way to protect Kluge and the ailing Bock from blame for these critical days of "discussion and investigation" (i.e., inactivity) is with the story that Kluge was engaged in field research. What actually happened is that as soon as news of the delays filtered through to Hitler, *via* army group headquarters, OKH and OKW—accompanied as it was by a good deal of verbal shuffling and dissimulation—he gave orders that the offensive was to be renewed along the entire front. It was the Führer's order, addressed personally to every man in the German Army, that made Kluge go over to the offensive on 2nd December, not the latter's consultation with junior officers and N.C.O.'s.

At the end of November other developments, at the extremities of the front, had been remarked at the *Wolfsschanze,* and there is no doubt that they made a deep impression on Hitler, confirming, if they did not initiate, his attitude in two vital respects—the offensive capacity of the Russians and the need for constant supervision of his senior commanders.

At the beginning of November the Russians had started an offensive against Leeb's positions in the Tikhvin-Volkhov bulge. This operation was intended to draw German reserves away from the central front and, on a longer view, to open the way for a complete relief of the beleaguered Leningrad garrison during the winter, but it was conducted in a manner reminiscent of the clumsiest battles of the summer and the Khalkin Ghol. Attacks were always frontal; the enemy was sought not at his weakest point, but where his resistance proved most stubborn; the operation was controlled directly from Moscow, and the commanders on the spot reacted with wooden orthodoxy under the brooding gaze of their "members of the military council" (commissars) and the NKVD. As the days passed, Russian losses piled up under the "categorical requirements" of the impatient *Stavka* while the Germans, from their well-sited defence positions, barely gave ground at all. All the predictions regarding Russian power in the offensive, to which the Finnish war had given urgency, were revived. It seemed as if the training and firepower of a German division would always be a match for a corps of the Red Army.

At the same time Timoshenko went over to the offensive in the extreme south, with results that were very different, for in this theatre Rundstedt's armies (in contrast to those of Leeb, which had been static for nearly three months) were dangerously overextended. Rundstedt himself had strong personal reservations about continuing the advance after the autumn rains started, and had gone on record with this opinion on at least three occasions. Whether Rundstedt's attitude affected the vigour with which the infantry divisions in his army group marched eastward is something for which there can never be direct evidence. Certainly it made no impression on Kleist, who had listened with substantially greater enthusiasm than his superior to Hitler's mid-August instructions to "clear the Black Sea coasts and master the Caucasus." The result was that while the 6th Army was blocked in front of Voronezh, the 17th was spread out between the Dnieper and the middle Don, and Manstein's 11th was removed from the main battlefield in its attempt to clear the Crimea. Kleist had continued to drive his weary Panzers east, across the Mius and into Rostov—"Gateway to the Caucasus" and the most easterly penetration achieved by the German Army in 1941. Kleist's own account of his operations is not always above suspicion,[5] particularly when relating to battles in which he was defeated—but his version of the Rostov setback (the first positive defeat suffered by the German Army in any theatre of the war up to that time) can be accepted. After the usual complaints about the weather and the fuel shortage Kleist said:

> My idea . . . was merely to enter Rostov and destroy the Don bridges there, not to hold that advanced line. But Goebbels' propaganda made so much of our arrival at Rostov—it was hailed as having "opened the gateway to the Caucasus"—that we were prevented from carrying out this plan. My troops were forced to hang on at Rostov longer than I had intended and as a result suffered a bad knock.

Whatever Kleist's intentions in Rostov, he had been singularly careless of his northern flank, screening this with a few battalions of satellite troops, mainly Italians and Hungarians. Timoshenko, in contrast, had built up three fresh armies out of local drafts and from the trans-Caucasian reserve and brushed the satellites aside on the first day (an experience from which, as will be seen, the Germans were singularly slow to take lessons). Kleist had to leave Rostov in such a hurry that forty tanks and a large quantity of heavy breakdown equipment was abandoned. When Kleist started to get into difficulties, Rundstedt had announced to OKH his intention of withdrawing to the line of the Mius River, but Hitler had forbidden this and, on Rundstedt's threatening resignation, had replaced him as commander of the army group with Reichenau. Just as Rostov was the first retreat by the German Army since 1939, so Rundstedt was the first senior commander to be summarily dismissed. They were omens of a new kind, but Hitler interpreted them in a significant way. The Russians, he believed, were not formidable in attack; this

[5] In particular, see p. 282 n.

was amply demonstrated by the fighting on the Volkhov. A resolute, unyielding defence was the answer—and if the professional soldiers, "trained" in the formal arts of tactics thought otherwise, then the remedy lay in the Führer's own hands. It is against this inauspicious background, and on the highly dangerous assumption that even if the German Army should exhaust itself the Russians were incapable of a serious counter-offensive, that the last attacks of 2nd and 3rd December were ordered.

The tables of strength shown in Greifenberg's staff appreciation still looked fairly impressive, and as every unit in the army group had orders to go over to the attack—even the hard-pressed Panzers protecting Hoepner's flank on the Moscow-Volga canal—there seemed a chance that the Russians' front would crack at one point, at least. But although the will was there, sheer physical difficulty made its fulfilment impossible. The advance started well enough at the various corps headquarters, but as it progressed downward and outward to the leading formations a strange numbness and imprecision are apparent, as at the extremities of a frostbitten limb. At some parts of the front that day the thermometer fell to 40 below and the breechblocks of rifles froze solid; oil in the sumps of tanks and trucks had the consistency of tar, and the drag on the dynamos made it impossible to start their engines; battery plates were warped, cylinder blocks were split open, axles refused to turn.

Bock himself could rise from his bed for only three or four hours a day. Less than a month before he had reminded his staff of the battle of the Marne, which was given up for lost when it might yet have been won. "Both opponents are calling on their last reserves of strength, and the one with the more determined will should prevail." Now he had lost heart. On 1st December he reported to OKH, ". . . it is only possible to gain local successes." Had Brauchitsch been completely fit, he might have intervened—though knowing what we do of his pliant character, it seems unlikely. As it was, he simply lay about gasping, his face an alarming bluish grey colour. "Great concern for the health of ObdH," noted Halder primly in his diary. On 4th December, OKW discussed the possibility "that the Commander-in-Chief might have to ask for his relief on account of his health." But by then the ingredients of disaster were piling up; the German offensive had burned itself out, and with it the whole striking power of the Wehrmacht; Zhukov's counterstroke was but twenty-four hours away.

On the last day of Kluge's attack the wind had begun to blow. Many of the infantry divisions had improvised "permanent" positions during their period of immobility and were reluctant to emerge from them to attack through a blizzard which cut visibility to fifty feet or less. Yet one, the 258th, did manage to penetrate the Russian positions during the short hours of the afternoon of 2nd December. Myth and legend have multiplied around this episode. Whether or not it is true that the Germans could "see the towers of the Kremlin reflecting the setting sun" or that they were brought to a halt by "Russian workers who poured out of their factories and fought with their tools and hammers,"[6] the fact remains that this was never a proper break-through, any more than the operation of which it was a part was a proper "offensive." Rather it was the last spasm in a desperate military convulsion which was nearly to prove fatal.

During the night of the 4th-5th December the whole of the "northwestern front" went over to the offensive, and by the 6th, Army Group Centre was under violent pressure along its entire length. The Russians were employing no fewer than seventeen armies[7] and in the new generation of commanders—Koniev, Vlasov, Govorov, Rokossovski, Katukov, Kuznetsov, Dovator—there were names that were to become and remain objects of dread to the German soldier for the duration of the war. Within days the three principal groups of Bock's command—Hoepner's armour, Kluge's, and Guderian's, had lost contact with one another, and it seemed as if the entire army group were about to disintegrate. The very suddenness of the transition from an offensive pattern where the flanks are protected by the momentum of the whole to a desperate defensive resulted in a fragmentation of the German position into a thousand combats fought in isolation by units with unusable vehicles, small arms

[6] An account of this, which rather dispells the legend of the "workers with hammers," has been given by Major General A. Surchenko. According to him, the break took place on the 5th Army sector, and as that had no reserves Zhukov allocated from front reserve the 18th Independent Rifle Brigade (commanded by Surchenko), 22nd and 23rd Ski Battalions, 140th and 136th Tank Battalions (totalling 21 tanks) plus 9 tanks from 5th Tank Brigade. The break-through was stopped by a tank battalion of the 20th Tank Brigade, and driven back by the mixed force allocated by Zhukov. Surchenko claims the operations lasted from 1st to 5th December. Two divisions of the Moscow Opolchenye (110th and 113th) defended the Narva River line, but it is not clear whether they were engaged.

[7] Owing to the smaller size of the Russians' division, their "armies" were seldom more than equivalent to a corps in the Wehrmacht.

frozen solid (the grenade was the only weapon which maintained its efficiency), half drunk on schnapps, frostbitten, and riddled with dysentery.

The "intestinal disorders" of which Bayerlein had been complaining in November were now rampant throughout the army; yet on days like 10th December, when Guderian recorded the temperature as falling to minus 63 degrees, it was death to squat in the open and "many men died while performing their natural functions, as a result of a congelation of the anus." Those who could still eat had to watch "the axe rebounding as from a stone" off the frozen horse meat, and the butter was being cut with a saw.

> One man was drawing his ration of boiling soup at the field kitchen could not find his spoon. It took him 30 seconds to find it, but by then the soup was lukewarm. He began to eat it as quickly as he could, without losing a moment's time, but the soup was already cold, and soon it would be solid.

There was no escape from this purgatory—save death itself. For that old soldiers' stand-by, the self-inflicted wound, besides being a capital offence, could mean only a slow death from exposure and gas gangrene. Some men took their own lives with a hand grenade held against the stomach, but even then the cold had the last word: ". . . the charred pulp . . . would be frozen solid in about half an hour." Not surprisingly, the special medal which was struck for those who took part in this first winter campaign in the East was known as the *Gefrierfleisch Orden,* the Order of the Frozen Meat.

Under the double impact of the blizzards and the ubiquity of the Russian attack, the perils of Army Group Centre mounted hourly. Kluge dared not withdraw his own divisions for fear of leaving the Panzers on his flanks in complete isolation, yet the extrication of the armour was proving virtually impossible; hundreds of tanks were abandoned in the drifting snow, and the crews retreated, fighting as infantry with their side arms; of the four divisions in Hoepner's *Gruppe* only one had a strength of more than fifteen tanks; on Christmas Eve, Guderian had less than forty runners in his entire command. German casualty returns for this period give some measure of the impact of the cold on those exhausted soldiers. Out of over 100,000 cases of frostbite no fewer than 14,357—over the strength of a division—were classified as "major" and requiring one or more amputations; there were 62,000 "moderate" (resulting in total incapacity, but without the necessity for amputation), and 36,270 "light" (in which the patient could be fit for action within ten days). Casualties from Russian action averaged at just under 3,000 per day. This was less than the figure at the height of the summer fighting, but its impact on the units concerned was very much greater, owing to the fact that the system of replacement drafts had broken down, together with the medical and supply services above divisional level.

As despair gripped the German foot soldier and a trembling paralysis afflicted his commanders, one man—the Führer—rose to the occasion. Disregarding the recommendations of OKH, too busy even to accept Brauchitsch's resignation, which the hapless C. in C. offered on 7th December, Hitler communicated directly with his army commanders from Rastenburg. His order of "No withdrawal" has been ridiculed as doctrinaire and amateurish. In reality, it was a principle which guided a ceaseless personal supervision of the development of the battle, a scrutiny of the reports, a complete mastery of the detail even of a regimental action. Hitler was the only person who could keep a tight enough rein on the separate commanders; to prevent them in their concern for their own armies to imperil those on their flanks, and to galvanise the Luftwaffe into maintaining an airlift to formations which had been cut off. By proceeding from the first principle that not an inch was to be yielded, Hitler won time for the concept of the *Igelstellen*—the net of defended localities—to be put into practice. Commanders who presumed to act on their own judgment soon found that Hitler's assumption of the title of Commander in Chief was no mere propaganda device. Hoepner, a trifle precipitate in pulling back the right flank of his *Panzergruppe,* was publicly cashiered; Kluge and Guderian had a race to the telephone to complain about each other, which the Panzer commander narrowly lost, and he, too, was dismissed. Thirty-five corps and divisional commanders were sent home in varying degrees of disgrace. Even Keitel stood in jeopardy. When Olbricht asked him how were the relations of OKW with the Führer, the miserable Field Marshal replied, "I don't know, he tells me nothing, he only spits at me."

We can now see that this winter crisis was no time for professional orthodoxy. Any attempt to withdraw from its positions, without fuel or serviceable vehicles, to retreat across the drifting snow fields at a rate that could not be more than three or four miles a day, would have resulted in the whole German Army being cut to pieces. Better to stand

and fight it out, relying on the innate tenacity and discipline of the German soldier. The Red Army was putting everything it had into this offensive—the few precious T 34's that had been saved from the autumn production; every man it dared to bring back from the Far East; every shell and bullet the factories had sent out. But it had no power to achieve, nor did the weather permit, a deep penetration in the manner of the summer battles. In the few cases where the Russians succeeded in surrounding their enemy they had neither the artillery to reduce them nor sufficient strength in the air to prevent their revictualling by the Luftwaffe. The Russians' recovery and their winter offensive of 1941 remain one of the most remarkable achievements in military history, but its dramatic quality should not obscure the essential limitations in material and talent which still confined the Soviet military machine. Once it had failed to carry the day in the first shock of assault, its chances of complete victory declined steadily along the immutable scale of relative force available.

In the net result, although the Russians savaged Army Group Centre without respite for three months, they never succeeded in any of the major encircling moves they attempted and their territorial gains were limited to a forty-mile belt at the approaches to Moscow. The Germans managed to hold on to Rzhev, Vyazma, and Orel.

But if Hitler's policy saved ground, it was profligate in its expenditure of talent. Crucified by the Russian winter, stripped of its most eminent commanders, the Wehrmacht had altered out of recognition from the days of June, and it was to carry the scars of the experience to its grave. As for Hitler, it was his finest hour. He had done more than save the German Army; he had achieved complete personal ascendancy over its ruling class. Yet this ascendancy did not ameliorate his dislike for the generals or his contempt for the professional *mystique* with which they surrounded themselves. Indeed, it aggravated the emotions to a degree which, sooner or later, had to precipitate a dangerous crisis. The Führer had become convinced, as he declared to Halder, that "This little matter of operational command is something that anyone can do."

BOOK II | Stalingrad

HITLER. The Russian is finished!

HALDER. I must say, it is beginning to look uncommonly like it.

15th July, 1942

ten | PLANNING AND PRELIMINARIES

In February 1942 the Russian offensive petered out. The temperature rose, the days lengthened, and to the Wehrmacht the end of its ordeal was in sight. The Red Army, although a few isolated successes remained to it, like the capture of Velikiye Luki on 15th February, was a spent force. The magnificent divisions of the Far Eastern Command were skeletons of their December strength, exhausted by three months' fighting in the worst winter for 140 years. More serious, as the pace of the attack slackened, the Russians had reverted to their old clumsy frontal tactics against the *Igelstellen,* so that by the end of the winter Zhukov's armies were in almost as parlous a state as those of his opponent—with the ominous corollary that his capital resources, in arms and trained men, were very much less.

For both sides the crucial problem was that of interpreting their enemies' intentions, and planning their own, for the full campaigning season that would follow the thaw. At OKH the question had come up for discussion as soon as it became plain that there was going to be a 1942 campaign—that is, during late November, when the "final" attack on Moscow started to run down. Blumentritt recalls that at that time

> A number of the Generals declared that a resumption of the offensive in 1942 was impossible, and that it was wiser to make sure of holding what had been gained. Halder was very dubious about the continuance of the offensive. Von Rundstedt was still more emphatic and even urged that the German Army should withdraw to their original front in Poland. Von Leeb agreed with him. While other generals did not go so far

> as this most of them were very worried as to where the campaign would lead. . . . But with the departure of von Rundstedt as well as von Brauchitsch, the resistance to Hitler's pressure was weakening, and that pressure was all for resuming the offensive.

Blumentritt does not fix these discussions at any specific date, and although the concept of the 1942 campaign got onto the planning boards at OKH in November 1941, it is more likely that Blumentritt's assessment is a symposium of opinions expressed to him by various commanders after he became Deputy Chief of the General Staff (under Halder) on 8th January, 1942, a time when the generals were competing with one another to produce the most pessimistic forecast. Certainly, once the front had been stabilised and it became possible to start accumulating a strategic reserve, the majority of professional opinion seems to have swung in favour of an offensive campaign that summer. The dispute centered around its scope. Could Russia still be knocked out of the war? Or was it wiser to limit the aims to reducing her potential to such an extent that she would no longer present a serious threat, to conduct an operation which would be, in terms of grand strategy, defensive?

In retrospect most of the generals who survived claim that they favoured a limited campaign and that anything more ambitious was a "gamble." Yet this is no more than another example (in which the Eastern campaign abounds) of the failure of the *Generalstab* to make correct appreciations at a global strategic level. It regarded—on its own admission—the summer campaign of 1942 as a narrow tactical problem in isolation from the world events which made it vital for Germany to win the war that year or be dragged under by the industrial weight of the coalition which was ranged against her.

The generals excuse themselves by complaining that they were never allowed to sit in on the economic conferences at which the requirements of wheat, manganese, oil, and nickel were discussed, and that the Führer "kept them in the dark" about this element of strategy. But this is demonstrably untrue. As will be seen, Hitler stressed the economic factor behind all his decisions on every occasion on which he argued with his military leaders. Nor were they as ignorant of these as they are prone to claim.

> There was a battle of opinion between Halder and him [Hitler]. The Intelligence had information that 600-700 tanks a month were coming out of the Russian factories, in the Ural Mountains and elsewhere.

> When Halder told him this Hitler slammed the table and said it was impossible.

Yet if Russian tank-production figures really were at this level, it was an argument for forcing rather than postponing the issue. It is clear that the generals either completely misunderstood Hitler or, as seems more likely, that they have since completely misrepresented him, as in Blumentritt's judgment that

> . . . he did not know what else to do—as he would not listen to any idea of a withdrawal. He felt that he must do something and that something could only be offensive.

In reality, Hitler had an absolutely clear idea of what he was going to do. He intended to smash the Russians once and for all by breaking the power of their army in the south, capturing the seat of their economy, and taking the option of either wheeling up behind Moscow or down to the oil fields of Baku. But instead of taking the staff at OKH firmly by the lapels and instilling this objective in them from the outset, the Führer was exceedingly circumspect—not to say tortuous—in the dissemination of his strategic ideas. In the result an operational plan was gradually evolved, but with Hitler and the General Staff of the Army assuming different objectives. The OKH plan accepted certain limitations of scope and ambition, while OKW—where Keitel and Jodl must be assumed to have been in some degree privy to Hitler's intentions—was always trying to graft onto it greater strength and a wider sweep. These differences were never reconciled, and their origins and history are important to an understanding of the course of the Stalingrad campaign and its disastrous climax.

The first draft plan prepared by OKH, in midwinter, when the Red Army was making a painfully formidable impression, was for a limited campaign in south Russia and consolidation of the line east of the Dnieper bend, which would safeguard the manganese deposits at Nikopol. The limitations of this plan were soon dropped in the euphoria which followed on the spring recovery, but the one "positive" measure for which it had provided—the reduction of Leningrad and a junction with the Finns—remained on the agenda, and was dutifully carried forward on each succeeding plan. This, as will be seen, led to a serious diversion of available force during the summer.

In April a more ambitious scheme was worked out. This involved the seizure of Stalingrad, and the isthmus between the Don and the Volga, "or at least to expose it [the city] to our heavy fire, so that it loses its importance as a centre of war industry and of communications." Granted the necessity of occupying the Donetz basin and protecting it with the "buffer" of steppe in the large bend of the Don, Stalingrad constituted an acceptable strategic long-stop—particularly when its actual capture was subject to the escape clause cited above. But for Hitler, Stalingrad was to be the first step only. His intention was to wheel north, along the line of the Volga, and cut the communications of the Russian armies defending Moscow while sending "scouting groups" still farther east, toward the Urals. But Hitler recognised that an operation on this scale would be possible only if the Red Army had suffered a defeat even more serious than those of the previous summer. The alternative was to seize Stalingrad as an anchor for the left flank while the mass of the Panzer forces wheeled south to occupy the Caucasus, cut off the Soviet oil supplies, and menace the borders of Persia and Turkey.

Halder claims that none of these ideas were communicated to OKH during the planning stage:

> In Hitler's written order to me to prepare an offensive in South Russia for the Summer of 1942, the objective given was the River Volga at Stalingrad. [We] therefore emphasised this objective and held only a protection of the flank south of the River Don to be necessary . . .

The eastern Caucasus were to be "blocked" and a mobile reserve was to be held at Armavir, to stand guard against Russian counterattacks south of the Manych. Halder remembered

> some critical remarks made just at that time about the lack of daring and initiative on the part of the General Staff. But Hitler did not connect them with the restriction of objectives south of the Don. Obviously he was not yet sure enough of himself to express his objections to the OKH order.

Far from being obvious, it must be reckoned highly unlikely that Hitler, who had ravaged the Army with his dismissals and conducted a successful winter campaign virtually singlehanded, was "not yet sure enough of himself" to impose his will on OKH. More probable is that he still hoped the Russians' strength might be broken before the Volga was reached, which would allow the fulfilment of the "large solution," the

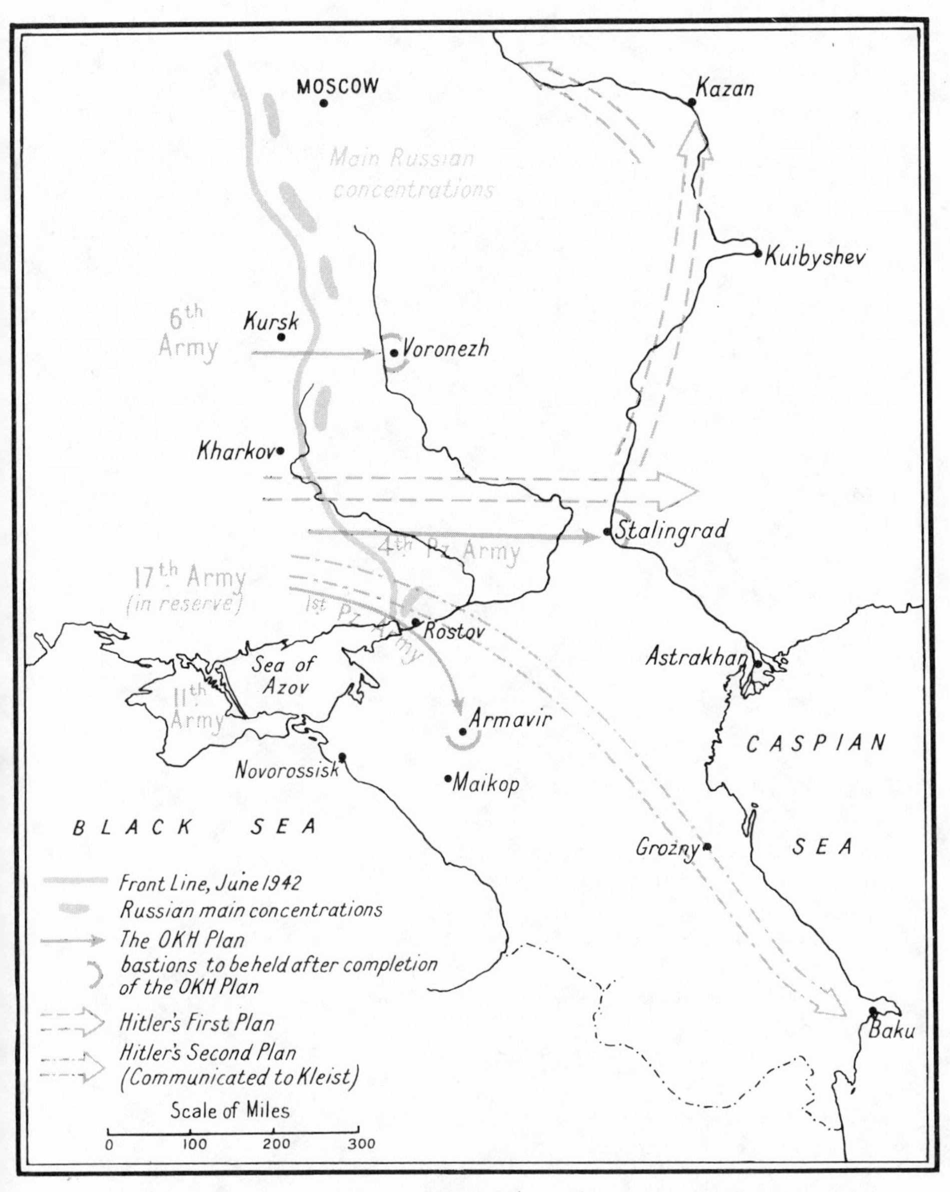

German Plans for 1942

northern sweep to Saratov and Kazan, and by leaving the sequel to the capture of Stalingrad in a planning vacuum he retained the option of a campaign in the Caucasus or a drive to the Urals. The result was that OKH went into the campaign believing that Stalingrad was the objective and that the forces in the Caucasus were to have merely a "blocking" role; while the conception of OKW, which Hitler was subsequently to communicate to some of the subordinate army commanders, was that the block was to be established at Stalingrad and the main forces committed either to the north or south. Things were further complicated by the fact that OKH continued to pay a kind of lip service to the idea of a campaign in the Caucasus. As long ago as the Orsha conference Paulus (who was Halder's deputy at the time) recalls the Chief of the Army Staff saying, ". . . when weather conditions permit we shall feel justified in making an all-out thrust in the south towards Stalingrad *in order to occupy the Maikop-Grozny area at an early date,* and thus improve the situation as regards our limited supplies of oil." Still more perplexing, Directive No. 41, which was issued in April 1942, included the "seizure of the oil region of the Caucasus" in the preamble concerning the general aim of the campaign, yet made no mention of this in the main plan of operations.

Not unnaturally, this duality was reflected in the organisation of the army group, which, although originally drawn up as one mass under Bock (now recovered from his winter ailment), had an intrastructure that would allow it to be split into two, Army Group B, under Weichs, and an A, which was to be placed under List, at that time G.O.C. in the Balkans. The B group was made up of the 2nd Army, the 4th Panzer Army, and the very strong 6th Army of Paulus, and was intended to do most of the fighting in the early stages. The A group looked at first sight like a reserve force, with its high proportion of satellite formations and its single German infantry army, the 17th, and under the scheme outlined in Directive No. 41 it was to run in tandem with, but slightly behind, Army Group B. But in his command List also had a whole Panzer army, the 1st, under the dashing and vigorous Kleist. And to Kleist the Führer had confided. As early as 1st April, Hitler had told him that he and his Panzer army were to be the instrument whereby the Reich would be assured of its oil supplies in perpetuity and the mobility of the Red Army would be crippled for good. Their discussion, Halder acidly commented, was a good example of the way

> in which Hitler succeeded in winning over the consent of leaders of lower rank by clearly misrepresenting those ideas which the OKH, as the superior organ, had rejected. It is characteristic of the conditions within the high leadership on the German side that I never came to hear of this discrepancy between the basic operational order of the OKH and the directions given personally to an army commander by Hitler.

The result of this "discrepancy" was that the commander of the largest single armoured force in the army group was to go into action with a private objective of his own. "Stalingrad," Kleist said after the war, "was, at the start, no more than a name on the map to my Panzer army."

Numerically German strength stood at about the same level for the campaign of 1942 as it had the previous year, and if the satellite armies are included the total of divisions surpasses the 1941 figure, for Hungary and Rumania had each increased its quota during the winter. In terms of equipment and firepower the average German division was better off, though only marginally so, and the number of Panzer divisions had been raised from nineteen to twenty-five. But in their quality and morale the Germans were already in a decline. No army could have passed through the experiences of that dreadful winter without suffering permanent damage; nor could it have felt the successive disappointments of apparent victory and cruel frustration which had alternated throughout the previous summer without a sense of futility and gloom. It was a feeling that spread right back to, and then recoiled from, the home front in the Reich. For the German nation "the war" meant the war in the East. The bombing, the U-boat campaign, the glamour of the Afrika Korps, these were incidentals when over two million fathers, husbands, brothers, were engaged day and night in a struggle with the *Untermensch.*

> Oh yes, we were heroes all right. Nothing was too good for us at home, and the press was full of our stories. The Eastern Front! There was something about the words, when you told people you were going there it was as if you had admitted some fatal disease. Everyone was so friendly, a sort of forced cheerfulness—but with a certain look in their eyes, that animal curiosity when you gaze on something condemned . . .
>
> And deep down so many of us believed it. In the evenings we used to talk of the end. Some slit-eyed Mongol sniper was waiting for each one of us. Sometimes all that mattered was that our bodies should get back to the *Reich,* so that our children could visit the graves.

The despair and fatalism which can already be detected in letters and diaries at this time were not nearly so widespread as they were to become in 1943, after the failure of *Zitadelle*. This was partly because relatively few units had been involved in the winter fighting and the practice of creating new divisions in preference to building up the old ones to full strength helped to limit the spread of defeatism at second hand. Nonetheless, the germ was there, ineradicable; and its effect was to be seen on many occasions in action at regimental and lower levels during the summer.

Those who went to the East passed already into a different world, separated by a gulf almost as wide as that which had existed in the Great War between the *permissionnaires* in Paris and the hills of Verdun. Once they had traversed the frontier of the occupied territories they were in a belt of country, up to five hundred miles across, where the septic violence of Nazism festered openly—no longer concealed beneath the trim roofs and *Gemütlichkeit* of suburban Germany. Mass murder, deportations, deliberate starvation of prisoner cages, the burning alive of school children, "target practice" on civilian hospitals—atrocities were so commonplace that no man coming fresh to the scene could stay sane without acquiring a protective veneer of brutalisation. One young officer, lately arrived in the East,

> received an order to shoot three hundred and fifty civilians, allegedly partisans but including women and children, who had been herded together in a big barn. He hesitated at first, but was then warned that the penalty for disobedience is death. He begged for ten minutes' time to think it over, and finally carried out the order with machine-gun fire. He was so shaken by this episode that [after being wounded] he was determined never to go back to the [Eastern] front.

The doctrine of *Befehlt ist befehlt*[1] could cut both ways on occasion. At the German frontier near Posen there were railway sidings often used

[1] Literally, "What is ordered is ordered." Since the war the Germans have rationalised their brutalities at three levels. First, that the end justifies the means, the end in this case being "the New Europe" (the term "Europe" being substituted for "Order" owing to the unhappy associations of the former concept in contrast with the prevailing bias toward European Union, etc.) Second, that the victims were Communists anyway, and therefore their liquidation was in defence of "freedom." And third, *Befehlt ist befehlt*—the obligation to obey orders relieves the individual of direct responsibility for his actions. These arguments can be heard with regularity at war crimes trials, and crop up periodically in print—though whether they were so carefully rationalised at the time is another question.

for the routing of troop trains and other, more sinister traffic from the East. Often freight cars jammed tight with Russian prisoners, or Jews rounded up for the extermination camps, would stand there, for days on end when the through traffic was heavy, "giving off a faint droning sound, the supplication of thousands of dying humans for air and water." Once the sidings were occupied by a Wehrmacht hospital train. It, too, had been sealed, and its markings removed as a protection against Partisans. The trucks should have been opened at Brest-Litovsk, but the movement orders were destroyed in an air raid and the train lost its identity. Soon, as it was shunted across eastern Poland, it assumed another, more fearful one. By the time it reached the sidings in Posen, over two hundred of the German wounded had died. The stationmaster and his staff had heard the cries for help from within, but had done nothing "as they thought it was a ruse . . . that the voices were those of German-speaking Jews."

Among other factors affecting the morale of the German troops was the failure to produce any really new weapons on a par with the T 34 or the *Katyusha* multiple rocket thrower. The infantry was going into action equipped almost exactly as it had been the previous summer, apart from an increase in the number of submachine gunners in certain companies. The Panzers had been subjected to a more thorough reorganisation (but this affected only those in the southern theatre. The "old" understrength formations in the northern and central army groups retained their original form in 1942).

The most important change in the composition of the Panzer divisions was the inclusion of a full-strength battalion of 88-mm. guns—still styled as an "anti-aircraft battalion" but in reality included because of the proved tank-killing powers of that famous gun. The motorcycle battalion was abolished, but one of the four infantry battalions (sometimes two, in cases of elite and SS Panzer formations) was re-equipped with armoured half-track personnel carriers, which greatly improved its effectiveness in difficult ground. The riflemen in these armoured carriers were designated *Panzergrenadier,* and this term soon spread to include all infantry attached to a Panzer division.

Of the tank battalions, their firepower had been increased by the long-delayed substitution of the L 60 50-mm. gun for the old 37-mm. in the PzKw III (although the effectiveness of this was diminished by the first batch being fitted with the L 42 gun in error; see p. 17) and by the L 48 75-mm. being fitted to the PzKw IV. At the same time the tank content of the divisions was raised by the addition of a fourth company to each tank battalion. However, this increase in strength was more nominal than real, as tank production in 1941 had amounted to a meagre 3,256 and less than 100 units were delivered in the first part of 1942. Losses in the 1941 campaign had totalled nearly 3,000 and the tables of strength had been further reduced by the withdrawal of many PzKw II's and I's to police and internal duties, these models being no longer of value in the rugged school of the Eastern front. Thus, although four companies were duly constituted in each battalion, very few were up to their required strength of twenty-two PzKw III's or IV's. In fact, at the start of the summer of 1942, German tank strength was numerically rather below that at the beginning of the 1941 campaign. But this was offset by a ruthless "starvation" of the units in the northern and central sectors, and a concentration of all the new armour in Bock's group, raising the Panzer density and the "force-to-space" ratio in the area selected for the attack.

If the Russian production of tanks had in fact been running at seven hundred per month, as Halder claimed, German prospects would have been bleak indeed. But two of the main centres of tank production, at Kharkov and Orel, had passed into enemy hands, as had the majority of the component factories in the Donetz basin. The KV factory at Leningrad was working at a reduced capacity, and the output was devoted to local requirements. The celebrated factories "in the Urals" (notably at Sverdlovsk and Chelyabinsk) were only beginning to get into their stride in the spring of 1942, and although the Soviet Official History claims a spectacular advance in tank-production rates in the later part of the year it seems unlikely that, in the first months at all events, it was any higher than the German, and Russian actual front-line strength was certainly inferior. In the carnage of that first summer the Red Army had lost its entire tank park of nearly twenty thousand. The economy had been reduced (taking 1940 figures as 100 percent) by nearly as disastrous a proportion: coal by 57 percent, pig iron by 68 percent, steel by 58 percent, aluminum by 60 percent, and grain by 38 percent. Ignoring the niceties of figures, which on either side are subject to some distortion, it would be safe to estimate the total Soviet industrial capacity as having been halved by German action in 1941. During the first months of 1942 a number of tanks from factories in Britain and the United States were delivered by the northern sea route to Murmansk and overland from Persia. What happened to these deliveries remains a mystery.[2] The Russians, very understandably, rejected the majority as being unfit for combat. (It is some commentary on the lag in Western tank design that the only model produced which would have been any use in the East, the Sherman, did not start coming off the production lines until, by Russian standards, it was already obsolete. First deliveries of the Sherman tank were not made until the autumn of 1942, by which time the T 34—to which it was inferior—had been in full production for eighteen months, and the T 34/85 and the Tiger were already laid down.) A few British infantry tanks, both Matilda and Churchill types, were used in the so-called "independent" (i.e., infantry co-operation) brigades, where they were acceptable because of their very thick frontal armour, and some American Honeys (a light, fast tank with a 37-mm. gun) were identified by Kleist in the later stages of the Caucasian fighting. But it would appear that the majority of the Western tanks were distributed among quiescent or non-combatant theatres like the Finnish front and the Far East, and played no more than an indirect role in releasing home-produced tanks for the more critical battles.

The severe shortage of armour and the evident clumsiness in handling large masses which the first generation of Soviet commanders had shown combined to influence the form of the new armoured units which were built up during the spring of 1942. These were uniform replicas of the *ad hoc* mixed groups which had been so successful in slowing up the German advance the previous November and had shown themselves adequate in the limited roles imposed by the conditions of the winter counteroffensive. Styled "armoured brigades," they consisted of two (sometimes three) tank battalions with mixed KV's and T 34's, a motorised infantry machine-gun battalion, a mortar company, and an anti-tank company—armed mainly with 75-mm. L 46's, although production of this gun was being phased out in 1942, and by the autumn of that year all anti-tank companies had the 76.2-mm. L 30's. "Cavalry" and

[2] Ogorkiewicz gives the totals as 5,258 from the United States, 4,260 from Britain, and 1,220 from Canada. The majority of these deliveries were made during the summer and autumn of 1942 and 1943, but it seems probable that at least 1,000 had been delivered by the start of the 1942 campaign.

"mechanised" brigades followed this pattern, but with the proportion of tanks to mounted horsemen and truck-borne infantry reversed, and often, owing to the shortage of armour, with the tank component left out altogether. These formations were intended for an offensive role, penetration and encirclement, but were in fact too light to be able to achieve much on their own, and the technique of co-ordinating them into corps was still not proceeding smoothly.

By the beginning of May the *Stavka* had accumulated about twenty of these new tank brigades. There were also a number of so-called "independent" brigades (independence in this case meaning separation from the orthodox armoured formations), which were placed under the command of the generals of particular divisions and used in close infantry support.

Although the majority of the Far Eastern units had been expended in the winter battles, the *Stavka* was able to draw again on this source in February and March, when the extent of the Japanese commitment in the Pacific and Indian oceans became apparent. In addition, there were about half a million reservists with some background training who had been recalled in the late autumn of 1941 but not yet committed to action. The mass of the "1921 and 1922 classes" were virtually untrained and without equipment, and would not be ready until the end of the year. For the time being, therefore, the Red Army was feeding drafts and new equipment into the organisation of old and battle-tried units, in preference to the creation of new armies (and in contrast to the practice of their opponent). The fruit of this policy, and of a truly ruthless discipline on the industrial front, where the factories were kept going throughout the twenty-four hours, often without heating and with windows and roofs blown in, was a strategic reserve of about thirty reconstituted rifle divisions, in addition to twenty armoured brigades of the type described above.

By the standards of the previous summer this was a negligible quantity —barely sufficient, it might have been thought, to bolster the remaining 160-odd formations that were strung out from Leningrad to Taganrog. Yet the thrashing they had meted out to the Germans during the winter, the poor state of individual German prisoners, and the evident superiority in certain types of equipment, notably tanks and gunnery, seem to have encouraged the Russians to believe that the Wehrmacht was in a sorrier state than the facts warranted. This belief was the stronger in proportion to the holders' distance from the battlefield, and persisted at the *Stavka* long after the disappointments of its attacks in March had convinced local commanders that the Germans were still formidable once they had recovered their confidence.

There is still no evidence regarding the strategic disputations which must have occupied the early spring in Moscow, and we do not know who, if anyone, on the *Stavka* opposed the plan for a triple offensive which was promulgated at that time. Certainly Stalin was in favour, and the result—a fruitless dispersal of force, which was barely adequate at the outset, and a ruthless prosecution of operations long after their futility had become apparent—shows all the marks of the dictator's personal interference.

Although the Soviet plan was based on a correct interpretation of the German objectives, it chose to counter them head on rather than by setting a trap of the kind which had worked so well in front of Moscow, in the somewhat dangerous hope that striking first would give the Red Army an advantage. Just as the Germans intended to subdue Leningrad during the summer of 1942, so Stalin was determined to relieve the town completely by breaking open the overland route between Tikhvin and Schlüsselburg; and the counterpart of Hitler's Caucasian ambitions was to be a sustained effort to reconquer the Crimea. Most important of all, and involving the commitment of practically the whole Soviet armoured reserve (and certainly all the new T 34 and KV formations), was to be a pincer attack by Timoshenko on the German positions before Kharkov. The town, fourth city of the U.S.S.R., was to be captured, and with it the whole communications system of the Germans in south Russia would be upset, and their capacity to stage an offensive in that theatre eliminated. The assumption of three distinct objectives, so widely separated that pressure against one could have no effect on the situation at another, was justifiable only where the attacker was much the stronger army. The Soviet miscalculation of the relative strengths brought disaster on all three projects and came near to damaging the Red Army beyond possibility of repair—at least during the summer of 1942.

The first of the *Stavka* spring offensives was launched in the Kerch Peninsula of the Crimea, on 9th April. The failure of Manstein's 11th Army to capture Sevastopol the previous autumn, and successful sallies by the garrison during the winter, had encouraged periodic attempts by the Russians to liberate the entire peninsula. On 26th December bridgeheads had been established at Kerch and Feodosiya, and although the latter had been bloodily eliminated by Manstein on 18th January, a strong Russian force had remained across the neck of the Kerch Peninsula, whence it made three separate and costly efforts to break into the Crimea proper, on 27th February, 13th March, and 26th March. On each occasion Russian strength was built up to a figure higher than the last, and each time it was not quite enough to break through Manstein's positions, which were themselves being reinforced. Finally, for the "Stalin offensive" in April, the *Stavka* had released a quantity of armour, five "independent" brigades. By that time, though, Manstein, too, had been heavily reinforced with a Panzer division (the 22nd) and a *lecht* division (the 28th), and the whole of Richthofen's 8th Air Corps of Ju 88's and *Stukas,* which were to be used in a renewed assault on Sevastopol. In the result the Russian strength was once again inadequate, the attack was stopped dead in three days, and within a month Manstein had cleared the whole Kerch Peninsula and could turn on Sevastopol. In the final count the Red Army lost over a hundred thousand in prisoners alone and over two hundred of their precious tanks. The total Russian commitment in the Crimea since Christmas Day (excluding the Sevastopol garrison) had come to over a quarter of a million men, but largely owing to the piecemeal way in which they had been fed into the mincing machine, they finished up without having achieved anything more than a minor diversion.

At least the Soviet attacks in the Crimea had given Sevastopol breathing space and drawn three German divisions down there. The second of the *Stavka* offensives was an unredeemed failure. Striking once again at the German positions on the Volkhov River, a strong column, including two fresh Siberian divisions and led by one of the most vigorous of the Red Army commanders, General Vlasov, achieved a temporary penetration. But it was soon to discover that in the May sunshine German confidence and tactical reflexes were very different than when the thermometer was 40 below. Unable to enlarge the flanks of his break-in, Vlasov was trapped in a narrow salient under steadily mounting pressure. He got no support from his "front" headquarters—only the habitual instructions to press the attack—and after five days of fierce fighting the Germans sealed off the neck of the Russian break-through and set about reducing the encircled divisions. Vlasov was so disgusted by the incompetence of the "front" command and by the useless sacrifices which his picked force had to endure that he refused to leave the pocket in an aircraft which was sent to fly him out. Together with his staff he was captured by the Germans and was later, as will be recounted, to play a strange role in the politics of the Eastern campaign.[3]

Now everything depended upon the centrepiece of the *Stavka* operations, Timoshenko's assault at Kharkov. Unfortunately, the Russian plan, besides being highly unimaginative and predictable, dovetailed fatally with a local offensive which Bock had ordered for almost the same date.

Bock's objective was the elimination of the "Lozovaya pocket," a salient which represented the high-water mark of the Red Army's advance during the winter, and which protruded into the German front to the southwest of the Donetz, at Izyum. At the beginning of May, Bock had withdrawn the German forces masking the western tip of the salient and replaced them with the 6th Rumanian Army; he then proceeded to concentrate Paulus to the north, between Belgorod and Balakleya, and Kleist to the south, at Pavlograd. It was intended that these two armies converge against the base of the Russian salient and cut it off, thereby straightening out the German line along the Donetz River before the main offensive was launched.

But at the very moment when Army Group South was making its dispositions Timoshenko, too, was on the move, and the bulk of the Russian armour was trundling into the very "pocket" which Bock intended to eliminate. The Soviet 9th Army, under General Kharitonov, backed by the 6th Army, commanded by Gorodnyanski, was to break out of the pocket and capture Krasnograd. Kharitonov was then to drive on Poltava, while Gorodnyanski swung north, toward Kharkov, and the northern pincer, composed of the Soviet 28th and 57th armies, attacked from the bridgehead at Volchansk.

If the Germans had struck first they would have had a severe shock, for Timoshenko had concentrated nearly six hundred tanks in the Lozovaya pocket—two thirds of his entire armoured strength—and OKH, which seems to have been quite unaware of the impending Russian attack, might have been forced into the gloomiest reappraisal of the Red Army's firepower and tank strength. As it was, Timoshenko forestalled Bock by about a week and started his offensive on 12th May.

In the north heavy fighting developed immediately as the two Soviet

[3] See p. 408.

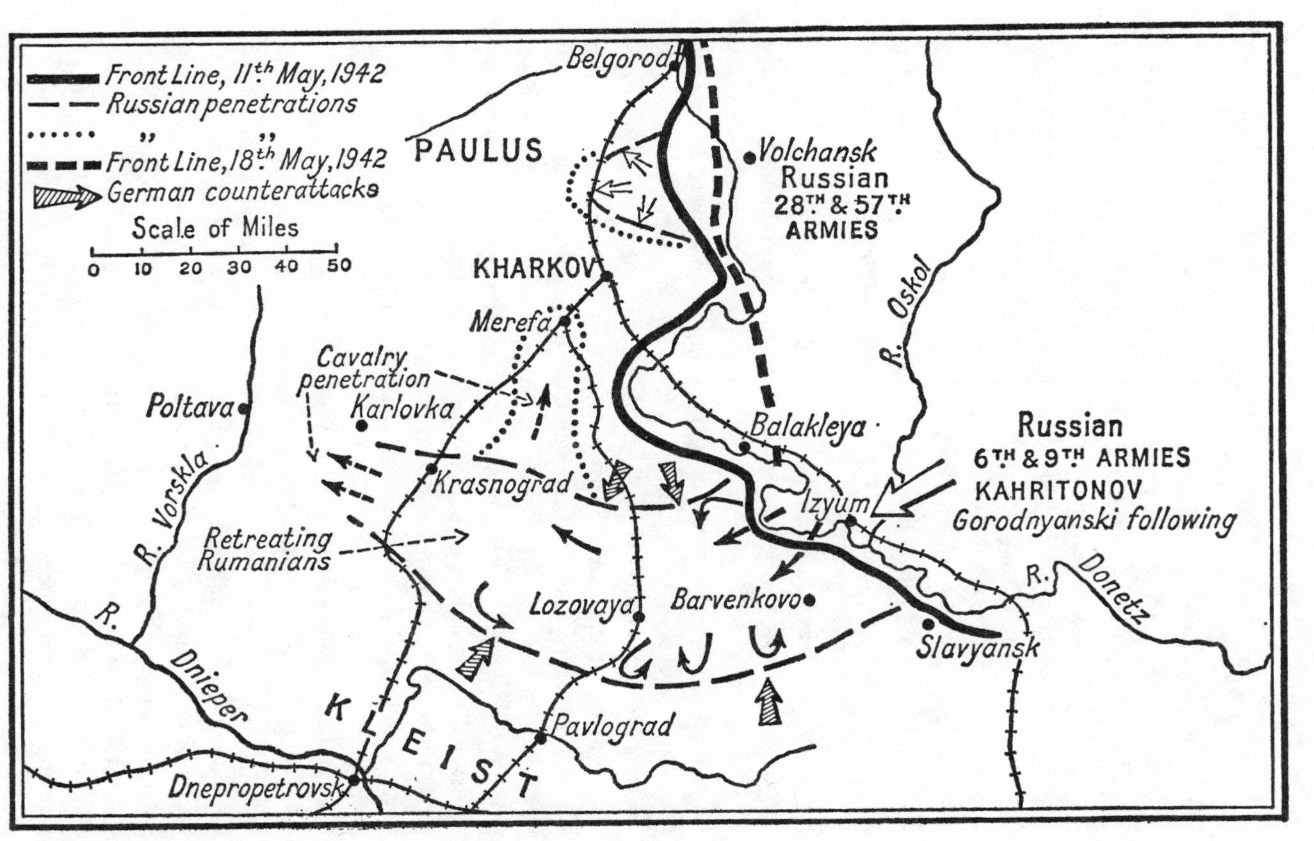

The Russian Defeat at Kharkov

armies became embroiled with Paulus and his fourteen fresh divisions, but in the south Kharitonov drove straight through the Rumanians and captured Krasnograd. For three days, as Gorodnyanski poured into the breach on the heels of the 9th Army, it must have seemed to Timoshenko that Kharkov was within his grasp. But on 17th May the first warning signs began to appear. The northern force, having driven Paulus back to the line of the Belgorod-Kharkov railway at the cost of heavy casualties, was unable to make further progress. Here there was no question of a break-through, the German line had been taken back, intact, as one piece. But in the south the 9th Army was still driving in a vacuum, and had reached Karlovka, west of Kharkov, and only thirty miles from Poltava. Yet all attempts to widen the breach southward from Izyum and Barvenkovo were proving fruitless against a determined resistance which was suspiciously strong in tanks.

The farther west the 9th and 6th armies drove, the farther they carried the mass of the Soviet armour from the danger point at Barvenkovo. And on 17th May the two armies diverged, with Gorodnyanski following his original instructions and turning north, toward Merefa and Kharkov. That evening was not a happy one at Timoshenko's headquarters. Prisoners captured in the southern sector had been identified as coming from the Panzers of Kleist's army, and mounting pressure against Kharitonov's flank confirmed this rapidly accumulating German strength there. Yet Russian armour was now strung out over seventy miles, without having brought the enemy to combat in anything greater than regimental strength. It was the first attempt at an armoured operation on the scale which the Germans had displayed the previous summer, and many weaknesses—the fragmentary nature of the brigade organisation, a shortage of supply vehicles, lack of anti-aircraft protection for the fuel convoys—were making themselves apparent. At midnight Timoshenko got through to the *Stavka* on the telephone in the hope of extracting some qualified permission for slowing down the offensive until he had cleared his flanks. Whether he spoke personally or not, we do know that his "political representative," N. S. Khrushchev, did. Stalin, however, would not come to the telephone, and sent Malenkov, who told Khrushchev that the orders remained in force and Kharkov must be captured.

At dawn on 18th May, Kleist's counteroffensive was unleashed against the south side of the breach and within hours had broken through to the confluence of the Oskol and the Donetz rivers at Izyum, narrowing the base of the Russian penetration to less than twenty miles. By evening Kharitonov had lost all control of his army, which was defending itself in a series of desperate but isolated battles; yet Gorodnyanski continued to press on northward, while the divisions protecting his rear crumbled away. Once again Timoshenko's headquarters approached the *Stavka,* and this time it was his Chief of Staff, Bagramyan, who spoke. Once again Moscow repeated its orders—the offensive was to be pursued to the end.

The end was truly not far off, for on the 19th, Paulus, who had transferred his two Panzer corps to his right flank, began to assault the northern side of the corridor, as it had now become, from the Donetz to Krasnograd. On 23rd May, Paulus' tanks met those of Kleist at Balakleya, and the noose was drawn tight. On 19th May the *Stavka* had at last relented, with a qualified order to Gorodnyanski to halt his advance, but it was then far too late to recover anything but debris from the encirclement. On the 20th, Timoshenko had sent his deputy, General Kostenko, into the pocket to salvage what he could, but less than a quarter of the 6th and 9th armies managed to escape and all left their heavy equipment on the west bank of the Donetz. Moscow admitted a loss of 5,000 dead, 70,000 missing, and 300 tanks destroyed. The Germans claimed 240,000 prisoners and 1,200 tanks. (The latter figure is certainly an overestimate, as Timoshenko's total armoured strength was 845, and although it is unlikely that any were rescued from the southern pocket, it is reasonable to suppose that the 28th Army managed to save some in the north.)

If the *Stavka* offensive had imposed a serious delay on the Germans' own plans, then it might be justified, even though failing in its principal objective—the capture of Kharkov. But under interrogation the German commanders maintained that the effect was negligible. Certainly it was bought at a crippling price, for at the beginning of June, while the German armies were reforming for their summer campaign, there were less than two hundred tanks left on the strength of the entire south and southwest fronts. The numerical ratio, from being five to one in the Russian favour the previous year, had now swung around to nearly ten to one against them.

On 28th June, under a sky heavy with foreboding, Bock's offensive broke like a clap of thunder. Three armies[1] split the Russian front into fragments on either side of Kursk, and Hoth's eleven Panzer divisions fanned out across hundreds of miles of open rolling corn and steppe grass, toward Voronezh and the Don. Two days later the southern half of the Army group went over to the attack below Kharkov, and Kleist took the 1st Panzer Army across the Donetz.

The Russians were outnumbered and outgunned from the start, and their shortage of armour made it difficult to mount even local counterattacks. Of the four armies which faced the German onslaught, one, the 40th, which took the full impact of Hoth's Panzer army, was broken up in the first forty-eight hours. The 13th, Golikov's flank guard on the "Bryansk front," was hastily folded back northward, opening a breach between the two theatres which was to widen daily. Two others, the 21st and the 28th—the latter barely recovered from its mauling at Volchansk in May—reeled back in a state of accelerating disorder, their command structure degenerating into independent combat at divisional, then at brigade, finally at regimental level. Without even the protection of mass, which had characterised the Red Army's deployment in the Ukraine in

[1] From north to south, the 2nd Army (Weichs); 4th Panzer (Hoth); and 6th Army (Paulus). Paulus, who was intended to establish the "block" at Stalingrad, had an unusually strong army of eleven divisions and a Panzer corps. It comprised the 29th Corps (General Obstfelder), 17th Corps (General Hollidt), 7th Corps (General Heitz), 30th Panzer Corps (General Stumme), and 51st Corps (General von Seydlitz).

1941, or of swamp and forest, which allowed small groups to delay the enemy in the battle of Moscow, these formations were at the mercy of the Germans. Polarising around the meagre cover of some shallow ravine or the wooden hutments of a *kolkhoz,* they fought out their last battle under a deluge of firepower against which they could oppose little save their own bravery.

> . . . quite different from last year [wrote a sergeant in the 3rd Panzer Division]. It's more like Poland. The Russians aren't nearly so thick on the ground. They fire their guns like madmen, but they don't hurt us!

The progress of the German columns could be discerned at thirty or forty miles' distance. An enormous dust cloud towered in the sky, thickened by smoke from burning villages and gunfire. Heavy and dark at the head of the column, the smoke lingered in the still atmosphere of summer long after the tanks had passed on, a hanging barrage of brown haze stretching back to the western horizon. War correspondents with the advance waxed lyrical about the "Irresistible Mastodon"—the *Mot Pulk,* or motorised square—which these columns represented on the move, with the trucks and artillery enclosed by a frame of Panzers. "It is the formation of the Roman Legions, now brought up to date in the twentieth century to tame the Mongol-Slav horde!"

During this triumphant period the philosophy of the *Untermensch* (subhuman) reached its peak, and every report and photograph from the advancing Nordic armies emphasised the racial inferiority of the enemy—"a mixture of low and lowest humanity, truly subhumans" . . . "degenerate-looking orientals." "This is how the Soviet soldier looks. Mongol physiognomies from the prisoner-of-war camps." The SS publishing house brought out a special magazine entitled, simply, *Untermensch,* made up of photographs showing the despicable character and appearance of the Eastern foe. "Whether under the Tartars, or Peter, or Stalin, this people is born for the yoke."[2]

No gift of psychiatric interpretation is required to see that this attitude was conceived as granting an unrestricted licence to ill-treat and exploit the "subhumans" as inclination directed—and inclination was the

[2] The author of this text, an SS Major Edwin Erich Dwinger, has been quoted elsewhere in this work. *Untermensch* ended with the assertion, "The *Untermensch* has risen to conquer the world. Defend yourself, Europe!" This is a sentiment which cannot be said to be entirely absent from West German circles to this day.

stronger because of the perverse effrontery with which these creatures resisted the will of their oppressors. "He fights when all struggle is senseless," complained one German journalist. "He fails to fight, or fights quite wrongly, when there is still a chance of success."

The Germans had not been slow to find legal as well as ideological justification for their treatment of the Russian soldier in captivity. The Soviet Union was not a signatory to the Geneva convention—therefore there was no obligation to apply even those minimum standards to the care of her nationals. To facilitate "administration" a special department of OKW, the *Allgemeines Wehrmachtsamt* (AWA), under General Reinecke, was charged with responsibility for the captives outside the immediate area of operations, which came under OKH. Overlapping both these authorities, as usual, was the SS, which had been granted general privileges in connection with "the liquidation of certain categories and the segregation of others."

As early as July 1941 OKH had issued the following instructions to the commanders of rear areas:

> In line with the prestige and dignity of the German Army, every German soldier must maintain distance and such an attitude with regard to Russian prisoners of war as takes account of the bitterness and inhuman brutality of the Russians in battle . . .

the instructions then proceeded to more detailed advice on the enhancement of "prestige and dignity" . . .

> . . . fleeing prisoners of war are to be shot without preliminary warning to stop. All resistance of the prisoners, even passive, must be entirely eliminated *immediately* by the use of arms (bayonet, rifle butt, or firearm).

Besides the direct infliction of violence the Germans virtually sentenced to death all prisoners who fell into their hands in the autumn and winter battles by stripping them of their magnificent greatcoats and astrakhan hats. Huddled together in "cages," often without shelter, much less heating, hundreds of thousands literally froze to death.

This, at least, had the effect of easing the "problem" of feeding them. General Nagel, of the economics branch of OKW (who merits a more conspicuous place than history so far has allowed him with his aphorism, "What matters is not what is true or false, but exclusively what is believed"), declared in a *Wirtschaftsaufzeichnung* of September 1941:

> In contrast to the feeding of other captives [i.e., British and American] we are not bound by any obligation to feed Bolshevik prisoners. Their rations must therefore be determined solely on the basis of their labour performance for us.

The result, as Goering laughingly confided to Ciano, was that ". . . after having eaten everything possible, including the soles of their boots, they have begun to eat each other and, what is more serious, have eaten a German sentry." One high-ranking SS officer sent a private report to Himmler, suggesting that about two million prisoners be shot "forthwith" so as to leave double rations for the remaining half, which would ensure that "real labour was available." But it was not shortage of food that caused the Russian prisoners to starve—simply the refusal of their captors to feed them.

Goering's jokes and the cold statistics of AWA must not be allowed to obscure the frightful horror of these prison cages. Dark compounds of misery and anguish, where the dead lay undisturbed in heaps for weeks on end, they were often ravaged by epidemics so virulent that no guard would enter—except with flame throwers when, in the interests of "hygiene," the dying and corpses were set alight together on their beds of verminous rags. The obsessive German passion for carving notches on their guns has left us with a specific record of the treatment accorded to their brave adversary. Recorded deaths in prisoner-of-war camps and compounds totalled 1,981,000. In addition to this there is the sinister heading of "Exterminations; Not accounted for; Deaths and disappearance in transit," with the horrifying total of 1,308,000. When these figures are augmented by the very large (but unverifiable) totals of men who were simply done to death on the spot where they surrendered, without ever passing through the prison cages, the new dimension of hatred and barbarism that the Eastern campaign was generating can be appreciated.

Following their defeat in the Kharkov battle during May the *Stavka* had been forced into a radical alteration of plan for the summer operations. The high concentration of armour identified with Kleist and Paulus indicated that the main weight of the German effort would be in the south, and this was confirmed by "Lucy." Yet "Lucy" had also forecast (with justification, as we now know) a renewal of the assault on Leningrad, and Moscow was still at the stage when "Lucy's" incredibly accurate information was suspect as laying the foundation for a gigantic trap. It had been decided therefore that the Red Army's reserves—such as they were—should be kept around Moscow to guard against a renewal of the offensive in the central area, whence they could be switched to Leningrad or the south once German intentions became apparent—for the configuration of the Soviet railway system as it had been left by the German gains of the previous year made it very much easier to send force out of Moscow to the flanks than to concentrate suddenly at the capital from the extremities of the front. Accordingly, Timoshenko had been ordered to hold the two "hinges" at either end of his front, at Voronezh and Rostov, and to allow—indeed, he had no option—the Germans to burst through the "gates" between, trading space for time across the Donetz basin and the large bend of the Don. When the German offensive broke on 28th June, its force, however, took the Russians by surprise. Paulus' Panzer divisions reached the Don on either side of Voronezh on 5th July, and at that time the *Stavka* still had no means of knowing whether they would force a crossing and swing northward to take Yelets and Tula, in the rear. A new "Voronezh front" was created forthwith out of the debris of Golikov's divisions and some of the meagre *Stavka* reserve and put under Vatutin, who was himself responsible directly to Moscow, and not to Timoshenko.

At this point Russian resistance, although still very patchy and disorganised, began to take effect on German planning. During the second week of July the only regions where there was any heavy fighting were at Voronezh and south of the Donetz, where the pit heads and slag heaps of the mining basin gave some protection for infantry trying to hold up armour. Between, in the broad corridor of land that separated the parallel course of the Don and the Donetz, over a hundred miles across at its narrowest point, the Red Army was barely functioning at all. A correspondent with the *Völkischer Beobachter* described how

> The Russian, who up to this time had fought stubbornly over each kilometre, withdrew without firing a shot. Our advance was only delayed by destroyed bridges and by aircraft. When the Soviet rearguards were too hard-pressed they chose a position which enabled them to hold out until night. . . . It was quite disquieting to plunge into this vast area without finding a trace of the enemy.

Bock wanted to "deal with" Vatutin before he extended his own flank too deeply into this yawning void, and proposed to use Weichs and a part of Paulus' army to do so. On a textbook appreciation this was, of course, absolutely correct policy. Furthermore, Bock had had personal experience in the summer of 1941 of the delays and frustrations that could follow from leaving large Russian forces undisturbed on his flank. Had Bock been allowed to do what he wanted, there can be little doubt that the whole course of the German campaign in south Russia (and therefore of the war itself) would have been very different. Yet he was prevented from doing so, and dismissed from his command, after a dispute of which the details remain obscure to this day. It does seem as if the disintegration (as it appeared) of the Russian forces in the Don corridor had come as a surprise to Hitler, as it had to many of his generals. At OKW the Führer was in a happier mood than at any time since the fall of France. On the telephone to Halder he showed none of the pettiness and apprehension which had characterised his inquiries the previous year. "The Russian is finished," he told his Chief of Staff on 20th July, and the latter's reply, "I must admit, it looks like it," reflected the prevailing euphoria at both OKW and the High Command of the Army. And following on this conviction OKW took two steps which were to have a radical effect on the development of the campaign. The first was a redirection of Hoth and the 4th Panzer Army; the second, the issuing of a new directive redefining the objectives of the army group.

Originally, under the terms of Directive No. 41, Hoth was to have led Paulus to Stalingrad, then handed over the "block" to the 6th Army and withdrawn into mobile reserve. But after the campaign got under way, Bock's anxiety concerning Russian strength at Voronezh had led him to recommend holding back the bulk of the 6th Army for an attack on the Russian position there and send Hoth on a dash to Stalingrad alone. OKW now decreed that Hoth was not to march on Stalingrad at all, but to swing southeast and "assist in the early passage of the lower Don"; Paulus could manage Stalingrad on his own—provided that the army group stood on the defensive from Voronezh to the Don bend. When Bock was dismissed, the two army groups within his huge "south" framework became independent, and were assigned independent—and diverging—objectives. Directive No. 45, dated 23rd July, promulgated:

> Army Group 'A' [under the command of List] is to advance southward across the Don, with the aim of taking possession of the Caucasus with its oil resources;
> Army Group 'B' [under the command of Weichs] is to attack Stalingrad, smash the enemy concentration there, take the town and cut off the isthmus between the Don and the Volga.

In spite of the progressive weakening of Russian resistance this order was received with some alarm at OKH as it amounted to a very considerable widening of the strategic scope of the operations. There was no longer any escape clause about "closing the Volga with gunfire," and the Caucasian objectives were no longer confined by Maikop and Proletarskaya, but included the whole oil-bearing region. By carrying through the change in two stages—first dismissing Bock and altering the order of priorities on the Don, then creating the two "new" army groups—Hitler had cleverly circumvented the opposition of OKH. Yet an interesting question remains: Why was there not a more concerted protest at the dismissal of Bock? The spokesmen of the conservative wing of the German Army have not been slow to attribute the blame for all the Wehrmacht's misfortunes to Hitler, and the fact that they keep silent over such a classic and far-reaching blunder as the diversion of Hoth and the failure to take Voronezh appears odd. Under interrogation, Blumentritt disclaimed any inside knowledge of the affair, and confined himself to asserting that ". . . there was never any intention of pushing beyond Voronezh and continuing this direct easterly drive. The orders were to halt on the Don near Voronezh and assume the defensive there as cover to the south-eastward advance—which was to be carried out by the 4th Panzer Army and the 6th Army."

The critical decision was plainly that the 4th Panzer Army should alter course. And here it seems that OKH *was* persuaded this was desirable—even though for a different reason from Hitler's. For Paulus in his papers clearly conveys the impression that the diversion of Hoth was first conceived with the possibility of cutting off the Russian divisions which were holding up Kleist's tanks and the 17th Army in the Donetz basin. Within a few days of the original order to Hoth, though, the Russians' resistance to Kleist in the Donetz basin had withered away and their troops were pulling out as fast as they could. The possibility of cutting off any large body was ruled out as it looked as though Kleist would arrive at Rostov no later than Hoth himself.

In the result the two Panzer armies arrived at the Don crossings together—a prodigious sledge hammer, as it had now become, to crush the tiniest of snails. For the Don crossings were virtually undefended. Timoshenko's troops had been hustled out of one position after another in the course of their retreat, and those who had not been trapped west of Rostov had already left the Don behind and were filtering up the valley of the Manych or making their way due east into the Kalmyk steppe, where the broken-up country and *balkas* of the Yergeni Hills would afford them some cover. Kleist, who was particularly free with comment on how operations in other theatres should have been conducted, claimed after the war, "The 4th Panzer Army . . . could have taken Stalingrad without a fight at the end of July, but was diverted to help me in crossing the Don. I did not need its aid, and it simply got in the way and congested the roads that I was using." A sergeant with the 14th Panzer Division has described how

> We got to the Don to find most of the bridges down, but very little sign of the enemy. The heat was stifling . . . the whole length of the right bank was smothered in dust clouds as more and more vehicles began to pile up there. Russian resistance was so slight that many of the soldiers were able to take off their clothes and bathe—as we had in the Dnieper exactly a year ago. Let us hope that history does not repeat itself!
>
> We were there for two days while the engineers were working. We suffered a good deal of attention from Russian aircraft, they would come over singly and in pairs at dusk, and at first light when the *Luftwaffe* was absent. In places Russian artillery was quite strong . . . they could be heard at night moving the guns into position, and would bombard us when the sun rose . . . its flat rays from the east showed our positions in detail, but made it difficult to spot their muzzle flashes.

As the two Panzer armies extended their flanks in an effort to find an intact bridge and a crossing point that would allow them to build up a bridgehead undisturbed, they soon overlapped and then became seriously intermingled. Kleist got across the Don with his light forces as early as 25th July, but traffic congestion and difficulty in bringing up fuel supplies prevented his getting many of his tanks across until 27th July. It was not until 29th July that Hoth got his first Panzers across at Tsimlyanskaya, and by then his orders were changed again—he was to send only one division southeast [3] to cover the gap between himself and Kleist, and lead the mass of the 4th Panzer Army northward through Kotelnikovo, across the Aksai River, to take Stalingrad on its unprotected southern side.

[3] This division, the 16th Motorised, was sent to Elista, where it remained, undisturbed by anything more violent than light patrolling activity for six months. As will be seen, its presence became a source of considerable friction between the various commanders concerned during the critical periods after the Russian counteroffensive. (See Ch. 14.)

Once he was across the Don, Kleist's advance accelerated at a tremenmendous pace. On 29th July he captured Proletarskaya (the original stop line on the old OKH plan); two days later he forced his way out of the Manych gorge and entered Salsk, sending one column down the Krasnodar railway to cover the left flank of the 17th Army and another directly across the steppe toward Stavropol, which fell on 5th August; Armavir on the 7th, and Maikop, where the first Russian oil derricks could be seen, on 9th August.

But things were turning out rather differently for Paulus in his march down the Don corridor. Although Russian resistance was negligible until the Germans reached the Chir River, the distance—over two hundred miles—and the fact that only Wietersheim's Panzer corps was completely mobile, meant that the 6th Army got badly strung out and would have little prospect of mounting a successful attack "off the march" if it should encounter serious resistance. On 12th July the *Stavka* had promulgated a new "Stalingrad front" [4] and was filling it up with divisions out of the Moscow reserve as fast as the railway system would allow. For three weeks a race of the kind familiar from the summer of 1941, between the attacking columns of the enemy and the urgent concentration of the defender's reserves, was on—and this time the Russians won, although by a very slender margin.

General Chuikov, who was ultimately to emerge as one of the three or four vital personalities who inspired and directed the battle of Stalingrad, was at that time in command of the reserve army which was dispersed around Tula. This force consisted of four infantry divisions, two motorised and two armoured brigades, and must have represented a substantial proportion of the remaining *Stavka* reserve. Some idea of the urgency (not to say confusion) and of the stress of their journey on the Russian railway system can be derived from their movement order, which specified detraining at no fewer than seven different stations. On his arrival Chuikov was given instructions so vague as to convince him that ". . . Front H.Q. obviously possessed extremely limited information about the enemy, who was mentioned only in general terms." And as these orders would have involved Chuikov's men in an immediate forced march of 125 miles, he protested:

> to carry them out in the time given was impossible, as parts of the Army which were to carry out these tasks had not yet arrived. The Chief of

[4] Commanded by Timoshenko until 22nd July, when Gordov was appointed.

> Staff replied that the instructions had to be carried out, but then he thought it over and proposed that I should call in to see him the following day.
>
> Next morning, however, he did not appear at H.Q., and no one was able to tell me when he would be there. What was I to do? I went to see the officer in command of the operations section at Front H.Q., Colonel Rukhle, and showing him that it was impossible to carry out the instructions according to schedule, asked him to report to the Front Military Council that the 64th Army could occupy its line of defence not earlier than July 23.
>
> Colonel Rukhle immediately, without reporting to anyone, with his own hand altered the date for the occupation of the line of defence from July 19 to July 21. I was astounded. How could the officer in command of operations, without the knowledge of the Commander, change the date of the operation? Who was in command of the Front?

It is plain from Chuikov's account that the "race" between Paulus and the Stalingrad front was more than a matter of concentration and deployment. The key problem was that of reviving the faltering morale of the Red Army. Could the arrival of the young commanders and fresh troops from the reserve armies restore cohesion to the battered flotsam of Timoshenko's old army, now being swept up by the rising tide of Army Group B in the Don bend? The Soviets' tactics of 1942 had been to retreat when their flanks were pierced, to give ground rather than lives, to avoid the ruinous encirclement battles of 1941. But these conditions—the long withdrawal across a burning homeland—are the most difficult in which to preserve morale, particularly among a relatively primitive and imperfectly trained body, as the bulk of the Red Army units were at this time. The vigour and heroism of the defence of Stalingrad are the measure of the revival which a few men, Chuikov, Khrushchev, Rodimtsev, Yeremenko, were to kindle in merely weeks; but equally there is evidence that all was not well with the Red Army in July 1942. Chuikov himself has described how on his first day at the front he was on a personal reconnaissance:

> I came across two divisional staffs . . . they consisted of a number of officers travelling in some three to five trucks filled to overflowing with cans of fuel. When I asked them where the Germans were, and where they were going, they could not give me a sensible reply . . . It was clear that to restore to these men the faith they had lost in their own powers and to improve the fighting quality of the retreating units would not be easy.

Of the 21st Army, on the right flank of the Stalingrad front, and the first control centre which Chuikov visited, he wrote:

> Army H.Q. was on wheels; signals and supplies were all mobile, in motor vehicles. I did not like such mobility. In everything here one could sense a lack of firm resistance at the front, a lack of tenacity in battle. It seemed as if someone were running after the Army H.Q., and in order to escape pursuit everyone, from the Army Commander downwards, was always ready to make another move.

Of Gordov (who was dismissed on the arrival of Yeremenko and Khrushchev):

> His hair was turning grey, and he had tired grey eyes which seemed to see nothing, and whose cold glance seemed to say, "Don't tell me about the situation, I know everything, and there's nothing I can do if that's how fate has turned out."

Between 25th and 29th July, while Hoth was milling about on the lower Don at Tsimlyanskaya, the 6th Army made an attempt to rush Stalingrad. The feeble resistance he had met up to that time encouraged Paulus to commit his divisions as they arrived on the battlefield instead of pausing for breath, and the result was that German and Soviet reinforcements were fed into battle at roughly the same rate, while the Russians had started the battle with a slight numerical advantage—for the battered 62nd Army (at that time under General Lopatin) had been ordered to stand and fight on the Chir. Paulus had a marked superiority in armour, supported at first by three, then five, then seven infantry divisions, and a long, untidy action followed in which the Russians were gradually levered out of the Don bend. But the 6th Army was so roughly handled that it no longer had sufficient strength to force the river unaided. Nor did it clear the Russians out of the loop of the river at Kletskaya, and the omission was to have a truly catastrophic sequel in November. At the time, Paulus did not have the strength to chase the Soviet infantry out of every little loop on the west bank, and these bridgeheads were soon to be forgotten in the all-consuming battle for Stalingrad. After the area had ceased to be an active sector of the front it was handed over to the Rumanians, and they did nothing about the bridgeheads, but remained on the defensive for the duration of their period in the line.

The unexpected strength of Russian resistance in the small bend of the Don had convinced Paulus that the 6th Army had no chance of forcing

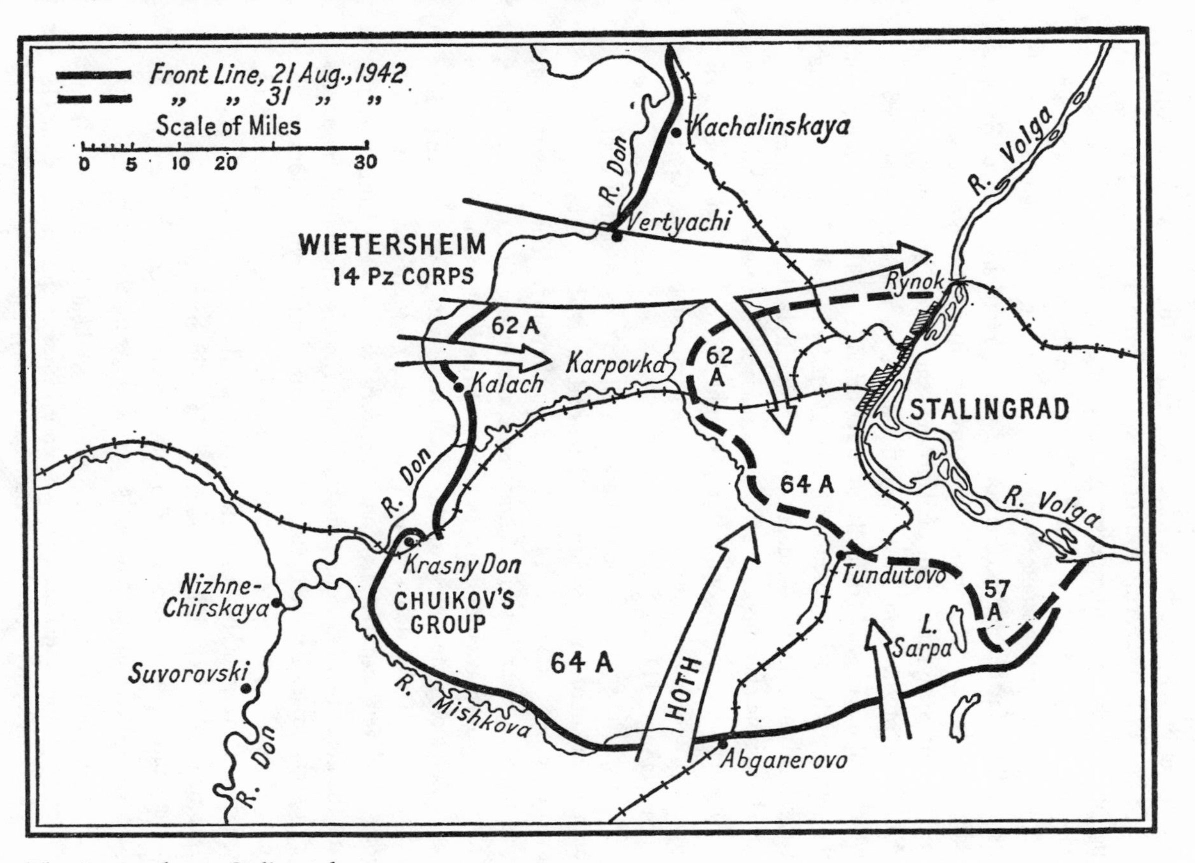

The Approaches to Stalingrad

a crossing by itself, and a lull followed during the first week of August while Hoth's Panzer army fought its way up from the south. During this period the balance of numbers began to swing back against the Russians, for the new 64th Army, which had played so important a role in stiffening the 62nd in its resistance to Paulus' first attack, was having to extend its left flank farther and farther to the west as Hoth approached. By 10th August the whole of Paulus' 6th Army was in position facing due east and all the army and divisional artillery had been brought up to the right bank of the Don. More significant, and a portent of the way in which Stalingrad was gradually to draw off all the offensive strength of the Wehrmacht, Richthofen's 8th Air Corps, which had been covering Kleist's advance in the Caucasus, was brought back to the Morozovsk airfield complex for support in the next attack on the town.

Another week passed while Hoth fought his way up from the Aksai, and then, on 19th August, the first serious attempt by the Germans to storm Stalingrad began.

Paulus, as the senior general, had over-all command of the operations, with Hoth subordinate to him, and he had evolved a conventional plan for a concentric attack with the armour on the wings. The Russian front was about eighty miles in circumference, but owing to its convex shape, from Kachalinskaya along the east bank of the Don and curving back to the Volga along the course of the Mishkova River, it was less than fifty miles across. It was defended by two armies, the 62nd and 64th, with a total of eleven infantry divisions, many of which were understrength, and the remains of different mechanised brigades and other fragments left over from earlier battles. Paulus had nine infantry divisions in the centre, two Panzer and two motorised divisions on his northern flank, and three Panzer and two motorised on his southern.

At first the attack did not prosper. Hoth, in particular, had difficulty in penetrating the positions of the 64th Army between Abganerovo and the Sarpa Lakes, and veterans of the 1941 fighting noted:

> The German tanks did not go into action without infantry and air support. On the battlefield there was no evidence of the "prowess" of German tank crews . . . they operated sluggishly, extremely cautiously and indecisively.
>
> The German infantry was strong in automatic fire, but . . . no rapid movement or resolute attack on the battlefield.
>
> When advancing they did not spare their bullets but frequently fired into thin air. Their forward positions, particularly at night, were beauti-

> fully visible, being marked by machine-gun fire, tracer bullets, often fired into empty space, and different-coloured rockets. It seemed as if they were either afraid of the dark, or were bored without the crackle of machine-guns and the light of tracer bullets.

The Germans certainly fought well enough later, and it may be that the initial caution came from the natural reluctance of soldiers who believe the war is over to expose themselves unnecessarily in the last flare-up. It is plain from diaries and letters at the time that this belief was universal:

> The company commander says the Russian troops are completely broken, and cannot hold out any longer. To reach the Volga and take Stalingrad is not so difficult for us. The Führer knows where the Russians' weak point is. Victory is not far away. [July 29th]

> Our company is tearing ahead. Today I wrote to Elsa, "We shall soon see each other. All of us feel that the end, Victory, is near." [August 7th]

On 22nd August, Wietersheim's 14th Panzer Corps succeeded in forcing a very narrow breach in the Russian perimeter at Vertyachi and fought its way across the northern suburbs of Stalingrad, actually reaching the steep banks of the Volga on the evening of 23rd August. It now seemed to Paulus, and to his superior, Weichs, that Stalingrad was within their grasp. For with Wietersheim ensconced on the Volga, with the railway bridge at Rynok in mortar range, the difficulty of supplying the Russian garrison, much less reinforcing it, seemed insuperable. During the day Seydlitz's 51st Corps followed Wietersheim into the breach, and it appeared that the whole of the 62nd Army could be rolled up from the north. That night the Luftwaffe was called upon to deliver the *coup de grâce.*

In numbers of aircraft employed, as in weight of explosives, the bombardment of the night of 23rd-24th August was the heaviest strike mounted by the Luftwaffe since 22nd June, 1941. The whole of Richthofen's air corps was used, together with all available Ju 52 squadrons and long-range bombers from airfields as far away as Orel and Kerch. Many of the pilots in Richthofen's corps made up to three sorties, and over half the bombs dropped were incendiaries. The effect was spectacular. Nearly every wooden building—including acres of workers' settlements on the outskirts—was burned down, and the flames made it possible to read a paper forty miles away. It was a pure terror raid, its purpose to kill as many civilians as possible, overload all the services, sow panic and demoralisation, to place a blazing pyre in the path of the retreating army—the pattern of Warsaw, Rotterdam, Belgrade, and Kiev.

With satisfaction Wilhelm Hoffmann, of the 267th Regiment, 94th Division, noted:

> The whole city is on fire; on the *Führer's* orders our *Luftwaffe* has sent it up in flames. That's what the Russians need, to stop them resisting . . .

But as 24th August came and went, and the 25th, and the days following, it became painfully clear that the Russians were determined to fight in front of, and if necessary in, Stalingrad. Wietersheim managed to keep open his corridor to the Volga, but had no success in widening it southward, while the Russian 62nd Army slowly withdrew along the Karpova and the railway line which ran parallel. Sheer weight of metal had allowed Hoth to force the 64th Army back to Tundutovo, but its front had held and the hopes of an orthodox Panzer break-through never approached fulfilment.

This was the second major German effort to be stopped dead in a month, and in its wake we can see the result—unpremeditated on either side—of the strange magnetism which Stalingrad was to exert on the two contestants. On 25th August the Regional Party Committee proclaimed a state of siege:

> Comrades and citizens of Stalingrad! We shall never surrender the city of our birth to the depredations of the German invader. Each single one of us must apply himself to the task of defending our beloved town, our homes, and our families. Let us barricade every street; transform every district, every block, every house, into an impregnable fortress.

On that very day the Führer and his entourage moved from Rastenburg to Vinnitsa, where his headquarters were to remain for the rest of 1942. Weichs was ordered to launch another attack and "clear the whole right bank of the Volga" as soon as Paulus' forces were ready, and on 12th September, the day before this "final" attack was to be launched, the two generals were summoned to the Führer's new headquarters, where Hitler repeated to them that ". . . the vital thing now was to concentrate every available man and capture as quickly as possible the whole of Stalingrad itself and the banks of the Volga." Hitler also told them that there was no need to worry about the left flank along the Don as the arrival of the satellite armies (which were to protect it) was proceeding smoothly.[5]

In addition he had allocated an additional three fresh infantry divisions (two of them came from the disbanded 11th Army of Manstein), which would be arriving in the 6th Army area within the next five days.

At almost the same moment that Hitler moved to Vinnitsa, the Russians, too (although undoubtedly without knowledge of this), gave recognition to the fact that the centre of gravity had shifted irrevocably southward, and that the war would be decided at Stalingrad. For Timoshenko was quietly removed from command and transferred to the northwestern front, and the same team that had evolved the battle-winning plan for the Moscow counteroffensive was moved to Stalingrad: Voronov, the artillery specialist; Novikov, chief of the Red Air Force; and Zhukov, the one commander in the Soviet Army who had never been defeated.

[5] Paulus claims in his "Papers," "Both General Weichs and I drew attention to the very long and inadequately held Don front, and the dangers inherent in the situation," though this assertion should be taken with a grain of salt as there is no confirmation of it from others who attended the conference of 12th September. Furthermore, the "dangers inherent" were largely imaginary at this period, while the Germans still had some reserves uncommitted and the Panzer divisions had not been ground down in three months of street fighting.

twelve | VERDUN ON THE VOLGA

The fighting in the Eastern campaign reflects the whole spectrum of military experience. The cold steel and ferocity of the cavalry charge differ little from the Middle Ages; the misery and privation of interminable bombardment in a stinking dugout recall the Great War. Yet the dominant characteristic of the Russian front is a composite one. Open warfare and manoeuvre alternate with bouts of vicious infighting in a manner that evokes both the Western Desert and the subterranean grapplings of Fort Vaux.[1]

Certainly the tremendous battle which was to be fought out in Stalingrad has its nearest parallel in the horrors of Falkenhayn's "mincing machine" at Verdun. But there are significant differences. At Verdun the contestants rarely saw one another face to face; they were battered to death by high explosives or cut down at long range by machine-gun fire. At Stalingrad each separate battle resolved itself into a combat between individuals. Soldiers would jeer and curse at their enemy across the street; often they could hear his breathing in the next room while they reloaded; hand-to-hand duels were finished in the dark twilight of smoke and brick dust with knives and pickaxes, with clubs of rubble and twisted steel.

At first, while the Germans were in the outskirts, it was still possible for them to draw advantage from their superiority in armour and air-

[1] The centrepiece of the French perimeter at Verdun. For an account of its siege (and probably the finest individual description of close combat in the Great War) see Alistair Horne, *The Price of Glory*.

craft. The buildings here had been of wood, and all had been burned down in the great air raid of 23rd August. Fighting took place in a giant petrified forest of blackened chimney stacks, where the defenders had little cover except the charred remains of the matchboard bungalows and workers' settlements that ringed the town. But as the Germans edged deeper into the region of sewers and brick and concrete, their old plan of operations lost its value. General Doerr has described how

> The time for conducting large-scale operations was gone for ever; from the wide expanses of steppe-land, the war moved into the jagged gullies of the Volga hills with their copses and ravines, into the factory area of Stalingrad, spread out over uneven, pitted, rugged country, covered with iron, concrete and stone buildings. The mile, as a measure of distance, was replaced by the yard. G.H.Q.'s map was the map of the city.
>
> For every house, workshop, water-tower, railway embankment, wall, cellar and every pile of ruins, a bitter battle was waged, without equal even in the first world war with its vast expenditure of munitions. The distance between the enemy's army and ours was as small as it could possibly be. Despite the concentrated activity of aircraft and artillery, it was impossible to break out of the area of close fighting. The Russians surpassed the Germans in their use of the terrain and in camouflage and were more experienced in barricade warfare for individual buildings . . .

If the battle had a tactical pattern it was one which revolved around the fate of the Volga ferries, the lifeline of the garrison. For although the Russians kept their heavy and medium artillery on the east bank, they were consuming small-arms ammunition and mortar bombs at a prodigious rate, and depended on the traffic across the Volga for many other services essential to the fighting spirit of the garrison, ranging from the provision of vodka to the evacuation of wounded. The slight curve in the course of the river and the numerous islets which obstructed the stream between Rynok and Krasnaya Sloboda made it very difficult to enfilade all the crossings even after guns had been installed on the right bank, and well-nigh impossible to do so at night, when the bulk of the traffic was on the move. The Germans were slow to realise this, and instead of putting all their energies into attacks at the extremities of the Russian position and working their way up and down the bank—a tactic which if successful would ultimately have left the garrison stranded on an island of rubble in the centre—they switched their effort to different points in the city, adopting the most extravagant method of simply battering away at one block after another. Each of the three major "offensives" launched during the siege was aimed at cutting across the thin strip of ground the Russians held and reaching the Volga at as many points as possible. The result was that even where they were successful in their aim, the attackers would find themselves stranded in a web of hostile emplacements, their access corridors too narrow to make the troops at their tip anything but a tactical liability.

If the Luftwaffe had been employed with single-minded persistence in an "interdiction" role (in the sense, that is to say, in which the term came to be understood in the West), the Volga ferries might have been knocked out. Certainly Richthofen, had he been properly directed, could have done more about the Russian 76-mm. batteries on the east bank, whose fire deterred the 6th Army from operating too close to the river. Yet the fact remains that while the Russians showed great skill and versatility in adapting their tactics as the battle wore on, Paulus mishandled it from the start. The Germans were baffled by a situation hitherto outside their military experience, and they reacted to it characteristically—by the application of brute force in heavier and heavier doses.

This bafflement extended from the senior commanders to the ordinary soldier. Hoffman (the diarist whose exultation at the 23rd August terror raid has already been quoted) reflects this attitude in the epithets he attaches to the defenders, ranging progressively through incredulity and contempt to fear, and then to self-pity.

September 1st: "Are the Russians really going to fight on the very bank of the Volga? It's madness."

September 8th: ". . . insane stubbornness."

September 11th: ". . . Fanatics."

September 13th: ". . . wild beasts."

September 16th: "Barbarism . . . [they are] not men but devils."

September 26th: ". . . Barbarians, they use gangster methods."

There is no further comment for a month on the quality of the enemy, and during this time the entries are filled with gloom at the plight of the writer and his comrades in arms.

October 27th: ". . . The Russians are not men, but some kind of cast-iron creatures; they never get tired and are not afraid of fire."

October 28th: "Every soldier sees himself as a condemned man."

When Paulus returned to his headquarters after the conference with Hitler on 12th September, H-Hour for his third offensive was imminent.

This time the 6th Army was deploying eleven divisions, of which three were Panzer. The Russians had only three infantry divisions, parts of four others, and two tank brigades. This drastic reduction in the defenders' strength was the result of Hoth's success in at last battering his way through to the Volga at Kuporosnoye, a suburb of Stalingrad proper, and thereby dividing the 62nd and 64th armies. Five days earlier, on 4th September, Hoth's tanks had split the 64th Army for the first time by reaching the Volga at Krasnoarmeisk, and the bulk of the Russian force, which had spent itself in six weeks of continuous fighting against the elite *Panzergruppe* of the whole German Army, was pinned down along a twelve-mile strip of the Stalingrad-Rostov railway embankment. The day after the 14th Panzer took Kuporosnoye, Chuikov was appointed to the command of the isolated 62nd Army. That night he crossed by boat from Beketovka, and after a nightmare jeep journey up the left bank of the river to report to Khrushchev and Yeremenko at "front" headquarters at Yamy, took the ferry at dawn from Krasnaya Sloboda across into the burning city.

Stalingrad had now been under continuous bombardment for twenty-four hours, as the whole of the 6th Army's artillery paved the way for Paulus' concentric assault. As their boat approached the landing stage, spent shrapnel and shell fragments were dropping in the inky water "like trout," and they could feel the air temperature several degrees hotter from the flames. Chuikov reflected:

> Anyone without experience of war would think that in the blazing city there is no longer anywhere left to live, that everything has been destroyed and burnt out. . . . But I know that on the other side of the river a battle is being fought, a titanic struggle is taking place.

Paulus had concentrated two "shock forces" with the intention of converging against the southern half of the town and joining at the so-called "central landing stage," opposite Krasnaya Sloboda. Three infantry divisions, the 71st, 76th, and 295th, were to move down from the Gumrak railway station, capturing the main hospital, to Matveyev Kurgan. An even stronger force, the 94th Infantry Division and the 29th Motorised, was to strike northeast from the Yelshanka mining suburb, backed by 14th and 24th Panzer.

Chuikov had only forty tanks left in action, and many of these were no longer mobile, but had been dug in as armoured fire points. He also had a small tank reserve of nineteen KV's, as yet uncommitted, but no infantry reserves whatever, for every man capable of carrying a gun had been sucked into the battle. Chuikov's predecessor, General Lopatin, had (allegedly) been convinced of the "impossibility and pointlessness of defending the city," and this feeling of depression

> had undoubtedly communicated itself to his subordinates . . . on the pretext of illness three of my deputies [for artillery, tanks, and army engineering] had left for the opposite bank of the Volga.

The defence problem was fourfold: first, it was essential to hold the flanks firmly anchored to the riverbank. Every yard of the steep Volga escarpment was precious to the Russians, who had tunnelled into it for depots, hospitals, ammunition dumps, fuel stores—even for garages for the *Katyusha* trucks, which would reverse out of their caves, fire a salvo, and get back under cover in less than five minutes. The northern flank below Rynok was the stronger of the two, for here the vast concrete edifices of the Tractor Factory and the Barrikady and Krasny Oktyabr were virtually indestructible. But at the southern end the buildings were less substantial and the ground was relatively open, undulating mounds of rubble and occasional patches of scorched heath dominated by a few towering grain elevators. Here, too, lay the shortest route to the central landing stage, along the course of the Tsaritsa rivulet; and to the nerve centre of the Stalingrad defence system, Chuikov's own command post, which was in a dugout known as the "Tsaritsyn bunker," sunk into the side of the riverbed at the Pushkin Street bridge.

The danger of concentrating his strength at the extremities was that Chuikov's very long west-facing front (it was over ten miles from Rynok to Kuporosnoye as the crow flies, and double that length along the "line") would be vulnerable to a concentrated assault on a narrow front, and in particular, that Matveyev-Kurgan, a grassy hillock of parkland that dominated the centre of the town, might be lost to the enemy before reinforcements could reach it.

Chuikov had sent urgent requests for infantry reinforcements to Yeremenko on 13th September, when Paulus had started his attack and he had learned during the night that the 13th Guards Infantry Division, a very strong unit under General Rodimtsev (who had started his experiences of street fighting in the Madrid University City, in 1936) would be sent over the river starting at dusk the following day. However, during the afternoon of 14th September, Paulus' central attack broke through the Russian front behind the hospital and the Germans of the

76th Infantry Division began to pour into the rear areas of the town, obstructed only by a few snipers.

> Lorry-loads of infantry and tanks tore through into the city. The Germans obviously thought that the fate of the town had been settled, and they all rushed to the centre and the Volga as soon as possible and grabbed souvenirs for themselves . . . we saw drunken Germans jumping down from their lorries, playing mouth organs, shouting like madmen and dancing on the pavements.

To deal with this break-through Chuikov used his last reserve of tanks, which meant bringing them up from the southern sector, itself under very heavy pressure, in daylight. His own staff officers and the bunker guard company were involved in the fighting, which raged all night. Infiltrating German soldiers got within two hundred yards of the Tsaritsyn bunker, and some managed to get heavy machine guns into position where they could fire on the central landing stage. Chuikov was now faced with the prospect of having his front once again broken into two pieces, yet to move any more troops from the southern end of the perimeter might lead to the collapse of the whole position there.

At this stage German tactics, though wasteful and unsophisticated, were highly abrasive against a defence stretched as thin as was that of the 62nd Army in the first days of Chuikov's command. They consisted of using tanks in packets of three or four at a time in support of each company of infantry. The Russians would never fire at tanks alone, but let them pass through into the field of fire of anti-tank guns and dug-in T 34's which were held farther back; so it was always necessary for the Germans to send infantry in first to draw the defender's fire. Once his position had been identified, the tanks would cover one another while they battered away at point-blank range until the building fell down. Where the houses were tall and substantial, this was a long and untidy business. Armour-piercing shot was useless—it would pass right through the walls, doing no more damage than a jagged hole about two feet across, yet to risk sending the tanks out with only high-explosive ammunition meant that they were at the mercy of any roaming T 34 that might come on the scene. Furthermore, although tank fire would gut the first two floors, the limited elevation of the turret often meant that unless the top story was set alight the rest of the building went undamaged.

> We would spend the whole day clearing a street, from one end to the other, establish blocks and fire-points at the western end, and prepare for another slice of the salami the next day. But at dawn the Russians would

> start up firing from their old positions at the far end! It took us some time to discover their trick; they had knocked communicating holes through between the garrets and attics and during the night they would run back like rats in the rafters, and set their machine-guns up behind some top-most window or broken chimney . . .

The tank crews were understandably reluctant to take their machines into narrow streets, where their lightly armoured rear deck could be penetrated by anti-tank rifles or grenades thrown from above. It was necessary to accompany each attacking force with teams of flame throwers so that the buildings could be burned down, but this was an extremely hazardous occupation as a single bullet could turn the operator into a flaming torch. Special rates of pay[2] were introduced, but it was still impossible to get sufficient volunteers without recourse to the punishment battalions.

However, during the first days of their September offensive the Germans enjoyed a superiority of three to one in men and over six to one in tanks, and the Luftwaffe held complete dominion of the daylight air. From 14th to 22nd September, while the 6th Army was relatively fresh and the Russians were defending with the remains of units which had been badly battered in earlier fighting, was the period of Stalingrad's greatest peril.

During the night of 14th September the whole defence front was creaking so badly that Rodimtsev's division had to be sent into action by battalion, immediately the men had formed up after disembarking from the ferries. The result was that it was dispersed over a wide area and many of the men were soon cut off in the strange wilderness of smoke and rubble in which they found themselves at daybreak. Yet even among these the stubborn refusal of the Russian to surrender while he still has ammunition played its part in dislocating the German advance. The account of a member of the 3rd Company of the 42nd Regiment, though its somewhat self-consciously heroic style may jar on Western ears, deserves to be quoted at length because of its relevance to the conditions of street fighting and the spirit of the defenders. At one point they found themselves cut off.

[2] Owing to Russian practice of putting captured *Flammenwerfer* operatives to death in a certain fashion, they were described in their paybooks as "Engineers, 1st Class," although drawing rates of pay even higher.

. . . We moved back, occupying one building after another, turning them into strongholds. A soldier would crawl out of an occupied position only when the ground was on fire under him and his clothes were smouldering. During the day the Germans managed to occupy only two blocks.

At the crossroads of Krasnopiterskaya and Komsomolskaya Streets we occupied a three-storey building on the corner. This was a good position from which to fire on all comers and it became our last defence. I ordered all entrances to be barricaded, and windows and embrasures to be adapted so that we could fire through them with all our remaining weapons.

At a narrow window of the semi-basement we placed the heavy machine-gun with our emergency supply of ammunition—the last belt of cartridges. I had decided to use it at the most critical moment.

Two groups, six in each, went up to the third floor and the garret. Their job was to break down walls, and prepare lumps of stone and beams to throw at the Germans when they came up close. A place for the seriously wounded was set aside in the basement. Our garrison consisted of forty men. Difficult days began. . . . The basement was full of wounded; only twelve men were still able to fight. There was no water. All we had left in the way of food was a few pounds of scorched grain; the Germans decided to beat us with starvation. Their attacks stopped, but they kept up the fire from their heavy-calibre machine-guns all the time. . . . The Germans attacked again. I ran upsairs with my men and could see their thin, blackened and strained faces, the bandages on their wounds, dirty and clotted with blood, their guns held firmly in their hands. There was no fear in their eyes. Lyuba Nesterenko, a nurse, was dying, with blood flowing from a wound in her chest. She had a bandage in her hand. Before she died she wanted to help to bind someone's wound, but she failed . . .

The German attack was beaten off. In the silence that gathered around us we could hear the bitter fighting going on for Matveyev Kurgan and in the factory area of the city.

How could we help the men defending the city? How could we divert from over there even a part of the enemy forces, which had stopped attacking our building?

We decided to raise a red flag over the building, so that the Nazis would not think we had given up. But we had no red material. Understanding what we wanted to do, one of the men who was severely wounded took off his bloody vest and, after wiping the blood off his wound with it, handed it over to me.

The Germans shouted through a megaphone: "Russians! Surrender! You'll die just the same!"

At that a moment a red flag rose over our building.

"Bark, you dogs! We've still got a long time to live!" shouted my orderly, Kozhushko.

We beat off the next attack with stones, firing occasionally and throwing our last grenades. Suddenly from behind a blank wall, from the rear, came the grind of a tank's caterpillar tracks. We had no anti-tank grenades. All we had left was one anti-tank rifle with three rounds. I handed this rifle to an anti-tank man, Berdyshev, and sent him out through the back to fire at the tank point-blank. But before he could get into position he was captured by German tommy-gunners. What Berdyshev told the Germans I don't know, but I can guess that he led them up the garden path, because an hour later they started to attack at precisely that point where I had put my machine-gun with its emergency belt of cartridges.

This time, reckoning that we had run out of ammunition, they came impudently out of their shelter, standing up and shouting. They came down the street in a column.

I put the last belt in the heavy machine-gun at the semi-basement window and sent the whole of the 250 bullets into the yelling, dirty-grey Nazi mob. I was wounded in the hand but did not let go of the machine-gun. Heaps of bodies littered the ground. The Germans still alive ran for cover in panic. An hour later they led our anti-tank rifleman on to a heap of ruins and shot him in front of our eyes, for having shown them the way to my machine-gun.

There were no more attacks. An avalanche of shells fell on the building. The Germans stormed at us with every possible kind of weapon. We couldn't raise our heads.

Again we heard the ominous sound of tanks. From behind a neighbouring block stocky German tanks began to crawl out. This, clearly, was the end. The guardsmen said good-bye to one another. With a dagger my orderly scratched on a brick wall: "Rodimtsev's guardsmen fought and died for their country here."

On 21st September both sides were prostrate with exhaustion. The Germans had cleared the whole of the Tsaritsa river bed and established their guns a few yards from the central landing stage. They had also carved out a large section, about a mile and a half square, of the built-up area behind the Stalingrad No. 1 Station, lying between the Tsaritsa and the Krutoy gully. Chuikov had been forced to move his headquarters out of the Tsaritsyn bunker to Matveyev-Kurgan, and with the central landing stage area neutralised, the garrison was now dependent on the factory ferries at the northern end of the town.

At this stage the Germans were perilously close to gaining control of the whole southern half of the city, up to the Krutoy gully at least, as only one Russian unit, the 92nd Infantry Brigade, was left fighting south of the Tsaritsa. But Hoth's forces were being seriously impeded by a

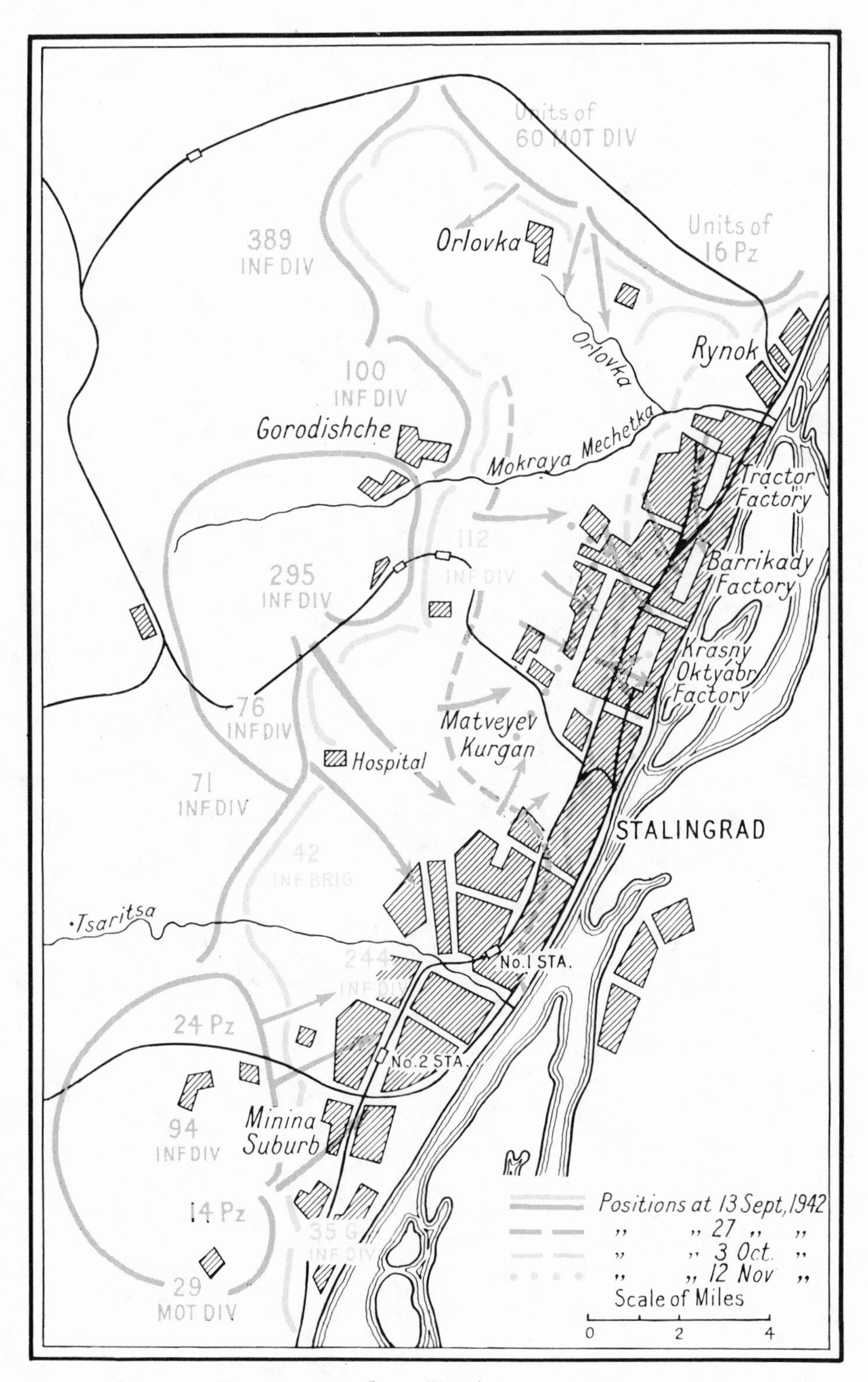

Street Fighting

number of isolated centres of resistance which had been left behind in the first armoured rush on 13th and 14th September. These were mostly centred around the giant grain elevators, and in one case we have available the diaries of men who took part on either side in a particular engagement.

First, the German:

> September 16th. Our battalion, plus tanks, is attacking the elevator, from which smoke is pouring—the grain in it is burning, the Russians seem to have set light to it themselves. Barbarism. The battalion is suffering heavy losses. There are not more than sixty men left in each company. The elevator is occupied not by men but by devils that no flames or bullets can destroy.
>
> September 18th. Fighting is going on inside the elevator. The Russians inside are condemned men; the battalion commander says: "The commissars have ordered those men to die in the elevator."
>
> If all the buildings of Stalingrad are defended like this, then none of our soldiers will get back to Germany. I had a letter from Elsa today. She's expecting me home when victory's won.
>
> September 20th. The battle for the elevator is still going on. The Russians are firing on all sides. We stay in our cellar; you can't go out into the street. Sergeant-Major Nuschke was killed today running across a street. Poor fellow, he's got three children.
>
> September 22nd. Russian resistance in the elevator has been broken. Our troops are advancing towards the Volga. We found about forty Russians dead in the elevator building. Half of them were wearing naval uniform —sea devils. One prisoner was captured, seriously wounded, who can't speak, or is shamming.

The "seriously wounded" prisoner was Andrey Khozyaynov, of the Marine Infantry Brigade, and his version conveys a remarkable impression of the character of the Stalingrad street fighting, where the individual courage and tenacity of a few soldiers and junior N.C.O.'s, often out of touch and given up for lost by their own high command, could effect the whole development of the battle.

> Our brigade was ferried over the Volga during the night of September 16 and at dawn on the 17th it was already in action.
>
> I remember that on the night of the 17th, after fierce fighting, I was called to the battalion command post and given the order to take a platoon of machine-gunners to the grain elevator and, together with the men already in action there, to hold it come what may. We arrived that night and presented ourselves to the garrison commander. At that time the elevator was being defended by a battalion of not more than thirty to

thirty-five guardsmen, together with the wounded, some slightly, some seriously, whom they had not yet been able to send back to the rear.

The guardsmen were very pleased to see us arrive, and immediately began pouring out jokes and witticisms. Eighteen well-armed men had arrived in our platoon. We had two medium machine-guns and one light machine-gun, two anti-tank rifles, three tommy-guns and radio equipment.

At dawn a German tank carrying a white flag approached from the south. We wondered what could have happened. Two men emerged from the tank, a Nazi officer and an interpreter. Through the interpreter the officer tried to persuade us to surrender to the "heroic German army," as defence was useless and we would not be able to hold our position any longer. "Better to surrender the elevator," affirmed the German officer. "If you refuse you will be dealt with without mercy. In an hour's time we will bomb you out of existence."

"What impudence," we thought, and gave the Nazi lieutenant a brief answer: "Tell all your Nazis to go to hell! . . . You can go back, but only on foot."

The German tank tried to beat a retreat, but a salvo from our two anti-tank rifles stopped it.

Enemy tanks and infantry, approximately ten times our numbers, soon launched an attack from south and west. After the first attack was beaten back, a second began, then a third, while a reconnaissance "pilot" plane circled over us. It corrected the fire and reported our position. In all, ten attacks were beaten off on September 18.

We economised on ammunition, as it was a long way, and difficult, to bring up more.

In the elevator the grain was on fire, the water in the machine-guns evaporated, the wounded were thirsty, but there was no water nearby. This was how we defended ourselves twenty-four hours a day for three days. Heat, smoke, thirst—all our lips were cracked. During the day many of us climbed up to the highest points in the elevator and from there fired on the Germans; at night we came down and made a defensive ring round the building. Our radio equipment had been put out of action on the very first day. We had no contact with our units.

September 20 arrived. At noon twelve enemy tanks came up from the south and west. We had already run out of ammunition for our anti-tank rifles, and we had no grenades left. The tanks approached the elevator from two sides and began to fire at our garrison at point-blank range. But no one flinched. Our machine-guns and tommy-guns continued to fire at the enemy's infantry, preventing them from entering the elevator. Then a Maxim, together with a gunner, was blown up by a shell, and the casing of the second Maxim was hit by shrapnel, bending the barrel. We were left with one light machine-gun.

The explosions were shattering the concrete; the grain was in flames.

We could not see one another for dust and smoke, but we cheered one another with shouts.

German tommy-gunners appeared from behind the tanks. There were about 150-200 of them. They attacked very cautiously, throwing grenades in front of them. We were able to catch some of the grenades and throw them back.

On the west side of the elevator the Germans managed to enter the building, but we immediately turned our guns on the parts they had occupied.

Fighting flared up inside the building. We sensed and heard the enemy soldiers' breath and footsteps, but we could not see them in the smoke. We fired at sounds.

At night, during a short lull, we counted our ammunition. There did not seem to be much left: one and a half drums for the machine-gun, twenty to twenty-five rounds for each tommy-gun, and eight to ten rounds for each rifle.

To defend ourselves with that amount of ammunition was impossible. We were surrounded. We decided to break out to the south, to the area of Beketovka, as there were enemy tanks to the north and east of the elevator.

During the night of the 20th, covered by our one tommy-gun, we set off. To begin with all went well; the Germans were not expecting us here. We passed through the gully and crossed the railway line, then stumbled on an enemy mortar battery which had only just taken up position under cover of darkness.

We overturned the three mortars and a truck-load of bombs. The Germans scattered, leaving behind seven dead, abandoning not only their weapons, but their bread and water. And we were fainting with thirst. "Something to drink! Something to drink!" was all we could think about. We drank our fill in the darkness. We then ate the bread we had captured from the Germans and went on. But alas, what then happened to my comrades I don't know, because the next thing I remember was opening my eyes on September 25 or 26. I was in a dark, damp cellar, feeling as though I were covered with some kind of oil. I had no tunic on and no shoe on my right foot. My hands and legs would not obey me at all; my head was singing.

A door opened, and in the bright sunlight I could see a tommy-gunner in a black uniform. On his left sleeve was a skull. I had fallen into the hands of the enemy.

The German offensive, which had opened so brilliantly, had reaffirmed in a few short weeks the Wehrmacht's power to make the whole world tremble, had carried the boundaries of the Reich to their farthest mark —was now, undeniably, stuck fast. For nearly two months the maps had remained unchanged.

The Propaganda Ministry affirmed that the "greatest battle of attrition that the world has ever seen" was taking place, and daily published figures which showed how the Soviets were being bled white. But whether the Germans believed them or not, the facts were very different. It was they, not the Red Army, who were being forced repeatedly to raise the ante. With the same coolness which stamped his refusal to commit the Siberian reserve until the battle of Moscow had already been decided, Zhukov was keeping the reinforcement of the 62nd Army to a bare minimum. In the two critical months from 1st September to 1st November only five infantry divisions were sent across the Volga—barely sufficient to cover "wastage." Yet in that period twenty-seven fresh infantry divisions and nineteen armoured brigades were activised from drafts, new material, and cadre skeletons of seasoned officers and N.C.O.'s. All were concentrated in the area between Povorino and Saratov, where their training was completed, and whence some of them were sent in rotation to the central sector for brief periods of combat experience. The result was that while the Germans were slowly running all their divisions into the ground with fatigue and casualties the Red Army was building up a formidable reserve of men and armour.

The feeling of frustration at being halted so near (as it seemed) to complete victory was soon compounded by a sense of foreboding which heightened as the weeks followed one another with the army always in the same position.

> The days were shortening again, you could definitely sense it. And in the mornings the air was quite cool. Were we really going to have to fight through another of those dreadful winters? I think that was behind our efforts. Many of us felt that it was worth anything, any price, if we could get it over before the winter.

While the spirits of the men alternated between frenzy and depression, recrimination and the clash of personalities enlivened the affairs of the army group at a higher level.

First to go were two tank generals, Wietersheim and Schwedler. The essence of their complaint was that the Panzer divisions were being worn out in operations to which they were completely unsuited, and that after a few more weeks of street fighting they would no longer have the ability to fulfil their primary task—that of engaging the enemy's armour in mobile battles. However, the dictates of military protocol do not permit corps commanders, however distinguished, to protest on broad strategic grounds, and each chose to complain on a narrower point of tactics.

General von Wietersheim commanded the 14th Panzer Corps, which had been the first unit of the 6th Army to break through to the Volga at Rynok in August. He certainly cannot be accused of timidity, for he had taken his corps across northern France in 1940 on Guderian's heels and been one of the very few officers in the German Army who had recommended pressing on across the Meuse. Wietersheim suggested to Paulus that the attrition from Russian artillery fire on both sides of the Rynok corridor was having such effect on his Panzers that they should be withdrawn and the corridor held open by infantry. He was dismissed and sent back to Germany, ending his military career as a private in the *Volkssturm* in Pomerania, in 1945.

General von Schwedler was the commander of the 4th Panzer Corps, and had led the southern arm of the counterstroke against Timoshenko's drive on Kharkov in May. His case is an interesting one in that he was the first general to warn against the danger of concentrating all the armour at the tip of a "dead *Schwerpunkt*" and the vulnerability to a Russian attack from the flanks.[3] But in the autumn of 1942 the concept of the Russians *attacking* was regarded as "defeatist," and Schwedler, too, was dismissed.

The next head to fall was that of Colonel General List, Commander in Chief of Army Group A. After the first rush across the Kuban, which had carried the 1st Panzer Army to Mozdok by the end of August, the front of the German advance had solidified along the contour lines of the Causasus range and the Terek River. Different factors accounted for this, notably the withdrawal of Richthofen's bombers to Stalingrad and a recovery by the defenders. Kleist has remarked:

> In the early stages . . . I met little organised resistance. As soon as the Russian forces were by-passed, most of the troops seemed more intent to find their way back to their homes than to continue fighting. That was quite different to what had happened in 1941. But when we advanced into the Caucasus the forces we met there were local troops, who fought more stubbornly because they were fighting to defend their homes. Their obstinate resistance was all the more effective because the country was so difficult . . .

[3] Practically every senior German commander in the southern theatre now likes to take the credit for this prescience. In fact, it seems to be due either to Schwedler or to Blumentritt. Blumentritt was sent on a tour of inspection of the Don front between Voronezh and Kletskaya and submitted a report to the effect that ". . . it would not be safe to hold such a long defensive flank during the winter." Goerlitz places the time of this inspection in early August, but Blumentritt himself under interrogation put it in September.

The result was that the first plan for the occupation of the oil fields was altered, and OKW directed List to force his way across the low Caucasus at their western end and capture Tuapse. Reinforcements, including three mountain divisions which would have been of great value to Kleist, were put into the 17th Army instead. Had this been successful, the Germans would have broken across the Caucasus at their lowest point, and by capturing Batum would have forced the Russian Black Sea fleet into internment and ensured the security of the Crimea and the compliant neutrality of Turkey. But in fact one difficulty after another supervened and, notwithstanding his reinforcement, List made little progress. In September, Jodl was sent as the OKW representative to List's headquarters to report "the Führer's impatience" and to try to get things moving.

But when Jodl returned he came with bad news.

> List had acted exactly in conformity with the Führer's orders, but the Russian resistance was equally strong everywhere, supported by a most difficult terrain.

Warlimont contends that Jodl answered Hitler's reproaches (and if he did, it was certainly for the first, and the last, time) by pointing

> to the fact that Hitler by his own orders had induced List to advance on a widely stretched front.

The result was "an outburst," and Jodl fell into disgrace.

> Further consequences were that Hitler completely changed his daily customs. From that time on he stayed away from the common meals which he had taken twice a day with his entourage. Henceforth he hardly left his hut in daytime, not even for the daily reports on the military situation, which from now on had to be delivered to him in his own hut in the presence of a narrowly restricted circle. He refused ostentatiously to shake hands with any general of the OKW, and gave orders that Jodl was to be replaced by another officer.

Jodl's replacement was never in fact appointed, and the OKW Chief of Staff soon came back into favour, having learned his lesson, which, as he confided to Warlimont, was that

> A dictator, as a matter of psychological necessity, must never be reminded of his own errors—in order to keep up his self-confidence, the ultimate source of his dictatorial force.

Nonetheless, the possibility of the appointment was duly communicated to that "other officer" concerned, with results which will be seen.

However, before that particular trail of private ambition and intrigue is followed up, there is one more dismissal which must be recorded, together with its effect on the running of Führer headquarters. Relations between Hitler and Halder had deteriorated steadily since the removal of the pliant ObdH, who had acted as an absorbent pad between Hitler's violence and the Chief of Staff's dry acerbity. Manstein, who had seen them together in August while passing through headquarters to take up his command at Leningrad, was "quite appalled" to see how bad their relations were. Hitler was abusive, Halder obstructive and pedantic. Hitler would taunt Halder with his lack of combat experience, in contrast to his own experiences in the front line in the Great War. Halder would mumble under his breath about the differences between professional and "untutored" opinion.

Matters came to a head over a quite minor point, relating to the central front. Many of the German commanders, particularly Kluge (whose responsibility it was), believed that the counteroffensive the Russians were expected to launch in the winter would be directed against Army Group Centre. Paradoxically, this was in part due to the Russians' practice of giving their new divisions a baptism of fire in the quiet central sector before taking them back into strategic reserve. The new divisions would be identified, then seem to disappear. Kluge, and Halder himself, formed the erroneous opinion that they were being accumulated behind the front where they had been identified instead, as was the case, of being sent south. At all events, a rather childish quarrel blew up between Hitler and Halder concerning the date on which one of these units was identified, and around it larger issues were invoked, notably the need (as Halder saw it) to reinforce Kluge and thus, indirectly, the overstretched condition of the Wehrmacht generally.[4] On 24th September, Halder was dismissed and Colonel General Kurt Zeitzler brought from the West to take his place.

[4] It is widely contended that Halder's dismissal arose out of his refusal to sanction further offensive operations at Stalingrad until the Don flank had been strengthened (Leyderrey, Goerlitz, Blumentritt, Halder himself, etc.). But the chain of argument by which this claim is rationalised seems pretty tenuous. *Vide* Warlimont's evidence to Liddell Hart:

> It was on this issue [Russian activity against Army Group Centre, at Rzhev] that the final clash between Hitler and Halder originated, which led to the latter's dismissal. (See Liddell Hart, *op. cit.*, 220, enlarged third ed.; London, 1956.)

The occasion of Halder's dismissal is of particular interest to historians of World War II because of a change which was inaugurated at that time in the administration of the daily Führer conferences. These conferences had become the medium through which the war was being run and the directives promulgated. For the old OKH apparatus had been in decay since Brauchitsch's dismissal, and Halder's real fault, in Hitler's eyes, was his shuffling attempts to reserve to OKH (and thereby to himself) certain of the old prerogatives of the *Generalstab* and his tacit reluctance to accept Hitler's "appointment" as Commander in Chief of the Army as anything but temporary. With the advent of Zeitzler, who had no memories of the time when OKH ran the Eastern campaign with Hitler no more than a petulant voice at the end of a bad telephone line, the centralisation of tactical as of strategic administration would be complete. The final step in the consolidation of the daily conferences as the prime mover in the executive process was the introduction of stenographers who faithfully recorded every word spoken by every participant. To the extent to which these records have survived, they are of enormous importance in showing how the war was run by the Germans, and where they impinge on the Eastern campaign they will be quoted at length in this work.[5]

One of those who benefited from the reshuffle at Hitler's headquarters was that loyal, well-meaning Nazi, General Schmundt (who will be remembered helping Guderian with his "problems" the previous summer,[6] and who will later be encountered in a less happy contest). Schmundt was promoted from the rather ill-defined post of Hitler's principal adjutant to that of head of the Army Personnel Office, where he enjoyed very considerable power in the field of posting and appointment. Paulus "felt that he ought to send Schmundt his congratulations."

Not long afterward Schmundt turned up at Paulus' headquarters, and the commander of the 6th Army launched straightaway into a long complaint about the condition of the troops, the shortages, the strength of the Russian resistance, the possible dangers if the 6th Army were to become too exhausted, and so forth. Perhaps he referred to the original text of Directive No. 41, which had limited his objective to bringing the Volga under gunfire—for he had certainly done *that*.

[5] The character and origins of this very important material are discussed at greater length in the Note on Sources.

[6] See pp. 93-94.

Schmundt, however, had the one answer which a reluctant commander can never resist. After some prefatory remarks about the Führer's desire to see the Stalingrad operations "brought to a successful conclusion," he broke the exciting news. The "other officer" under consideration for the post of the Chief of the OKW staff was none other than Paulus himself! It was true that General Jodl's actual standing-down was hanging fire at the present time, but Paulus had been "definitely earmarked" for a more senior post, and General von Seydlitz would take his place as head of the 6th Army.

Paulus may have been a good staff officer; as a commander in the field he was slow-witted and unimaginative to the point of stupidity. Equally certain, as can be seen from the course of his career up to and after capture, he had a keen awareness of the sources of power, or to put it bluntly, he knew what was good for him. With the news from Schmundt of what extra was at stake, he threw himself into preparations for a fourth offensive with a special enthusiasm.

This time Paulus had decided to strike head on at his adversary's strongest point—the three giant erections of the Tractor Factory, the Barrikady, and the Krasny Oktyabr, which lay in the northern half of the city, ranged one after another a few hundred feet from the Volga bank. This was to be the fiercest, and the longest, of the five battles which were fought in the ruined town. It started on 4th October and raged for nearly three weeks.

Paulus had been reinforced by a variety of different specialist troops, including police battalions and engineers skilled in street fighting and demolition work. But the Russians, though now heavily outnumbered, remained their masters in the technique of house-to-house fighting. They had perfected the use of "storm groups," small bodies of mixed arms—light and heavy machine guns, tommy gunners, and grenadiers usually with anti-tank guns—who gave one another support in lightning counterattacks; and they had developed the creation of "killing zones," houses and squares, heavily mined, to which the defenders knew all the approach routes, where the German advance would be canalised. "Experience taught us," Chuikov wrote:

> . . . Get close to the enemy's positions; move on all fours, making use of craters and ruins; dig your trenches by night, camouflage them by day; make your build-up for the attack stealthily, without any noise;

carry your tommy-gun on your shoulder; take ten to twelve grenades. Timing and surprise will then be on your side. . . .

. . . Two of you get into the house together—you, and a grenade; both be lightly dressed—you without a knapsack, and the grenade bare; go in grenade first, you after; go through the whole house, again always with a grenade first and you after. . . .

. . . There is one strict rule now—give yourself elbow room! At every step danger lurks. No matter—a grenade in every corner of the room, then forward! A burst from your tommy-gun around what's left; a bit further—a grenade, then on again! Another room—a grenade! A turning—another grenade! Rake it with your tommy-gun! And get a move on!

Inside the object of attack the enemy may go over to a counter-attack. Don't be afraid! You have already taken the initiative, it is in your hands. Act more ruthlessly with your grenade, your tommy-gun, your dagger and your spade! Fighting inside a building is always frantic. So always be prepared for the unexpected. Look Sharp! . . .

Slowly and at a tremendous price the Germans inched their way into the great buildings, across factory floors; around and over the inert machinery, through the foundries, the assembly shops, the office. "My God, why have you forsaken us?" wrote a lieutenant of the 24th Panzer Division.

We have fought during fifteen days for a single house, with mortars, grenades, machine-guns and bayonets. Already by the third day fifty-four German corpses are strewn in the cellars, on the landings, and the staircases. The front is a corridor between burnt-out rooms; it is the thin ceiling between two floors. Help comes from neighbouring houses by fire escapes and chimneys. There is a ceaseless struggle from noon to night. From storey to storey, faces black with sweat, we bombard each other with grenades in the middle of explosions, clouds of dust and smoke, heaps of mortar, floods of blood, fragments of furniture and human beings. Ask any soldier what half an hour of hand-to-hand struggle means in such a fight. And imagine Stalingrad; eighty days and eighty nights of hand-to-hand struggles. The street is no longer measured by metres but by corpses . . .

Stalingrad is no longer a town. By day it is an enormous cloud of burning, blinding smoke; it is a vast furnace lit by the reflection of the flames. And when night arrives, one of those scorching, howling, bleeding nights, the dogs plunge into the Volga and swim desperately to gain the other bank. The nights of Stalingrad are a terror for them. Animals flee this hell; the hardest stones cannot bear it for long; only men endure.

thirteen | THE ENTOMBMENT OF THE 6TH ARMY

By the end of October the Russian positions at Stalingrad had been reduced to a few pockets of stone, seldom more than three hundred yards deep, bordering on the right bank of the Volga. The Krasny Oktyabr had fallen to the Germans, who had paved every metre of the factory floor with their dead. The Barrikady was half lost, with Germans at one end of the foundry facing Russian machine guns in the extinct ovens at the other. The defenders of the Tractor Factory had been split into three.

But these last islets of resistance, hardened in the furnace of repeated attacks, were irreducible. The 6th Army was spent, as raddled and exhausted as had been Haig's divisions at Passchendaele exactly a quarter of a century before; and on a purely military assessment, the concept of another "offensive" in the town was unthinkable. Had Army Group B the strength, the correct course would have been to strike at Voronezh and lever the Don front away by starting from its northern end. But it had not the strength; the whole Wehrmacht was desperately overextended on a front which had nearly doubled in length since the start of the summer campaign. It was in that peculiarly dangerous position of being the weaker army, which had nothing save "initiative" to compensate for its material inferiority. Once the momentum was lost its perils would become acute, and once the reserves were exhausted the momentum would be lost.

This rationalisation could, of course, be made to serve two arguments. The first, the obvious one, dictated immediate withdrawal; losses should be cut and a deep "winter line" occupied many miles in the rear, on the

Chir River, or perhaps even the Mius. The alternative case, and one with which soldiers can often be convinced, was the familiar "lesson" of Waterloo and the Marne—that "The last battalion will decide the issue." The Germans, who had seen their own strength sucked into the inferno for week after week, could not admit to themselves that the Russians were not suffering attrition at the same rate. To many, and especially to Hitler, the parallel with Verdun was irresistible. Once a place assumes a symbolic importance its loss can destroy the defenders' will regardless of its strategic value. In 1916, Falkenhayn's mincing machine had been turned off when another month would have destroyed the whole French Army. At Stalingrad it was not only the Russian will, but the whole world's assessment of Germany's power which was at stake. To withdraw from the field of battle would be an admission or defeat which, though it might be acceptable to a detached and calculating military professional, was unthinkable "in the cosmic orientation of world political forces," as Schwerin von Krosigk might have put it.

Hitler's attitude could have been changed (though it is by no means certain) if he had been getting accurate intelligence reports instead of the highly misleading figures Paulus was sending in. But the 6th Army, from an understandable desire to justify its own further reinforcement and to emphasise the weight it was carrying, tended to report whole Russian divisions where only regiments or even battalions existed, by assuming the presence of the "parent" division once one of the subordinate formations had been identified. Owing to the number of *ad hoc* miscellaneous units Chuikov had cooped up in isolated parts of the city, this habit resulted in an estimate of the Russian strength five or six times greater than the true figure. Besides inducing the Germans to believe that they were wearing down the Red Army at a faster rate than they themselves were suffering, this delusion also ruled out the possibility of a Russian counteroffensive for lack of reserves. Another serious error, for which Weichs and Paulus must share responsibility, was the failure to supervise the Rumanian forces on the flanks. It was bad enough trying to protect these vulnerable positions with units which were underequipped and had already shown themselves inferior to the ordinary Russian infantry—still less prudent to neglect co-ordination of intelligence and reconnaissance on their front, and to ignore the periodic warning signals which the Rumanians sent in.

For these Rumanian divisions were quite unsuited to independent front-line operations against the Red Army. They were organised along the lines of a French infantry division of World War I (and relied heavily on French equipment captured by the Germans in 1940). There was only one anti-tank company to each division, and they had been equipped with the obsolete 37-mm. gun. After repeated requests from the army commander, General Dumitrescu, they received an allocation of German 75-mm. guns in October—six per division! Ammunition of all kinds was short, and there were no modern anti-tank or anti-personnel mines. The Rumanians were also short of rations and winter clothing; a German visitor at the beginning of November noticed that ". . . the building of defences was being neglected in favour of large dug-outs for the command posts and shelters for men and animals."

This weakness, and the fact that the Rumanians were not really in position along the Don at all, but opposite a whole series of Russian bridgeheads, some of which were as much as ten miles deep, made their sector an obvious place for the counteroffensive. Indeed, as winter approached, its prospects became a common subject for speculative conversation. "The one consolation was that the whole of this Eastern campaign had been based on such seemingly impossible improvisations and that, somehow, the impossible had always been achieved." What no one seems to have realised, from Paulus' staff all the way up to the headquarters of OKW at Vinnitsa, was the sheer weight of the impending Russian attack. The first positive indication of what was brewing had not come until as late as 29th October, when a report from Dumitrescu to Weichs listed:

1. marked increases in the number of Don crossings in the Russian rear
2. statements from deserters
3. continuous local attacks, "the sole object of which must be to find the soft spots and to pave the way for the major attack."

After some rather dilatory measures to confirm this, chiefly by means of air reconnaissance (which was itself becoming increasingly difficult in the deteriorating winter weather), Paulus went to army group headquarters at Starobelsk with a report which grossly underestimated the strength of the Russian concentration.

Paulus' figures were "positively identified" at Kletskaya: "three new infantry divisions with some tanks thought to be concentrated in the area; one new armoured, one new motorised and two new infantry formations." At Blynov, "two new infantry formations with a few tanks." On this assessment, of course, the Soviet offensive would be no stronger than many the Wehrmacht had shaken off in the past. Even as late as 12th November, exactly a week before the storm broke, Richthofen (admittedly a habitual optimist) recorded in his diary, after a personal aerial reconnaissance of the Russian bridgeheads:

> Their reserves have now been concentrated. When, I wonder, will the attack come? At the moment there appears to be a shortage of ammunition [this because the Russian artillery was holding its fire so as not to give away its position]. Guns, however, are beginning to make their appearance in the artillery emplacements. I only hope that the Russians won't tear too many big holes in the line!

The majority of staff officers in Army Group B were still more concerned with the preparations for the "last heave" at Stalingrad. Richthofen claims that even Zeitzler agreed with him, ". . . if we can't clear up the situation now, when the Russians are in real difficulty and the Volga is blocked with ice floes, then we shall never be able to." The Chief of the General Staff would surely have held a very different opinion had he known that the Russians, far from being "in difficulties," had concentrated over half a million infantry, 900 new T 34's, 230 regiments of field artillery, and 115 regiments of *Katyushas* on an attack frontage of under forty miles—a higher density of man and firepower than had been achieved in any previous battle of the Eastern campaign.[1]

While the Germans gathered their strength for a last attempt on the rubble islets of Stalingrad and over their shoulders Zhukov's armies moved stealthily into position, an uneasy quiet would descend on the town for hours at a time.

> Sometimes a silence more disturbing than the roar of explosions spread over the town, which seemed like a dead place. But it continued to watch, although no-one could distinguish night from day any longer. Even in the short periods of calm, each factory, each destroyed house, observed everything with intensity. The piercing eyes of the snipers spied upon the slightest movement of the enemy. The supply units, loaded with mines and shells . . . hastened along the ditches which zig-zagged between the ruins. From the height of the highest floors artillery observers were on the watch. In the cellars the leaders bent over

[1] The armour was split up into four tank corps, three mechanised corps, and fourteen "independent" tank brigades. The attacking armies were the 5th Tank Army, part of the 1st Guards Army, and the 21st Army. In the south the 57th Army, the 51st Army, and the resuscitated 64th Army. (See maps generally for dispositions.)

> maps, orderlies tapped on their machines, papers circulated, the soldiers were given directions. The miners, busy at their dangerous work, dug galleries and tried to find out those of the enemy.

Local actions at company level would break out all the time as each side constantly attempted to improve its position. A German tank would appear at the corner of a street; slowly it would swing around and grind cautiously toward the Russian-held buildings, iron hatches closed tight, the crew trembling with the anticipation of combat. The Russian infantrymen would watch it pass, trembling, too, while they waited for the rest of the German force to show its hand. Another tank appears at the street corner; it halts there, and follows its compatriot's progress with the gradual traverse of a still silent turret. Then suddenly, an explosion. A Russian 76-mm. anti-tank gun at the eastern end of the street opens fire; the range is less than fifty yards, but it seems to have missed, and at once the whole scene becomes animated in a storm of noise and pain. The German tank accelerates desperately in reverse, the cover tank fires instantaneously at the Russians' gun smoke; at the same time a section of German infantrymen, armed with submachine guns and grenades, rise from the rat runs in the rubble, where they have been crawling, and empty their magazines at the anti-tank gun. As they do so Russian snipers who have been lying motionless for hours in the eaves of skeletal buildings, high on the ledges of tottering façades, pick them off one by one. If the action does not escalate, with each side calling in more and heavier weapons in support, it will soon die away, leaving only the wounded exposed to view, crying in agony where they lie, until dark.

These "quiet" days were dominated by the sniper. It was an art at which the Russians excelled. Individual marksmen of particular skill soon became known, not only to their own side but also to the enemy, and the Russian ascendancy became so marked that the head of the snipers' school at Zossen, Standartenführer SS Heinz Thorwald, was sent to Stalingrad in an attempt to restore the balance. One of the crack Soviet snipers was set the task of catching him. He has described how

> The arrival of the Nazi sniper set us a new task: we had to find him, study his habits and methods, and patiently await the right moment for one, and only one, well-aimed shot.
>
> In our dug-out at nights we had furious arguments about the forthcoming duel. Every sniper put forward his speculations and guesses arising from his day's observation of the enemy's forward positions. All sorts of different proposals and "baits" were discussed. But the art of the

sniper is distinguished by the fact that whatever experience a lot of people may have, the outcome of an engagement is decided by one sniper. He meets the enemy face to face, and every time he has to create, to invent, to operate differently. There can be no blue-print for a sniper; a blue-print would be suicide.

I knew the style of the Nazi snipers by their fire and camouflage and without any difficulty could tell the experienced snipers from the novices, the cowards from the stubborn, determined enemies. But the character of the head of the school was still a mystery for me. Our day-by-day observations told us nothing definite. It was difficult to decide on which sector he was operating. He presumably altered his position frequently and was looking for me as carefully as I for him. Then something happened. My friend Morozov was killed, and Sheykin wounded, by a rifle with telescopic sights. Morozov and Sheykin were considered experienced snipers; they had often emerged victorious from the most difficult skirmishes with the enemy. Now there was no doubt. They had come up against the Nazi "super-sniper" I was looking for. At dawn I went out with Nikolay Kulikov to the same positions as our comrades had occupied the previous day. Inspecting the enemy's forward positions, which we had spent many days studying and knew well, I found nothing new. The day was drawing to a close. Then above a German entrenchment unexpectedly appeared a helmet, moving slowly along a trench. Should I shoot? No! It was a trick: the helmet somehow or other moved unevenly and was presumably being held up by someone helping the sniper, while he waited for me to fire.

"Where can he be hiding?" asked Kulikov, when we left the ambush under cover of darkness. By the patience which the enemy had shown during the day I guessed that the sniper from Berlin was here. Special vigilance was needed.

A second day passed. Whose nerves would be stronger? Who would outwit whom?

Nikolay Kulikov, a true comrade, was also fascinated by this duel. He had no doubt that the enemy was there in front of us, and he was anxious that we should succeed. On the third day, the political instructor, Danilov, also came with us to the ambush. The day dawned as usual: the light increased and minute by minute the enemy's positions could be distinguished more clearly. Battle started close by, shells hissed over us, but, glued to our telescopic sights, we kept our eyes on what was happening ahead of us.

"There he is! I'll point him out to you!" suddenly said the political instructor, excitedly. He barely, literally for one second, but carelessly, raised himself above the parapet, but that was enough for the German to hit and wound him. That sort of firing, of course, could only come from an experienced sniper.

For a long time I examined the enemy positions, but could not detect his hiding place. From the speed with which he had fired I came to the conclusion that the sniper was somewhere directly ahead of us. I continued to watch. To the left was a tank, out of action, and on the right was a pill-box. Where was he? In the tank? No, an experienced sniper would not take up position there. In the pill-box, perhaps? Not there either—the embrasure was closed. Between the tank and the pill-box, on a stretch of level ground, lay a sheet of iron and a small pile of broken bricks. It had been lying there a long time and we had grown accustomed to its being there. I put myself in the enemy's position and thought—where better for a sniper? One had only to make a firing slit under the sheet of metal, and then creep up to it during the night.

Yes, he was certainly there, under the sheet of metal in no-man's-land. I thought I would make sure. I put a mitten on the end of a small plank and raised it. The Nazi fell for it. I carefully let the plank down in the same position as I had raised it and examined the bullet-hole. It had gone straight through from the front; that meant that the Nazi was under the sheet of metal.

"There's our viper!" came the quiet voice of Nikolay Kulikov from his hide-out next to mine.

Now came the question of luring even a part of his head into my sights. It was useless trying to do this straight away. Time was needed. But I had been able to study the German's temperament. He was not going to leave the successful position he had found. We were therefore going to have to change our position.

We worked by night. We were in position by dawn. The Germans were firing on the Volga ferries. It grew light quickly and with day-break the battle developed with new intensity. But neither the rumble of guns nor the bursting of shells and bombs nor anything else could distract us from the job in hand.

The sun rose. Kulikov took a blind shot; we had to rouse the sniper's curiosity. We had decided to spend the morning waiting, as we might have been given away by the sun on our telescopic sights. After lunch our rifles were in the shade and the sun was shining directly on to the German's position. At the edge of the sheet of metal something was glittering: an odd bit of glass or telescopic sights? Kulikov carefully, as only the most experienced can do, began to raise his helmet. The German fired. For a fraction of a second Kulikov rose and screamed. The German believed that he had finally got the Soviet sniper he had been hunting for four days, and half raised his head from beneath the sheet of metal. That was what I had been banking on. I took careful aim. The German's head fell back, and the telescopic sights of his rifle lay motionless, glistening in the sun, until night fell . . .

For the 6th Army's last offensive both tactics and organisation had been revised. The Panzer divisions had already virtually lost their identity with their tanks split up in company packets to support the infantry. Four more pioneer battalions had been flown into the city, and these were to be used as spearheads to four separate thrusts aimed at completing the fragmentation of the defenders' position. The last "rectangles," as they were called, would then be pulverised by concentrated artillery fire. The old, wasteful house-to-house technique, where one building could consume a whole company in its stairways, balconies, attics, and corridors, was employed only as a last resort. On both sides the infantry had gone underground: cellars, drains, saps, tunnels—these made the contours of the battlefield. Only the tanks crept slowly about on the surface, watched from their precarious nests by the snipers.

Paulus' attack, which started on 11th November, was as misguided and as hopeless as had been the last winter offensive of Army Group Centre against Moscow twelve months before. Within forty-eight hours it had degenerated into a series of violent personal subterranean battles without central direction. Many small groups of Germans managed to cover the last three hundred yards to the Volga, but having arrived at the river, they would themselves be cut off by Russians moving back across the narrow corridors they had opened. For four more days fighting of a desperate ferocity flared and slackened between these isolated groups. Prisoners were no longer being taken, and the combatants had little hope of personal survival. Filled with alcohol and benzedrine, bearded, exhausted from days without sleep or relief, they had lost all sense of motive and purpose save the ultimate obsession of close combat—to get at one another's throats.

By 18th November total exhaustion and shortage of ammunition imposed a lull. During the night the crackle of small arms and the thud of mortars died down and each side began to take in its wounded. Then, as dawn came to lighten the smoke clouds, a new and terrible sound overlaid the dying embers of the battle in Stalingrad—the thunderous barrage of Voronov's two thousand guns to the north. Every German who heard this knew that it presaged something quite outside his experience.

> Although the Russian artillery had always been quite good they would not keep up their fire for long periods at a time. But that day [19th November] was different . . . continuous drum-fire since dawn with the scream of a *Katyusha* discharge every minute or so.

At nine thirty on the morning of the 19th the sound was swollen by the guns of Tolbukhin, Trufanov, and Shumilov as they debouched

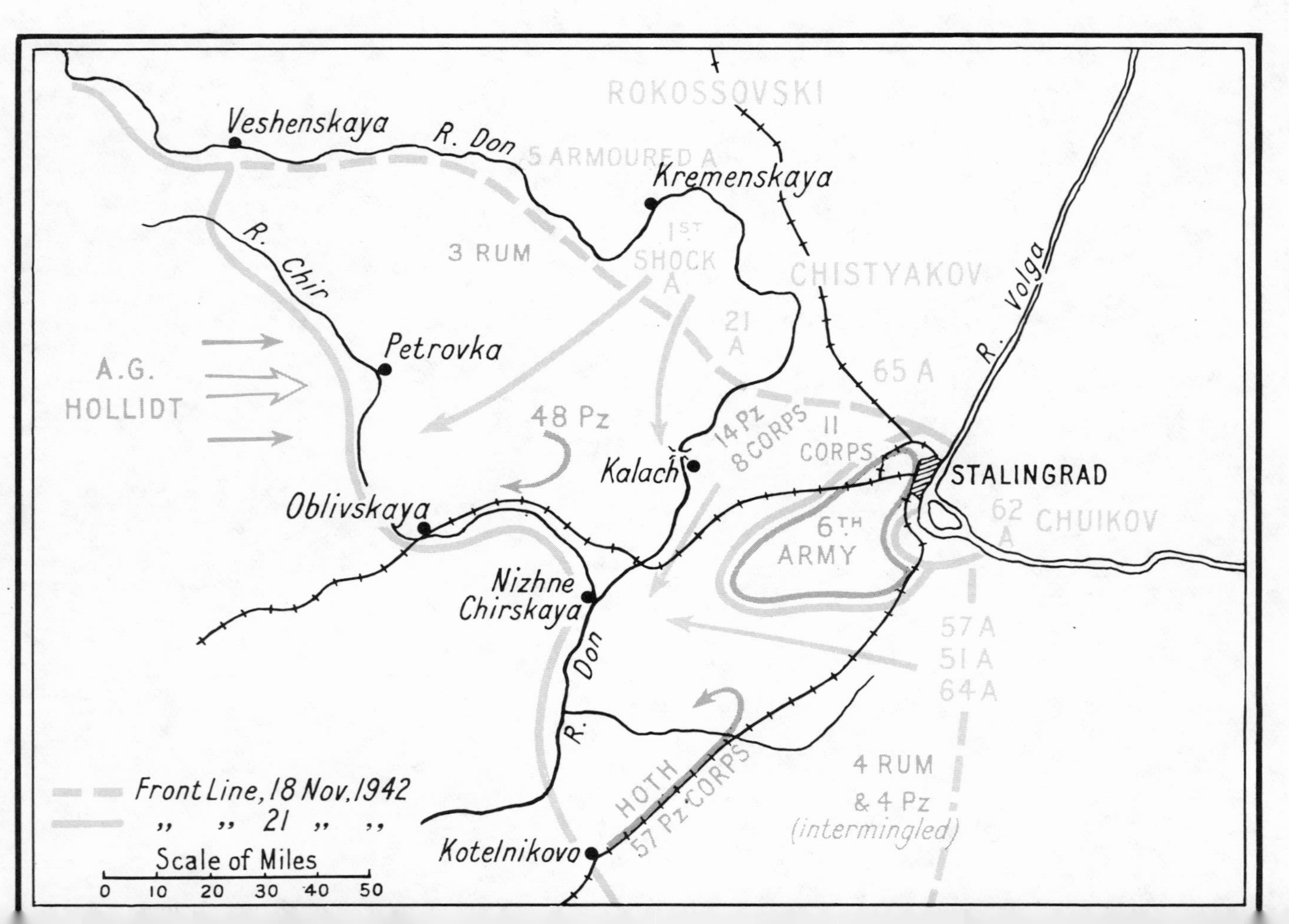

The Stalingrad Encirclement

from their positions to the south, and the scale of the Red Army's counterblow, together with the threat it posed to the whole German position, began to percolate to the officers of the 6th Army.

Paulus had already taken two steps to "deal with" the Russian threat, following his (disastrously inaccurate) estimate of Russian strength and intentions formed on 9th November. The Rumanian Army had been reinforced by a close support group under Colonel Simons, and the 48th Panzer Corps had been moved into the small bend of the Don as mobile reserve. Simons' group consisted of a *Panzergrenadier* battalion with an anti-tank company and a few pieces of heavy artillery. The Panzer corps itself had barely the strength of a division, and 92 out of its total tank strength of 147 were Czech 38-T's, manned by Rumanians. The 14th Panzer, with an additional 51 Mark IV's, had also been put into the corps, but had been so disorganised by its experiences in street fighting that it was not fully extricated by the time the Russian offensive started.

For three days, from 19th November to the evening of the 22nd, the German and Rumanian front broke up along a length of over fifty miles in the north and thirty in the south. Into the breach Zhukov was pouring six armies, flattening the resistance of a few defiant pockets, brushing aside the paltry efforts of Simons' group and the debilitated 48th Panzer corps. The 6th Army staff went sleepless for two nights as they struggled to regroup the precious Panzers and pull back their infantry from the smoking maze of Stalingrad to protect the collapsing flanks. In Paulus' rear confusion was absolute; the western railway from Kalach had already been cut by Russian cavalry in several places; the sound of firing came from every direction, and periodically broke out between Germans going up to the front and ragged groups of Rumanians in leaderless retreat. The huge bridge at Kalach, over which every pound of rations and every bullet for the 6th Army passed, had been prepared for demolition, and a platoon of engineers were on duty there all day on 23rd November in case the order to destroy the bridge should come through.

At half past four that afternoon tanks could be heard approaching from the west. The lieutenant in charge of the engineers thought at first that they might be Russians, but was reassured when the first three vehicles were identified as Horch personnel carriers with 22nd Panzer Division markings; assuming that it was a reinforcement column for Stalingrad, he instructed his men to lift the barrier. The personnel carriers halted on the bridge and disgorged sixty Russians, who killed most of the engineer platoon with tommy guns and made the survivors prisoners. They removed the demolition charges, and twenty-five tanks from the column passed over the bridge and drove southeast, where that evening they made contact with the 14th Independent Tank Brigade from Trufanov's 51st Army. The first tenuous link in a chain that was to throttle a quarter of a million German soldiers had been forged, and the turning point in World War II had arrived.

fourteen | THE ADVENT OF GENERAL VON MANSTEIN

When the tanks of the Russian 26th Armoured Corps captured Kalach and joined the infantry that had driven up from the south, they achieved something greater even than the spectacular victory which was promised by the isolation of the 6th Army. For this brilliant stroke marked, in its every aspect—its timing, its concentration, and the manner in which it exploited the enemy's own disposition—a complete and final shift in the strategic balance between the two contestants. From this time on the Red Army held the initiative, and although the Germans were to try on many occasions (and to succeed on some) to reverse this balance, their efforts turned out to be no more than tactically significant. From November 1942 on, the posture of the Wehrmacht in the East was fundamentally a defensive one.

This reversal can be attributed to a number of interlocking factors. First, unwarranted German overconfidence. It was this that lay behind the idiotic dispositions, with the weakest formations at the front responsible for the most vital sectors. (The most vital, that is, if the main battle should alter character and become defensive.) Second, erratic leadership, chiefly but by no means entirely the product of Hitler's interference. This led to a confusion of objectives and their priority. And finally (as a function of this), the emotional obsession with Stalingrad which caused the *Schwerpunkt* to become fatally entangled in a web of street fighting, imposing on the whole army a static process of attrition which was severer than that suffered by its enemies—and to which it was less suited.

In essence, though, the miscalculation of the Germans went deeper than this. The hard fact was that they were attempting too much. They were relying entirely on superior leadership and training to compensate them for material deficiencies. Having failed to destroy the Red Army in 1941—and having experienced its recuperative powers that winter—they had embarked on a campaign that stretched their own resources to the uttermost, ignoring at their peril the inexorable laws of time and distance, of numbers and firepower.

The defeat at Stalingrad shocked the whole nation, and its tremors, echoing back from the mass of the people, were registered at OKW. The concept of ultimate defeat, though still having no substance, lengthened as a shadow. The effect of this shock is most marked in the character of the leadership. This is what gives a peculiar interest to 1943. For the first six months of that year the conduct of the war in the East enjoyed a greater degree of professional direction than at any other time before or afterward. It is almost as if Paulus, by his sacrifice, had bought for his colleagues on the General Staff a final chance to redeem their fortunes.

The two men Hitler selected as architects of recovery were Manstein and Guderian. These brilliant soldiers had conceived, independently of each other, the principles on which the campaign must be fought—a reversion to the mobile defence, a war of manoeuvre, luring the enemy forward to entrap and destroy him in the manner of Tannenberg or Galicia-Tarnow. Only thus (they believed) could Germany redress the balance of material which tilted against her and recover the strategic ascendancy. On their early successes and subsequent frustrations—owing as much to the jealousy and intransigence of their professional colleagues as to Hitler's interference—the history of this period hangs.

As this break in the continuity of the campaign in the East is more a matter of change in the strategic balance and in the character of the German leadership than the ending of a particular battle or calendar period, the easiest point at which to take up the narrative is on 20th November, when Manstein was ordered to travel to the headquarters of Army Group B at Starobelsk.

Manstein had been more or less inactive during October, establishing his headquarters at Vitebsk, in preparation for playing a "special role" against the Russian offensive which was expected against Army Group Centre.[1] Now the order from OKH, "for the purpose of stricter coordination of the armies involved in the arduous defensive battles to the

[1] For Manstein's career after his promotion to Field Marshal, and his relations with Hitler, see Ch. 16.

south and west of Stalingrad," directed Manstein to form a new army group in the Don bend, at the junction of Army Groups A and B. Under command he would take the 4th Panzer Army, 6th Army, and the 3rd Rumanian Army. The task of the new group was somewhat optimistically framed as "to bring the enemy attacks to a standstill and recapture the positions previously occupied by us."

Owing to the bad weather which had spread over the whole of central Russia—low ceiling, snow blizzards, and temperatures around 20 below zero—Manstein and his staff had to make the journey by train. Leaving Vitebsk at seven o'clock on the morning of 21st November, they made their first stop that evening at Orsha. Here they found Kluge waiting on the platform with his Chief of Staff, General Wöhler. The commander of Army Group Centre radiated gloom. The latest information from OKH, he told Manstein, was that there were two Soviet tank armies involved, "in addition to a great deal of cavalry—in all, some thirty formations." [2] As to the possibilities of restoring the situation, "You will find it impossible to move any formation larger than a battalion without first referring back to the Führer."

Whatever Manstein may have thought of Kluge's foreboding, he can have derived little comfort from a detailed scrutiny of the units which were to compose the strength of his new responsibility. The 6th Army, being surrounded, was unusable in an operational sense. Moreover, Paulus' divisions, representing as they did the head of the *Schwerpunkt,* had been under the closest supervision and direction of OKW, and Hitler maintained direct control of operations through a liaison officer who was attached to Paulus' headquarters with his own signals section. Of the remainder, their real strength belied the grandiose "corps" and "Army" titles the situation map attributed to them. The 3rd Rumanian Army had taken the full shock of the Russian attack from Kremenskaya and, with the exception of two divisions in the west, had been annihilated. The 48th Panzer Corps, the mobile reserve in the Don bend, had, after some hesitation, been committed to a counterattack.[3] It had run

[2] In fact, this was a serious underestimate. By 25th November, OKH had identified 143 "formations." This term included divisions referring to infantry (the Russian infantry division was, of course, smaller than the German; see Ch. 2), and brigades when referring to armour and cavalry. Kluge's gloomy reference to cavalry is further evidence of the impression made on the German Army by the skilful way in which the Russians used this anachronism on their home ground.

[3] The corps consisted of two divisions. One, a Rumanian armoured division equipped mainly with captured French tanks, and largely untrained. The German compo-

head on into a renewed advance by the 2nd Guards Tank Army and been cut to pieces. Finally, the 4th Panzer Army, on the southern wing, had itself been severed by the southern arc of the Russian pincer. The bulk of its tank strength had been swept up into the Stalingrad pocket, and the remainder, concentrated in the Kotelnikovo area, consisted mainly of service and communication troops, with one intact Rumanian division. The only full-strength German unit, the 16th Motorised Division, was based at Elista, 150 miles from the Don, and carried the critical responsibility of guarding the junction between Army Group A and the right wing of the main front.

These forces were manifestly incapable of serious resistance if the Russian effort should change direction and turn westward—or, still more perilous, southward to the Sea of Azov and across the communications of the extended Army Group A. And the notion of their "recapturing positions previously held by us" was an absurdity. Army Group Don had, excluding the 6th Army, little more than the fighting strength of a corps—and was spread over two hundred miles. Manstein's first task, therefore, was to collect sufficient strength under his own hand to give him, at least, certain tactical options. From his railway coach he had teleprinted OKH that ". . . in view of the magnitude of the enemy effort, our task at Stalingrad could not be merely a matter of regaining a fortified stretch of front. What we should need to restore the situation would be forces amounting to an army in strength—none of which, if possible, should be used for a counter-offensive until their assembly was fully complete." The disorganised state of the railway system and the fact that Partisan activity had made large stretches of line unusable prolonged the journey to Starobelsk over three days and two nights. On the morning of 24th November, while his train was waiting in Dnepropetrovsk for the last stage, Manstein was handed a teleprint from Zeitzler which promised "an armoured division and two or three infantry divisions."

nent, the 22nd Panzer, was understrength, and many of its tanks had been immobilised by mice eating the electric insulation. (This very odd excuse is confirmed by two independent sources: Mellenthin 166, and Leyderrey 94.) Rather than pose the many questions of maintenance and service which this curious statement leaves unanswered, Manstein confines himself to admitting that the division "was obviously not up to standard from a technical point of view." Nor is the affair illuminated by its sequel. The corps commander, General Heim, was immediately court-martialled by a tribunal presided over by Goering, and sentenced to death. But he was later rehabilitated on the ground that "his forces had indeed been too weak for the task confronting them."

But by this time the situation had worsened to a degree that made reinforcements on that scale of small account.

In the three days following their penetration of the Rumanian corps, the Russians had moved thirty-four divisions across the Don, twelve from Beketonskaya bridgehead and twenty-two from Kremenskaya. Their tanks had turned westward, defeating the 48th Panzer Corps and probing dangerously into the confusion of stragglers, service and training units, and mutinous satellites who milled about in the German rear. Their infantry had turned east, digging with feverish energy to build an iron ring around the 6th Army. Zhukov kept the whole of the Stalingrad pocket under bombardment from heavy guns sited on the far bank of the Volga, but for the first few days he had exerted only a gentle pressure upon the surrounded Germans. The Soviets' intention was to probe in sufficient strength to be able to detect the first signs of their enemy's actually striking camp, but to avoid any action which might precipitate this. For them, as for Paulus, these first hours were vital. All night on the 23rd and during the morning of 24th November, men and tractors hauled and struggled with battery after battery of 76-mm. guns across the frozen earth. By that evening, when Manstein finally arrived at Army Group B, Russian firepower on the west side of the pocket had trebled. Over a thousand anti-tank guns were in position in an arc from Vertyatchi, in the north, around to Kalach, then eastward below Marinovka, and to join the Volga at the old Beketonskaya bridgehead. The extrication of the 6th Army could no longer be an extempore affair—undignified and costly, perhaps, but conducted under conditions of confusion and mobility. It would have to be a set-piece and one, according to Clausewitz, which "habitually presents exceptional difficulty," that of "Sortie to Relief."

Thus when Manstein finally descended from his train at Ştarobelsk, he found that the situation, which had been unpromising when he had started out from Vitebsk, was now highly critical.

At Army Group B a certain atrophy was apparent. Weichs and his Chief of Staff, General von Sodenstern, were responsible for seven armies, of which three were "allied" and a fourth had a high non-German content.[4] Their front extended for more than 250 miles. Sodenstern's pessi-

[4] Prior to the formation of Army Group Don, Weichs's command included, from north to south, the 2nd Army, 1st Hungarian, 8th Italian, Army Detachment Hollidt, 3rd Rumanian, 6th Army, and 4th Panzer.

mism seems also to have effected the accuracy of his estimates because he told Manstein that the 6th Army had, "at most," two days' supply of ammunition and six days' rations. Although Manstein had expressly telegraphed before leaving from Vitebsk that the ". . . Sixth Army be instructed to withdraw forces quite ruthlessly from its defence fronts in order to keep its rear free at the Don crossing at Kalach," professional etiquette had obliged him to send this message down the cumbersome chain of command that stretched through Army Group B. He was "unable to discover" whether these instructions were ever passed on to Paulus.

Whether it was that he found the atmosphere at Army Group B unsatisfactory, or for whatever reason, Manstein remained there only a few hours. He had brought with him the majority of his old 11th Army staff, and was allocated the quartermaster organisation originally set up for Marshal Antonescu.[5] On the evening of the 24th, therefore, the Field Marshal and all his attendant personnel climbed back again into their train and set off on a further twenty-four-hour journey to Novocherkassk, the site selected for the headquarters of Army Group Don.

However, before leaving Starobelsk, Manstein had had a long telephone conversation with Zeitzler. Their discussion seems to have been confined to the plight of the 6th Army, but is important as representing the first impact of a cool and rational intellect upon the problem (as distinct from a series of conflicting reflexes, triggered by alarm and emotion or rigid professional orthodoxy), and also in view of the recriminatory atmosphere which has pervaded the whole issue of Paulus' encirclement and reduction ever since.

Manstein maintained that a breakout to the southwest (i.e., down the left bank of the Don) was "probably still possible even now." To leave the army at Stalingrad any longer constituted an extreme risk, in view of the ammunition and fuel shortage.[6] But although the bulk of the Panzer forces would probably get through, there was a risk that the infantry, leaving its prepared positions around and in the town, might be destroyed on the open steppe.

[5] Just before the Russian counteroffensive broke, it had been mooted that the Marshal might be given an army group. Colonel Eberhard Finkh, one of the ablest transport officers in the German Army (executed in 1944 for his suspected complicity in the 20th July plot; he was Deputy Chief of Staff to Kluge), had been placed in charge of the quartermaster general's branch.

[6] It should be remembered that at the time of this conversation Manstein was basing his assessment on the faulty estimate made to him by Sodenstern.

Nevertheless, since he considered that the best chance for an independent breakout had already been missed, Manstein's view was that ". . . it was preferable from the operational point of view at the present time to wait until the projected relief groups could come to the army's aid." He would be able to launch this relief operation with the forces due to arrive at the beginning of December. "To achieve real effect, however, it would require a steady flow of further reinforcements, as the enemy would also be throwing in powerful forces on his own side." An isolated breakout by the 6th Army might still be necessary "if strong enemy pressures were to prevent us from deploying these new forces."

Manstein claims that he concluded the conversation by emphasising that unless the delivery of supplies could be guaranteed ". . . one could not risk leaving Sixth Army in its present situation any longer, however temporarily."

The whole question of Stalingrad and the fate of the 6th Army is so clouded with guilt in the German mind that in holding an inquiry twenty years after, it is almost impossible to find any "witness" who has told the whole truth. In his account of this conversation Manstein makes no mention of having used the same basic strategic argument, which he prints later on in his memoirs, as a "reflection," namely that ". . . at the same time as the extricated elements of Sixth Army might have been joining Fourth Panzer Army, the entire enemy siege forces would have been released. With that, in all probability, *the fate of the whole southern wing of the German forces in the east would have been sealed—including Army Group A.*" Indeed, he goes further, and says, ". . . the latter consideration played absolutely no part in shaping our appreciation of 24th November." Why not? one may well ask. Are we really to believe that this fundamental strategic truth neither occurred to nor was expounded by the leading military intellect on the German side?

But because the 6th Army never escaped, and because had it attempted a breakout in November *some* of its soldiers would have escaped, no responsible person will now admit to having advised against it. Instead, the usual semi-articulate conspiracy which aims to saddle Hitler with the responsibility for every defeat in the field which the German Army sustained has evolved a conventional fiction—to the effect that the Army was "prevented" from breaking out by Hitler's expressly denying them this course of action.

The facts are as follows: The capture of the Kalach bridge and the junction of the Russian 21st and 51st armies took place on 23rd November. This effectively blocked the 6th Army's last escape route—but by then the ring had already hardened around it, over at least three quarters of its circumference. It was not until that day (23rd November) that Paulus made any request for freedom to manoeuvre. Then, in a message sent *directly to Hitler* at OKW headquarters, he said that all his corps commanders "considered it absolutely imperative that the army should break out to the southwest. To raise the forces needed for such an operation, he would have to shift certain formations around inside the army and, for the purpose of economising in troops, take his northern front back on a shorter line." Why was this request sent straight to OKW? The proper course was for Paulus to communicate with Weichs at Army Group B. Furthermore, Paulus had been Oberquartiermeister I at OKH throughout the previous winter. He was painfully familiar with the Führer's attitude to "shortening the front" under enemy pressure (an attitude, let it be remembered, which had worked remarkably well during that critical period). He must have known in advance what the answer would be.

There is also another point which should be made. Why did Paulus wait nearly four days before asking for permission to redistribute his forces? The 6th Army had felt its flanks severed on 19th November. Ordinary prudence—let alone the rigid training of the General Staff—should have led to an immediate adjustment in co-ordination with Army Group B. This at least would have avoided the situation where the 48th Panzer Corps and 3rd Motorised Division (striking westward toward Kalach from Stalingrad) were defeated in detail. For these two moves, already jeopardised by their inadequate strength, lost any advantage they might have derived from converging axes by starting off at an interval of nearly twenty-four hours.

Paulus' delay in asking for instructions has additional importance. His message to OKW was dated 23rd November. Even if immediate assent to a withdrawal had been given, the earliest by which the 6th Army could have been moved into a "ram" formation for breaking out would have been 28th November.[7] By that time the Russian concentrations

[7] This was "The view taken at Army Group Headquarters" (Manstein 302). It is possible that it erred on the side of pessimism (but see p. 271). There was no question of the encircled troops' simply leaving their emplacements and doing an about-turn. Their whole order of battle was inverted, with the majority of the tanks (which would have to spearhead the drive southwest) distributed in close support at the eastern side of the pocket. The bulk of the artillery was also sited offensively

would have been so strong that the result would in all probability have been the same as in February—namely, total annihilation. Had a certain debris finally got through, its arrival would have been small compensation for the release of all the investing forces which would have been free to strike at Rostov and aggravate the precarious situation of Army Group A.

In making this assessment we are greatly helped by what we now know of the Russians' strength and intentions at that time. The most important point to remember and the most difficult, in the light of their subsequent scale of operations, is that Zhukov's aim was strictly limited —and greatly affected by his experiences of the previous winter. In December 1941 the Russians had fought like a boxer who, after flooring his man for a count of eight, charges in again and smothers his opponent with blows no single one of which is deadly. These exertions sapped their strength and gave their dazed opponent time to recover. This time the *Stavka* had but one primary aim, the isolation and destruction of the 6th Army. If that was achieved it was confident that the offensive power of the Wehrmacht would be broken, and that it would never have cause to dread the onset of the summer campaigning season again. The whole operation was deliberately confined to the quadrilateral—less than a hundred miles square—bounded by Stalingrad and the eastern corner of the Don bend. Into this area the Russians had concentrated seven of the nine reserve armies which had been built up for the winter campaign, and limited their scope to take maximum advantage of their qualities—mass, surprise, and (once the trap had been closed) resolution in defence —and to throw the least strain on their weaknesses. Zhukov knew that the standard of training, and the initiative of commanders at the lower level, would make too deep and ambitious penetration with his armour a risky affair.[8] He knew, too, that many of the corps and even the army

---

in the east. It would have had to be redistributed in a screen along the northern side.

[8] Mellenthin (182), writing of the battles against the 5th Tank Army (the elite of all the new Soviet armies) in December, when it was first probing to the west of the Don, says, "The foolish repetition of attacks on the same spot, the rigidity of artillery fire, and the selection of the terrain for the attack betrayed a total lack of imagination and mental mobility. Our Wireless Intercept Service heard many a time the frantic question: 'What are we to do now?' . . . On many occasions a successful attack, a break-through, or an accomplished encirclement was not exploited simply because nobody saw it."

commanders had neither the flexibility nor the imagination for a "general intention." At all costs those wasteful and repetitive attacks which had characterised the fighting in the Rzhev salient the previous winter had to be avoided. And so each phase of that critical first week was meticulously worked out; every task and objective was covered three or four times over. Zhukov was determined to dig his two thousand guns around the 6th Army in an unbreakable chain, and determined also that no other opportunity, however temptingly it beckoned, would be allowed to distract him from this.

In fact, though, the Russian blow had been delivered with such strength that the whole German front was broken at its apex and threatened with disintegration. There is no doubt that the primary factor in the recovery was Zhukov's determination to avoid the risks of mobile warfare, and his refusal to commit the mass of his army westward until the 6th Army had been eliminated. Thus the decision (if it can be called that) to make Paulus defend Stalingrad as a hedgehog position meant that the weight of the Russian offensive was tied back between the Volga and the Don, and time, both to reshape its front and to organise a relieving force, was granted to the German High Command at the very moment when it must have seemed to run out.

During the first days of December, Manstein worked frantically to assemble sufficient strength for an effort to relieve the 6th Army. His responsibility was effectively divided into three separate areas, of which Paulus' command was by far the strongest[9] in terms of numbers. During the first desperate week after the Russian break-through Army Group Don had been holding its front with a ragged mixture of "*Ad hoc* units

[9] Surrounded at Stalingrad were:

Headquarters and entire command organisation of the 6th Army.

Headquarters staff of five army corps (4th, 8th, 11th, 14th Panzer, and 51st).

Thirteen infantry divisions (44th, 71st, 76th, 79th, 94th, 100th Jäger, 113th, 295th, 305th, 371st, 376th, 389th, and 397th).

Three Panzer divisions (14th, 16th, and 24th).

Three motorised divisions (3rd, 29th, and 60th).

One anti-aircraft division (9th).

In addition there were a number of special engineer units which had been brought in to help in the street fighting with their expert knowledge of demolition, two depleted Rumanian divisions, and a Croat regiment. There must also have been about 8,000-10,000 German B-echelon troops, making a total (excluding satellites) of 220,000-230,000 Germans.

formed from non-combat units, headquarters staffs, Luftwaffe troops and Army personnel who had been on the way back to their parent units from privilege or sick leave. These 'emergency units' lacked cohesion, seasoned officers and weapons (especially anti-tank protection and artillery), and most of them had little or no battle experience or training in close combat." But as has been seen, it was no part of Zhukov's plan to drive westward until he had really tightened his lock on Stalingrad, and as the days passed, the thin screen gradually acquired a stiffening of men and firepower. The Germans even managed to hold their bridgehead at Nizhne-Chirskaya, at the confluence of the Chir and the Don.

It was into this area, the flat plain that lay southwest of the Chir, that Manstein directed the first of his reinforcements. The remains of the 48th Panzer Corps were folded back due south from Veshenskaya as an anchor on which the line of the northern Don might still be held and a new corps headquarters formed to the southeast, into which by 4th December three fresh divisions[10] had been moved.

One of these, the 11th Panzer, was probably the finest armoured formation on the Eastern front. Its commander, General Balck, was a leader of Rommel's calibre—though his antithesis in appearance. Photographs show him a slight, almost stooping figure, with a detached expression. Only the eyes, hard and alert, betray his restless energy. Balck was particularly ruthless with his subordinates,[11] and every officer of the division was of his mould. The 11th Panzer had been in OKH reserve since October, and had its full complement of tanks and assault guns.

Another very strong division, the 6th Panzer, had been entrained from the West on 24th November, and was scheduled for redeployment in Army Group Don by 8th December. Furthermore, an additional two infantry divisions (the 62nd and 294th), another Luftwaffe field division, and a mountain division were put into Army Detachment Hollidt.[12]

In spite of the strength he was accumulating below the Chir, and the fact that the bridgehead at Nizhne-Chirskaya was only twenty-five miles from the western tip of the Stalingrad siege front, Manstein formed the

[10] These were the 11th Panzer, 336th Infantry, and 7th Luftwaffe Field Division.

[11] Balck wrote, "We were fortunate that after the hard fighting in previous campaigns all commanders whose nerves could not stand the test had been replaced by proven men. There was no commander left who was not absolutely reliable."

[12] To eliminate "command difficulties and obstructions" the Rumanian headquarters staff had been dispersed and Army Detachment Hollidt enlarged and made responsible for the Chir River front.

view that it would be dangerous to rely on the forces there for a relief thrust. He considered that the Russians would regard this as the obvious direction from which to expect such a move, and that they had the ability to double or treble their forces on the left bank of the Don at a few hours' notice. In addition there was the potential threat from the long northern flank, stretching along the upper Chir and up to the junction with the 2nd Hungarian Army and Weichs's boundary.

Manstein therefore decided that, if it was possible from an operational point of view, the main thrust should be delivered by Hoth with an invigorated 4th Panzer Army, and that the 48th Panzer Corps and the Hollidt detachment should restrict their activities to a show of force designed to draw off Zhukov's mobile reserve once Hoth began his approach march. If and when Hoth's columns drew level with the bridgehead at Nizhne-Chirskaya, the 48th Panzer Corps would attempt to cross the Don. Ideally this would result in Paulus' being offered two alternative routes for the extrication of his garrison, and the possibility of dividing his forces between them.

To reinforce Hoth, Manstein decided to bring the 6th Panzer Division down through Rostov and to use the whole of the 57th Panzer Corps, which (after a considerable back and forth of telegrams between Rastenburg, Novocherkassk, and the headquarters of the headless Army Group A, in whose command they had been) was sanctioned by OKW. While he was waiting for these forces to get into position, a reconnaissance group was sent out of Elista by the commander of the 16th Motorised Division in a wide sweep across the steppe to the southwest of the Volga. This force was relatively small, consisting of two motorcycle companies, a few half-tracks with towed 50-mm. anti-tank guns, and eleven Mark III's. But after a three-day sortie they were able to confirm that Hoth's open right flank was safe and, even more important, that there was no immediate danger of a Russian drive to cut off the forces in the Caucasus.

While waiting for the 4th Panzer Army to gather strength, Manstein found things beginning to go wrong on the Chir. Zhukov had, as has been seen, taken his tanks out of the line within three days of completing the encirclement at Kalach, but after less than a week of rest and refit elements of the 5th Guards Tank Army began to be identified between Nizhne Kalinovski and Nizhne-Chirskaya. On 7th December two armoured brigades[13] got across the river still farther west, and penetrated

[13] From the 1st Armoured Corps of the 5th Guards Tank Army.

nearly twenty miles before nightfall, placing themselves deep on the flank of the new 336th Infantry Division, which had itself just moved into position.

Fortunately for the Germans, Balck's 11th Panzer was driving up from Rostov during the day at almost the same speed as the Russian tanks (fewer in number) were moving south. That evening the heads of the two columns collided at the huge "State Farm No. 79," just north of Verkhne-Solonovski, and exchanged fire until darkness fell. The Russians then went into leaguer among the farm buildings, but Balck, with characteristic energy, led the mass of his tanks around in a wide arc to the west and north, leaving only his engineer battalion and a few 88-mm. guns to mask the Russian position. This feat, over snow-covered unmapped ground and coming after a two-day forced march, brought its reward. Ten hours later the Panzers were astride the Russian approach route. At first light they could see a long column of Russian trucks, infantry sent down to reinforce the armoured break-through, driving serenely along, nose to tail. They charged down and ran parallel with the column, destroying it with machine-gun fire at ranges of about twenty yards, in order to conserve their armour-piercing ammunition. After the destruction of the infantry, Balck's tanks continued to drive south on the road the Russian motorised column had been following and arrived at the state farm just as the T 34's were moving off (also in a southerly direction) to attack what they mistakenly believed to be the weak left flank of the 336th Infantry Division. The Russian tanks faltered as they caught the first shells from Balck's screen of 88's, and at that moment they were set upon by the Panzers in their rear. Their two brigades kept up the fight throughout the day, but by its end were practically annihilated, losing fifty-three tanks. Only a few managed to slink away under cover of darkness. They lay low in the frozen gullies which intersected the area, and were to emerge and cause some trouble in the days that followed.

There was no resting on the field of victory for the 11th Panzer. Almost simultaneously with their crossing of the Chir in the north, the Russians had started up a succession of attacks against the Nizhne-Chirskaya bridgehead, and the division turned west to restore the situation there. During the two days that followed a sequence of small bridgeheads and crossings were made against the position of the 336th Division and it became plain that the Russians were now mounting their strength in earnest against the Chir position, both as a spoiling move against any concentration of a relief army and with the more ambitious purpose of capturing the airfields at Tatsinskaya and Moravichin, which were the Ju 52 bases for the "airlift" to Stalingrad.

The German strength was inadequate to maintain a static defence along the whole length of the Chir, whose sinuous course almost doubled the apparent front. Yet the infantry, although fresh, had neither the equipment nor the weapons for a flexible and mobile defence. Only the 11th Panzer had the means for *"Klotzen, nicht Kleckern."* The Chief of Staff of the 48th Panzer Corps wrote an appreciation of Russian bridgehead technique at that time:

> Bridgeheads in the hands of the Russians are a grave danger indeed. *It is quite wrong not to worry about bridgeheads and to postpone their elimination.* Russian bridgeheads, however small and harmless they may appear, are bound to grow into formidable danger-points in a very brief time and soon become insuperable strong-points. A Russian bridgehead, occupied by a company in the evening, is sure to be occupied by at least a regiment on the following morning and during the night will become a formidable fortress, well equipped with heavy weapons and everything necessary to make it almost impregnable. No artillery fire, however violent and well concentrated, will wipe out a Russian bridgehead which has grown overnight. Nothing less than a well-planned attack will avail. The Russian principle of "bridgeheads everywhere" constitutes a most serious danger, and cannot be over-rated. There is again only one sure remedy which must become a principle: If a bridgehead is forming, or an advanced position is being established by the Russians, attack, attack at once, attack strongly. Hesitation will always be fatal. A delay of an hour may mean frustration, a delay of a few hours does mean frustration, a delay of a day may mean a major catastrophe. Even if there is no more than one infantry platoon and a single tank available, attack! Attack when the Russians are still above ground, when they can be seen and tackled, when they have had no time as yet to organise their defence, when there are no heavy weapons available. A few hours later will be too late. Delay means disaster; resolute, energetic and immediate action means success.

However, Knobelsdorff, the new commander of the 48th Panzer Corps, decided that the most important task was to preserve his own bridgehead at Nizhne-Chirskaya. At the evening conference of 10th December he restrained Balck from setting off once again with his "fire brigade," and that night the 11th Panzer occupied itself with moving into position for a counterattack against the Russians who had broken the defence perimeter. On the following morning the German bombardment started, and carried extra weight from the whole of the 336th Division's artillery and some heavy mortars which had been brought from the west to help in breaking up the Russian positions in Stalingrad and had been discovered in the railway sidings there. The tanks were to go in during the afternoon, and it was intended to pull them out as the light failed and leave the infantry to mop up during the night. Balck himself was not sanguine about the prospects of a frontal attack, and understandably reluctant to allow his division to get bogged down in the maze of small islets, frozen creeks, and fire-swept *balkas* that interlaced the area where the two rivers joined. Then, just as the leading regiment was about to leave its starting line, a message came up from General Lucht[14] that his front had been penetrated at Nizhne Kalinovski, and also at Lissinski (about midway between there and Balck's present position).

The tanks had their engines running, and the fire of the barrage had already slackened. After the briefest of consultations Balck and Knobelsdorff decided that the attack should be called off and the tanks sent north to deal with the new emergency. Both commanders agreed that the strength of the Germans' artillery fire and their evident preparations for a counterattack would be sufficient to deter the Russians for a few days.

Once again, therefore, the 11th Panzer spent the night in marching to a new battlefield, and once again it went into the attack, unprepared and unreconnoitred, at dawn. The Russian force was a mixed group of tanks, cavalry, and a few gun crews with 76-mm.'s. The horses had slipped away into the steppe under a full moon, but many of the tanks were still in leaguer when the Germans attacked and the 76-mm.'s had not yet been dug into the frozen earth. By midday the bridgehead had been eliminated, and that afternoon the 11th Panzer covered the fifteen miles to Nizhne Kalinovski, where the second breach had been reported. As at Lissinski, it went straight into the attack with its leading regiment ("Our engines have been warm, and our gun barrels, too, ever since we arrived on the Chir," wrote a lieutenant of the 115th *Panzergrenadier* Regiment).

But this time the Russians were in greater strength. They had got nearly sixty T 34's across the river, and two companies of these had swung eastward during the morning toward the sound of firing from

[14] Commanding the 336th Division.

the Lissinski battle. This screen took the first shock of the 11th Panzer's thrust, and by the time the Germans had struck the main mass the tanks were "hull down" and prepared. The 11th Panzer made little impression that evening, and in the morning its first attack was silhouetted against the rising of a wintry sun. Heavy fighting all day took its toll of the exhausted Germans. Machines broke down, crews had hardly the strength to lift shells into the breech. When night fell, the division was reduced to half its November strength, and compelled to do the one thing which Balck most dreaded—to settle down and dig in, in a position of containment. After moving by night and fighting by day for over a week, the 11th Panzer had ground to a standstill.

As the precious days slipped past and the Russian build-up along the Chir gathered weight, Hoth was suffering agonising delays in his efforts to concentrate the main relief column at Kotelnikovo, in the south.

The 57th Panzer Corps, so reluctantly ceded by Army Group A, had set out two days behind schedule. But in the Caucasus a thaw had begun, making the roads impassable. The corps struggled back to its railhead at Maikop and entrained, but there were not enough flatcars for carrying the tanks, and some of them had to be left behind. Nor was any of the "heavy army artillery" which Zeitzler had promised loaded—allegedly for the same reason. The 17th Panzer, which Manstein had repeatedly requested from OKW reserve, was sent first to Voronezh, then back to its original concentration area, and did not entrain for Rostov until ten days after it had originally been asked for. OKW was no more co-operative in allocating a division from Army Group A to garrison Elista (this would have released the 16th Motorised, a full-strength division which had hardly fired its guns since September, and was only forty-eight hours from the 4th Panzer Army concentration area).

Manstein could see that the Russians' transfer of forces to the west of the Don was accelerating, and knew that it must be a matter of days before their tanks started to reappear in strength in the south. He therefore decided to push Hoth forward the moment the detrainment of the 57th Panzer Corps was completed. The planning of the operation, which went under the name of "Winter Tempest," offered Hoth two alternatives. The first, or "large solution," was that of an independent thrust directly at the siege perimeter, aimed at a point just west of Beketonskaya. The "small solution," to be implemented if Russian strength below the Volga bend became insuperable, was a thrust up the left bank of the Don, making junction with the 48th Panzer Corps at the Nizhne-Chirskaya bridgehead and then swinging east to the "nose of Marinovka." In either case, on receipt of the code signal *Donnerschlag* (Thunderclap), the 6th Army was to rupture the siege perimeter and advance with its mobile elements to meet the approaching relief force. Hitler had given express instructions to Paulus that, as well as making the breakout in a particular section, the 6th Army must continue to hold its existing positions in the pocket.

But it appears that Manstein was not allowing this restriction to worry him unduly, for he wrote that it was obvious this would not be possible in practice, "for when the Soviets attacked on the northern or eastern fronts [the Army] would have to give way step by step. In the event, undoubtedly, Hitler would have had no choice but to accept this fact, as he did on later occasions. (Not that we could say so in the operation order, of course, as Hitler would have learnt of it through his liaison officer at Sixth Army Headquarters and immediately issued a countermand.)" [15]

This was the key element in all the calculations—the handling of the 6th Army. Because, short though it was of fuel and ammunition, exhausted by the strain of fighting without relief, this force was the largest single concentration of the German Army in the East. It had been the very spearhead of the summer offensive. It contained some of the finest divisions.[16] These men, the flower of the Wehrmacht, were desperate. Their qualities, and their desperation, would put a diamond head on the drill.

It is still not clear how close to unanimity Paulus and his corps commanders were on the desirability of attempting a breakout. Indeed, the army commander's ambivalence reflects, as well as magnifies, that of Man-

[15] Manstein 323-24. It is perfectly possible that this rather abstruse device—explained by Manstein in memoirs written some twelve years later—carries the subtle implication which he attributed to it. But from his writing *at the time* it is clear that Manstein was still of two minds about the right course for the 6th Army to adopt. For in a long appreciation to Zeitzler (which as a matter of course would have been seen by Hitler) he had written, "Should [the 6th Army] be left in the fortress area, it is entirely possible that the Russians will tie themselves down here and gradually fritter away their manpower in useless assaults."

[16] As, for example, the 29th Motorised, which had led Guderian's drive across White Russia the previous year. See Ch. 4.

stein himself. In a letter to Manstein, dated 26th November, he spoke of ordering a breakout "in an extreme emergency," and concluded by saying that he regarded Manstein's appointment as a guarantee that "everything possible is being done to assist" the 6th Army. But neither of these terms is very precise, and at the time when this appreciation was drafted Paulus was still trying to consolidate his new perimeter. It seems likely that by "an extreme emergency" he was thinking of failure to do this, or the collapse of the perimeter under a concentric assault. At all events, there is no evidence of a comprehensive battle plan having been drawn up by his staff for getting the army into an assault formation *at any time,* either during the first crisis or in accordance with the Winter Tempest plan.

Manstein had no means of knowing what was in Paulus' mind. They communicated seldom—in contrast to the direct link between Paulus and Hitler. He was compelled to rely (until the last stages of the battle, when a short-wave radio link was established) on reports in longhand, delivered "by hand of officer." General Schulz, the Chief of Staff at Army Group Don, and Colonel Busse, the Chief of Operations, flew into the pocket at different times to try to establish a closer contact, and to brief the army commander on the plans for raising the siege. How successful they were in this is not known, but each returned (according to Manstein) with the over-all impression "that Sixth Army, provided it were properly supplied from the air, did not judge its chances of holding out at all unfavourably." In other words, there was a substantial body of opinion, to put it no higher, that preferred holding to sortie.

Manstein himself knew that time was running out. The Russian redeployment, his own weakness, the threat of a drastic strategic reversal in some other sector of the front—these things made it impossible to postpone the relief attempt any longer. On 10th December he sent word to Paulus that the attack would be started within twenty-four hours, and on the 12th, Hoth crossed the starting line, with the 23rd Panzer leading. Winter Tempest had begun.

The column was spearheaded by the 57th Panzer Corps,[17] with parts of two Luftwaffe field divisions, and its flanks protected by the re-formed remnants of the 4th Rumanian Army. At the rear a mass of vehicles of every kind, trucks of French, Czech, and Russian manufacture, English

[17] The corps was commanded by General Kirchner, and comprised the 6th Panzer (Rauss) at full strength and 23rd Panzer (Vormann), which had only about thirty runners.

Bedfords and American GMC's captured during the summer, agricultural tractors towing carts and limbers—pressed into service by the resourceful Colonel Finkh—waited with three thousand tons of supplies which were to be run through the corridor to revictual the 6th Army.

During 13th and 14th December the attack made good progress. The Russian force guarding the approach route was the 51st Army, but it had been reduced to about half its strength[18] since the break-through in November. Three tank brigades had been switched across to the attack on the Nizhne-Chirskaya bridgehead, and some of the artillery had been placed on the siege perimeter.

Against light opposition the Panzers rolled steadily forward, making about twelve miles a day. The ground was hard, iced over with a light covering of snow. At first sight it seemed completely flat, without ground elevation or cover of any kind. But in fact the whole terrain was split and crisscrossed by a network of deep and narrow gullies (not unlike the *wadis* of North Africa) into which the snow had drifted. Here lay groups of Russian infantry, sometimes up to a battalion in strength and with a full complement of heavy weapons. The cavalry kept its horses there during the day, sheltered from the freezing winds, and rode out at night when the air was still to harass the German flanks with mortar and machine-gun fire. At times—usually in the evening or at first light—isolated packets of T 34's would attack the columns, forcing a halt for a few hours. An iron-grey sky, with an overcast ceiling at five hundred feet, had grounded the Luftwaffe, and Hoth had no means of knowing whether with each clash he might not have run head on into a full-scale counterattack. Between ten and fifteen miles in the rear engineers struggled to keep the great soft "tail," with its eight hundred loaded trucks, from lagging too far behind the armoured carapace.

By 17th December the leading tanks of the 6th Panzer had reached the Aksai. The river was seventy feet broad, frozen hard enough to carry a foot soldier, but too thin for a tank. There were two bridges, at Shestakovo and Romashkin, where the railway from the Caucasus crossed the river. During the night gunfire from the siege front, thirty-five miles to the north, could be heard.

[18] The actual strength of the 51st Army at the start of Winter Tempest was four rifle and four cavalry divisions, with one tank and one motorised brigade forward and one tank and one rifle brigade in reserve. The tank brigade in reserve (117th) had been badly mauled in an ineffective spoiling attack against the 4th Panzer Army positions at Kotelnikovo on 8th December.

At his headquarters at Stary-Sherkatsk, Zhukov was receiving twice daily reports of the progress of Hoth's column. To say that he was regarding its approach with equanimity is too facile—particularly in the light of the constant tendency on the part of the Russian commanders in general and the *Stavka* in particular to overestimate German capabilities, which persisted right up until the last days of the war. But the only measures taken locally to deal with the threat were the despatch of about 130 tanks, one mechanised and one tank brigade, and two infantry divisions (each with full tank and artillery complement) to defend the Aksai crossings. Whether this was because he was a prisoner of his own planning rigidity or whether (as many German writers have suggested) Zhukov was in fact regarding the thrust by the 4th Panzer Army as a perfectly timed push on the "swing door," much as Schlieffen had intended the French to thrust into Alsace, we have no direct means of knowing.

What we do know is that the Russians were determined not to be distracted from their main prize—the 6th Army. Once they had tightened the noose at Stalingrad and began to concern themselves with the purpose of breaking up the German relief attempts, they began to redeploy along the Chir, which would seem to indicate that they expected this threat to come from the most obvious quarter—the bridgehead at Nizhne-Chirskaya. The real essence of the Russians' strategic planning, when they felt themselves to be secure at Stalingrad, was their second blow, whose purpose, even more ambitious than the isolation of the 6th Army, was nothing short of the disintegration of the German southern wing. Here it was they who had rejected the obvious course. A blow down the east bank of the Don would, it was felt, "be too constricted by the lay of the land, be vulnerable on both flanks and open to the danger of a pincer movement by the enemy at Rostov and in the Caucasus. The thaw, being unpredictable in the Black Sea region, would have restricted mass operations." (That this consideration was wholly valid is apparent from the difficulties experienced by the 57th Panzer Corps in travelling north.)

It seems probable, though its official historians do not admit as much, that the *Stavka* felt a certain uneasiness about the central sector, which had lain dormant throughout the year, and believed that a blow at the junction of the southern and central sectors would give its forces more elbowroom and serve to draw off any German reserves that might be accumulating there. To this end it had concentrated two army groups, under Generals Golikov and Vatutin, and put into them its last three reserve armies.[19]

The area selected for the attack—a stretch of front thirty miles on either side of the Don bridgehead at Verkhni Mamon—was defended in the main by Italians.[20] There was only one German division (the 298th) in the combat zone, and two battalions of another (the 62nd) at Kantemirovka. The mobile reserve (27th Panzer) was a weak unit, being re-equipped with repaired and reconditioned tanks from the workshops at Millerovo. The ice on the Don was so thick that the Russian tanks could cross at will, and a thick fog covered the battlefield during the day, heightening the panic and confusion of the luckless Italians.

That evening, as the first coherent reports began to come into Manstein's headquarters, it was plain that something very serious indeed had happened. The immediate responsibility was not his, for the attack had fallen against the right flank of Army Group B, but a glance at the largest-scale map showed the threat the new Russian thrust carried, both to Army Group Don and to every man in the Caucasus. In a telephone conversation that night Weichs told Manstein that he had "committed" (*übergeben*) the whole of the 27th Panzer at the western end of the Russian break-through, but had "as yet no news of its fortunes." (Two days later it was reduced to eight runners.) Weichs also asked that Army Detachment Hollidt pull back and westward, in order to take over some of the shattered flank of his own army group.

During these last critical days Manstein's problem resembled more and more that of a three-dimensional chess player, with each board showing a losing game. On that same day when the Italians broke below Voronezh, his whole position on the lower Chir started to crumble. While the 11th Panzer had been crouching, hull down, around the Nizhne Kalinovski bridgehead, the Russians had put in four infantry divisions against the weak foothold to the east of the Don at Nizhne-Chirskaya and driven the Germans back to the west bank. That night they made

[19] These were the 6th (Golikov), which was primarily an infantry army, and the 1st and 3rd Guards (Vatutin), each of which contained one tank corps and one mechanised corps. If all units on these two "fronts" were up to strength at the start of the offensive—which seems likely—this would have given an armoured strength of about 456 T 34's.

[20] The 8th Italian Army (General Gariboldi). The Italian area overlapped with that of the 2nd Hungarian Army in the north, and also included a few weak remnants from the 3rd Rumanian Army.

two more crossings in force on either side of Lissinski, and the following morning threw an independent armoured brigade and an entire motorised corps (the 94th) against the 7th Luftwaffe Field Division[21] at Oblivskaya. Once again Balck roused the weary 11th Panzer and started west to tackle the most serious of the new penetrations. But it was now plain that any idea of a supporting move by the 48th Panzer Corps to assist Hoth's relief attempt was out of the question. Sheer weight of numbers was levering Army Detachment Hollidt out of the Chir bulge, and its complete evacuation was in sight.

The only glint of light came from the far eastern tip of the front. On the morning of 18th December, Manstein received a message from Hoth saying that the 17th Panzer had arrived in the line and concentrated. This meant that the 4th Panzer Army now contained three Panzer divisions[22] and their supporting elements, and was substantially stronger than any of the Russian forces that had so far been identified against it. If it could prise open the ring around Stalingrad and release the eleven divisions of the 6th Army, the whole balance of strength might still, if not be reversed, at least be brought nearer parity. Manstein knew that the Russians' efforts on all other fronts would weaken if they believed the main prize was slipping from their grasp. But to press forward with Hoth when his own northeastern flank was disintegrating along its entire two-hundred-mile length remained a gigantic gamble. One, moreover, whose responsibility was solely Manstein's. Neither OKH nor Hitler, nor even Paulus himself, showed any great enthusiasm for the plan or appreciated the urgency of the situation.

And here lay the rub. The 4th Panzer Army could not possibly hack its way over the whole distance, right up to the ruins of the city, by itself. Paulus, too, had to co-operate, to concentrate the mass of his two hundred thousand men against a single point of the siege perimeter and blast a way through. Yet when he was asked to do this, or what his plans were, Paulus was evasive.

In the light of this unresponsive attitude, on 18th December, Manstein

21 Of the Luftwaffe field divisions Balck said, "After a few days they were gone—finished—in spite of good mechanical equipment. Their training left everything to be desired, and they had no experienced leaders. They were a creation of Hermann Göring's, a creation which had no sound military foundation—the rank and file paid with their lives for this absurdity." (Mellenthin, 180.)

22 One, the 17th, at full strength. One, the 6th, less combat losses (not severe) since 8th December. One, the 23rd, with about thirty runners.

sent a message directly to Zeitzler at OKH, requesting him "to take immediate steps to initiate the breakout of 6th Army towards 4th Panzer Army." And that evening the chief intelligence officer of Army Group Don, Major Eismann, was sent into the pocket to recount Manstein's views on how the operation should be conducted.

No feat of imaginative power is required to visualise the dramatic tension that surrounded this visit. Eismann was one of the last visitors to enter the pocket from the outside world—at least while there was still hope of relief. During the night he drove from Novocherkassk to Morozovsk, taking off from the airstrip there in a Fieseler Storch an hour before dawn.

Eismann touched down at Gumrak at ten minutes to eight on the morning of 19th December, and was immediately taken to Paulus' headquarters. Those present, in addition to Paulus and his Chief of Staff, Schmidt, were two corps commanders and the Chief of Operations and the Quartermaster General of the 6th Army. Eismann put Manstein's views with as much force as he could muster, but Paulus went no further than to admit being "not . . . unimpressed." Paulus then proceeded to emphasise "the magnitude of the difficulties and risks which the task outlined to him would imply." After a moment or two the Chief of Operations and the Quartermaster General said their piece. Each began with a general résumé of the difficulties, and echoed—though perhaps with less enthusiasm—the views of their chief. But at the end, when they came to express their own personal opinions, both said that ". . . in the circumstances it was not only essential to attempt a breakout at the earliest possible moment but also *entirely feasible*."

However, the last word was spoken by Paulus' Chief of Staff, Major General Arthur Schmidt. Schmidt was an ardent Nazi and a forceful personality. There is no doubt that he exercised a considerable influence on Paulus, playing the part of a "Party conscience," always standing at the General's elbow. "It is quite impossible to break out just now," he told Eismann. Such a solution would be "an acknowledgement of disaster. Sixth Army," he claimed, "will still be in position at Easter. All you people have to do is to supply it better."[23]

[23] Many of the 6th Army General Staff officers, as will be seen, had second thoughts about their loyalty oath after a few months in Russian prison camps. But not Schmidt. He refused to recant, and was sentenced to twenty-five years' hard labour. See p. 288, *et seq*.

The conference dragged on during the day. At intervals gunfire shook the building. During the afternoon an egregious meal was served. To an audience at the same time gloomy and resentful Eismann tried to explain that the breakout was necessary "from the point of view of operations as a whole." As to the airlift, ". . . although the Army Group was doing everything in its power to maintain supplies, it was not to blame when the weather brought the airlift to a virtual standstill, nor was it in a position to produce transport machines out of a hat." Paulus remained unconvinced. Indeed, his attitude seems to have hardened during the day, for he finally dismissed Eismann with the assertion that the breakout was "a sheer impossibility," and that in any case, the surrender (i.e., the evacuation) of Stalingrad was forbidden by order of the Führer.

Before Eismann returned, on 19th December, Manstein was brought the news that Hoth had forced the Aksai line and penetrated as far as Mishkova. When his intelligence officer told him of Paulus' refusal to cooperate, Manstein at first thought of dismissing both him and Schmidt and replacing them with members of his own staff or promoting corps commanders from within the pocket. But time was very short, and the likelihood of getting such an appointment sanctioned by OKH—still less by Hitler—without a long back and forth of communication, threatening his own resignation, and so forth, must have deterred him.

That afternoon, at 2:35 P.M., Manstein teleprinted to Zeitzler, stating "I now consider a break-out to the south-west to be the last possible means of preserving at least the bulk of the troops and the still mobile elements of Sixth Army." He waited until six o'clock that evening, and then, still not having secured a reply from OKH, teleprinted directly to Paulus, "Sixth Army will begin Winter Tempest attack as soon as possible," and, "It is essential that operation Thunderclap should immediately follow Winter Tempest attack."

During the next twenty-four hours Paulus came back several times over the very high-frequency link with Novocherkassk. First he said that to regroup for the attack would take at least six days; then that the regrouping itself would entail serious, possibly prohibitive, risks in the northern and western sectors of the front. Then again, that ". . . the general debility of the troops and the reduced mobility of units following the slaughtering of horses for food made it most unlikely that such a difficult and risky undertaking—particularly when carried out under conditions of extreme cold—could possibly succeed." Finally, when all these protestations had been patiently, or firmly, or brusquely, set aside, Paulus played his trump card. It was impossible to move within the terms of the order, he declared, as this entailed an advance of thirty miles and he had fuel only for twenty. (Apart from the fact that units always have more fuel in hand than they care to admit in official returns, on Paulus' own admission he could have covered the same distance by reducing the number of vehicles engaged by 30 percent. The very fact that such an objection could be put forward as real at such a moment of crisis showed that Paulus had, in reality, no intention of moving.)

In the meantime the other "boards" in Manstein's game were showing a progressive deterioration. The withdrawal of Detachment Hollidt was becoming more and more precipitate, and within days threatened to uncover the airlift bases which were supplying the pocket. And Hoth was reporting a sudden stiffening of resistance. A new Russian tank corps (the 13th), together with an infantry division and an independent tank brigade, had been identified opposite him in the 51st Army.

On the afternoon of 21st December, Manstein spoke on a direct telephone line to Rastenburg, in a final effort to persuade Hitler that the whole of the 6th Army must break camp and drive southward, without avail. "I fail to see what you are driving at," was all Hitler would say. "Paulus has only enough gasoline for fifteen to twenty miles at the most. He says himself that he can't break out at present."

Here, then, was the log jam in the Stalingrad problem. The army commander pleaded that whatever the technical requirements, he was bound by an order of the Führer. The Führer refused to rescind his order on the grounds that anyway, the army commander was raising technical objections. All Manstein's efforts had been useless, and now the risks he had run were coming home to roost. The cream of his armour was poised right out in the steppe, at the eastern extremity of his front, weighted down by a huge and vulnerable convoy of supply vehicles. The mass of his infantry was written off—gone for good, nearly a quarter of a million men. And along the whole of his northeastern flank, for nearly two hundred miles, Army Group Don was retreating in disarray, virtually cut off from its neighbours, exhausted and outnumbered by five or six to one. In eighteen months of triumph, disappointment, and fluctuating fortunes this moment was for German arms the nadir.

BOOK III | Zitadelle

GUDERIAN. My Führer, why do you want to attack in the East at all this year?

HITLER. You are quite right. Whenever I think of this attack, my stomach turns over.

We are in the position of a man who has seized a wolf by the ears and dare not let him go.

Mellenthin, 14th May, 1943

fifteen | CRISIS AND RECOVERY

The first days of 1943 opened—as had 1942—with the German Army in dire peril. But in that first winter its predicament had arisen largely out of accident and miscalculation. In 1943 the causes were more serious, and more fundamental.

For over half its length, nearly six hundred miles, the front had solidified. From the frozen Baltic, around the siege perimeter at Leningrad, due south to Lake Ilmen, and across the pine forests of the old Rzhev salient, and then down to Orel, the German front had hardly altered in twelve months. Permanent emplacements of logs and earth sheltered the soldiers; reinforced concrete protected guns whose field of fire traversed enormous mine fields, laid during spring and summer, while the earth was soft. In these positions the "garrison" had a comfortable enough time. Fuel was plentiful, clothing adequate, mail was delivered regularly. Its situation is comparable to that of the Western front in World War I between St. Mihiel and the Swiss frontier. Its bitterest enemies were the terrible cold and the huge bands of Partisans who roamed the desolate terrain, usually on horseback, and came out of the freezing night to attack lonely German billets far behind the lines. The front itself was often quiet for days at a time. The Germans used it as a rest area for worn-out divisions, the Russians as a training ground for new ones.

It was to the south, where the three great rivers of the Ukraine flowed into the Black Sea, that the campaign was being decided. Here, six months before, the Germans had deployed the flower of their Army, and here it was now in headlong retreat. It had failed to force an issue in its prime. How, weakening daily, could it avoid annihilation?

For Manstein, as he considered this problem in the first week of January 1943, there was not one gram of comfort. The forces for which he had responsibility were broken into three separate groups, each too far, and too preoccupied with its own perils, to render the others mutual support. With Paulus gone, German strength in south Russia was halved. The 6th Army had at most a few weeks of diminishing effectiveness to distract what it could of Russian strength. To the southeast, still deep in the Caucasus, Army Group A lingered on, outside the scope of Manstein's direct command and alarmingly vulnerable to Russian encirclement. Manstein's own units, in Army Group Don, had taken such a battering since November as to be hardly recognisable. Corps and divisions had lost their identity; shot-up Panzers, anti-aircraft and Luftwaffe remnants, had polarised around a few energetic commanders—Hollidt, Mieth, Fretter-Pico, who gave their names to *Gruppen* responsible for stretches of front up to a hundred miles long.

Nonetheless, the Germans' inferiority was not so great as they, and the majority of Allied observers, believed it to be at the time. Many of the factors present the previous winter had recurred—men and machines had worn themselves down in the exertions of the summer battles; winter equipment was still inadequate, for mobile warfare at least; the tenacity and resilience of the Russian soldier had again been underestimated —and these factors were transient. As the Germans fell back on their railheads, if and when they could gain time to breathe, when the temperature moderated, then they might still expect their situation to improve. The Russians now definitely had the stronger army—whereas in the winter of 1941 they had never achieved more than a local numerical superiority and owed their victory simply to the toughness and bravery of the Red Army man and his personal ascendancy over the individual German when the thermometer was 20 below. But equally the Russians had inherited many of the weaknesses of the previous period. They had brought two and a half million men into uniform since the outbreak of war. They had lost over four million trained soldiers. A ruthless standardisation of equipment—two types of trucks, two tanks, three artillery pieces—had allowed them to raise production rates in spite of losing two thirds of their factory space. But of leaders to handle the new army there was a desperate scarcity. Some were too cautious, others too headstrong, all compensated for lack of experience with blind obedience to orders from above. The result was that tactical flexibility and speed in exploitation were far below the German standard. Only the artillery, some of the cavalry, and a very few of the tank brigades truly merited the "Guards" accolade that was being so liberally dispensed. The real problem for the Red Army had become one of adaptation: the change-over from a defensive stance, where its rugged courage and fortitude had carried the day, to the more complex structure of an offensive pattern, where the initiative and training of even the smallest units could be of vital importance.

In their advance from the left bank of the Don the Russians reverted to the techniques they had employed the previous winter, advancing on a broad front with groups of mixed arms, maintaining constant pressure at the price of diluting the concentration which would be needed for a deep and narrow thrust. While Manstein was compelled to hold his army as close as possible to Stalingrad, Russian tactics were effective. But once Paulus had been abandoned and the Germans could start using the space of the Don bend for manoeuvre, it was possible to delay the Russian advance at a much lighter cost in men and equipment. Ironically, it was after it had forfeited all hope of relief that the Stalingrad garrison played its biggest part in the recovery of fortune by its comrades. For throughout January, Zhukov was pursuing the Germans with his left hand tied behind his back. Over half the Soviet infantry and 30 percent of its artillery were still in the Stalingrad area at the conclusion of the siege.

It seemed, then, that Manstein could continue to run risks on his left flank to achieve what had now become his most urgent aim—the extrication of Army Group A from the Caucasus and the redeployment of the fresh units of which it was composed. The most powerful force in Army Group Don was still Hoth's 4th Panzer Army, in spite of its casualties on the Aksai, and Manstein decided to leave it in position, granting Hoth the discretion to pull back gradually toward Rostov while keeping open the line of retreat for Army Group A. Such an operation was just feasible, given the obvious decline in Russian offensive vigour and the ability of Paulus to hold out for at least one more month.

It soon became evident that there were other obstacles to this course of action, and they threatened to be more intractable than the enemy in the field. The first stumbling block was to be found in that familiar German handicap—duality and contradiction in the chain of command. Army Group A had been commanded by Kleist since October. Nominally it was of equal status to Army Group Don. Both received their orders from OKH and did not communicate directly, other than on particular and incidental problems at divisional level. In addition, the army group, like the 6th Army, was one of those which were subject to a great deal of personal attention from Hitler, and so received separate (and sometimes conflicting) orders from OKW.

Hitler did not at this time envisage a complete evacuation of the Caucasus—he thought in terms of contracting the front, to leave a "balcony" from which a later operation against the Russian oil fields could be mounted. He believed that it was important to hold on to Novorossisk, thereby confining the Russian Black Sea fleet to one port (Batum); that even if Rostov should fall it would still be possible to supply Kleist across the strait from the Crimea; and that German presence in the Kuban as well as at Kerch would seal off the Sea of Azov. Hitler's argument was at a strategic level. At the start of the campaign he had said, ". . . if we do not capture the oil supplies of the Caucasus by the autumn, then I shall have to face the fact that we cannot win this war." The projected "balcony" offered him the chance of one more lunge at the oil fields if the main front could be stabilised, and from a defensive aspect, the Kuban protected the Crimea, which in turn protected the Reich's only source of natural oil—the Rumanian fields at Ploesti.

Manstein and Kleist, in contrast, viewed things from a narrower tactical (not to say personal) standpoint. Manstein did not want Kleist, and his staff, and his independent command structure. What he wanted were Kleist's fresh and intact divisions—particularly the 1st Panzer Army—and he wanted them under his own control. Kleist, not unnaturally, was very reluctant to see them go. He had no taste for seeing his vigorous and undefeated army reduced in size and strength to a mere appendix. In the result, although the subordination of Army Group A to Manstein was "under consideration at OKH for some time," nothing more was heard of the idea.[1]

[1] It is not hard to read between the lines of Manstein's comment on the command dilemma:

> . . . Generally speaking, it is not a good thing to put an army or army group under a headquarters of equal status. In the present critical situation, however, *this would probably have had its advantages*—provided, of course, that no strings were attached. Any possibility of interference by Hitler *or of Army*

The result of this quadripartite division of opinion and attitude—between Manstein, OKH, Hitler, and Kleist—was that no positive order was ever sent to Army Group A as a whole. Instead it was moved about piecemeal, one division or two at a time, a course of action which gave the maximum scope for muddle and procrastination.

It has already been shown how these delays affected the striking power of Hoth's relief force in December.[2] At that time it had been a question of moving two divisions; in January, when the whole future destination of the army group was under consideration, the inertia became leaden. The army group staff contended that "a substantial period of grace" would be required if a large-scale evacuation were intended. Immovable weapons would have to be lifted from their emplacements, reliefs arranged, stores packed and despatched, horses brought up to the line, wounded moved from the hospitals . . . Then again, what were the specific movement orders? Was the army group to move west, or northwest? Or was it to divide? Was it to take up intermediate positions en route, and if so which? Could the positions and times be clearly scheduled? Was it to halt on the Kuma line? In default of positive instructions to the contrary it would halt on the Kuma line. To reach the Kuma line the army group would require twenty-five days, as "in the interests of getting the equipment out" the withdrawal would take place "sector by sector."

When Manstein finally succeeded in getting one division (SS Viking) given to Hoth, he had to supply most of the fuel to move it. The 1st Panzer Army staff also pleaded fuel shortage for its continued delay in relieving the 16th Motorised at Elista. Kleist made inordinate demands on the railway system, claiming that 155 trains would be required to move the army back to the Don, and that when this had been done he would need 88 more to stock up the Kuban bridgehead. On 18th January, Manstein, who had threatened to resign ten days earlier, was still grumbling, ". . . whether First Panzer Army will be withdrawn towards Rostov or into the Kuban is anybody's guess." By now, so much time had been consumed that it was too late to move the infantry—except across the strait of Kerch—and a very good division from the 1st Panzer Army, the 50th (Mountain), also had to be left behind. On the last day of

Group A's invoking his decisions in opposition to my own had to be expressly barred. Hitler, however, was unwilling to accept my conditions, and Army Group A consequently remained autonomous. [My italics.]

[2] See p. 264.

January, the 13th Panzer, for which the door was still being held open at Rostov by Hoth's battle-weary formations, was reallocated to Army Group A, turned around, and sent back to the Kuban. The result of this was that there were nearly 250,000 men (over 400,000 if satellites are included) fit, well equipped, but virtually immobilised, at the extreme southern end of the Eastern front.

Army Group A was kept in being, its area to include the Crimea as well as the Kuban, and Kleist was made a Field Marshal "for his achievements in conducting the retreat." [3]

[3] Kleist has given his own version of the events of January 1943, under interrogation by Liddell Hart at the end of the war, subsequently published in *The Other Side of the Hill,* pp. 230-31. (Published in the U.S. under the title *The German Generals Talk.*) His account is extremely foreshortened and inaccurate, takes the traditional avoiding action by blaming Hitler and, where factual, runs counter to the other evidence—namely that of Manstein and the OKH movement orders. However, in fairness to the Field Marshal, it should be cited:

> When the Russians were only forty miles from Rostov, and my armies were 390 miles east of Rostov,[a] Hitler sent me an order that I was not to withdraw under any circumstances. That looked like a sentence of doom. On the next day,[b] however, I received a fresh order—to retreat, and bring away everything with me in the way of equipment. That would have been difficult enough in any case, but became much more so in the depths of the Russian winter.[c]
>
> The protection of my left flank back to the Don from Elista had originally been entrusted to the Rumanian Army Group under Marshal Antonescu.[d] Antonescu himself did not arrive on the scene, thank God! Instead the sector was placed under Manstein, whose Army Group South included part of the Rumanian forces. With Manstein's help we succeeded in withdrawing through the Rostov bottleneck before the Russians could cut us off . . . Even so, Manstein was so hard-pressed that I had to send him some of my own divisions[e] to help in holding off the Russians who were pushing down the Don towards Rostov. The most dangerous time of the retreat was the last half of January.[f]

The following comments suggest themselves:

[a] These distances never coexisted. For example, from the Sal, south of Kotelnikovo, to Rostov, is 130 miles. From Rostov to the Kuma is under 300.

[b] The date of these two "orders" is not given, and there is no other record of their having emanated from OKW.

[c] The climate prevailing was unseasonably mild, in an area which has the mildest winter weather in all Russia.

[d] Presumably Kleist is referring to the order of battle prior to the start of the Russian counteroffensive on 19th November, 1942. There was no Rumanian army incorporated in Manstein's dispositions here (i.e., after December 1942).

[e] See text above.

[f] If by "the retreat" Kleist refers to the withdrawal of the 1st Panzer Army (the 17th Army remained in the Kuban), this was not even authorised by OKH until "the last half of January." (*Vide* Manstein, quoted above.)

Hoth had fought a clever and flexible battle, with forces which had been in almost continuous action for two months, many of them for far longer, and his tanks were completely worn out by their constant marching and countermarching. Nonetheless, it must not be taken as a detraction of Hoth's skill and the endurance of his soldiers to repeat that the main burden throughout the month of January was being carried by the beleaguered divisions of Paulus' 6th Army. This should always be borne in mind when the argument of the "useless sacrifice" at Stalingrad is expounded.

After the failure of the relief attempt at Christmas, Zhukov had stepped up the rate of transfer of armour and mechanised forces from the Stalingrad perimeter, but he continued to retain a total strength there of nearly half a million men, a ratio of two to one against the Germans. In contrast, the forces deployed between the Don and the Sal, toward Rostov, were by no means as strong as the Germans believed at the time, or claimed subsequently. Only the 2nd Guards Army was properly balanced, with tank and self-propelled artillery brigades; four of the infantry corps had been badly mauled in the fighting on the Aksai; and the two fresh infantry corps, the 51st and 28th, did not have the mobility to keep up with their prey in the open conditions which characterised these battles.

The real moment of crisis had come and passed on 8th January, when the Russians presented a demand for the surrender of the 6th Army, and it had been rejected. The appeal had been signed by Rokossovski and Voronov, and offered

> Honourable surrender . . . sufficient rations . . . care for the wounded . . . officers to keep their weapons . . . repatriation after the war to Germany or any other country.

Hitler was still in daily touch with Paulus by short-wave radio, and the army commander would not consider surrender without the Führer's permission. Nor is there evidence that any but a tiny proportion of the rank and file thought seriously of taking advantage of the Russian offer. "We did not have much faith in Russian promises." "Anything was better than Siberia." "We all knew 'Ivan' too well; one never knew what he would do next, promises or no promises." This was the typical reaction, although by that time the beleaguered army was suffering miseries which would have impelled any Allied commander to surrender, on humanitarian grounds alone. Some German authorities even attribute to the 6th Army more altruistic motives: ". . . we were surrounded by three Russian armies which would be free for other operations if we capitulated . . ." And there was always the hope—for man must have hope, however slender—that they would be relieved.

Until 10th January the Russians had mounted no serious attacks against Paulus' perimeter, but had been content to maintain harassing fire from their immensely superior artillery and conduct local operations aimed at paving the way for the final assault. Throughout December and the first week in January conditions within the perimeter got worse and worse.

> Only twenty to thirty cartridges were distributed daily to each man, with the order to use them solely to repulse an attack. The ration of bread was reduced to 120 and then 70 grams—a slice only! Water came from melted snow. Because of a lack of potatoes a kilogram box had to make do for fifteen men. There was no meat; we ate our horses at Christmas.

The 6th Army's minimum requirements of supplies of all kinds had been estimated at 550 tons. The round trip from the airfields at Tatsinskaya and Morozovsk involved a flight time of nearly three hours—excluding that spent loading and unloading—so, with only one sortie per day likely, this meant that a force of 225 Ju 52 aircraft would have to be serviceable every day. In fact, there were never more than eighty Junkers operational at a time. Their efforts were supplemented by two squadrons of Heinkel III's (which had a capacity of only 1.5 tons), but the largest amount ever brought into Stalingrad in one twenty-four-hour period was 180 tons, on 14th December. After Christmas, when Tatsinskaya and Morozovsk had been overrun, the nightly average fell to about 60 tons.

Virtually no gasoline was issued—until the very end the meagre supplies were being hoarded for a breakout, and the army's tanks and self-propelled artillery were dug into permanent positions in the frozen rubble. The men were too weak to dig fresh emplacements or communication trenches; when forced out of their old positions they would simply lie in the ground behind heaped-up snow "parapets," numb with cold and the inevitability of death. To be wounded might be lucky, more often it was a stroke of hideous misfortune among comrades too exhausted to lift a man onto a stretcher; where medical services had no anaesthetic other than artificially induced frostbite.

While the landing ground at Pitomnik was still usable, some of the seriously wounded were flown out from there in the returning transports, but as the days passed fewer and fewer pilots dared risk touching down on the cratered runway. The Heinkels, with their weaker undercarriage, confined their mission to making low-level drops. Many of the Junkers broke up on landing or were destroyed by Russian artillery fire. Others, with their take-off runs perilously reduced by craters and wreckage, had to leave half empty or, more harrowing still, lighten their load by turning out "passengers" who were already aboard. This led to some ugly scenes:

> There were about thirty of us on the 'plane, mostly wounded, with stretcher cases piled on top of each other all over the floor. There were also some people, couriers and the like, who were quite unharmed—the sort of people who always, it seems, get themselves out of the tightest scrape by the use of their wits. We started trundling across the ground at an ever-increasing speed, with clouds of snow blowing back from the propellers; at intervals one wheel would drop in a crater with a terrible crash. Then to our horror the engines cut and we could feel the brakes coming on. The pilot turned round and started taxi-ing back . . .
>
> A Lieutenant of the *Luftwaffe* came through and said that we could not get airborne because of the ground, and that we would have to shed about 2,000 kilos . . . twenty men would have to get out. At once there was the most terrific din, everybody shouting at once, one man claimed that he was travelling by order of the Army Staff, another from the SS that he had important Party documents, many others who cried about their families, that their children had been injured in air raids, and so on. Only the men on the stretchers kept silent, but their terror showed in their faces . . .

Sometimes the wounded would have to wait for days, huddled around stoves in matchboard shanties at the edge of the airfield or in the "safety" of open trenches, where they would freeze to death overnight. At others shortage of fuel and transport meant that they never got to Pitomnik at all. Then the prospect of an aircraft returning empty was an unbearable temptation, and at the last moment there would be attempts to rush it. On New Year's Day 1943 it was promulgated that nobody was allowed to board an aircraft, for whatever purpose—even unloading and ground servicing—without a written permit from the 6th Army Chief of Staff. This led to further delays, particularly for the seriously wounded. Many men were shot out of hand trying to force their way on board, and their corpses lay about in the snow. There were at least two cases of soldiers being taken aloft hanging on to the undercarriage or tail wheel in desperation; they fell to their death in a matter of minutes.

Others found more ingenious ways of getting back to Germany:

> I had taken out a case of medical supplies to the advanced dressing station at Dmitriyevka. It was in a warehouse with the roof open to the sky in places from shell-fire. It was absolutely crammed with wounded and most of them were in a bad state, dead and dying together, crying and praying aloud. . . . an orderly told me that they were going to be flown out . . . just then a *Katyusha* salvo fell in the street and calls from some more wounded took him and the Doctor outside. I went over to a part of the building where the men were quiet. They were so badly injured they were unconscious and some of them had already died. I turned one of them off his stretcher . . . I fired three shots through my left foot and lay down. I lost consciousness . . . it was dark and the pain was frightful . . . there were no lights in the warehouse . . . I kept telling myself, "It will be an hour, a few hours, and then the flight." Two days passed and the blood round my foot froze solid, but I dared not call for attention . . . two of the men near me died. Then—Morning of Joy! They started to move us . . .

But the corporal's elation was short-lived. The wounded were taken to a casualty clearing station for inspection and the issue of "flight permits." There a doctor found powder burns on his skin and decided the wound was self-inflicted—a capital offence on the Eastern front. He lay in the cellars of the GUM department store for a fortnight before being captured, in agony from frost gangrene, and the Russians saved his life by amputating the leg at the hip.

Following the rejection of the surrender ultimatum on 10th January, the Russians opened an all-out offensive. Throughout the night their artillery ravaged the lines of the 6th Army, and with the icy dawn, the first moves in a great concentric assault began.

In that inferno most of the Germans must have shared Colonel Selle's judgment: "The cover of the tomb is closing upon us." But a rumour spread that because his surrender terms had been rejected Zhukov had given orders that no prisoners were to be taken, and many detachments fought literally to the last round and then committed suicide. (Suicides had become so common in the preceding period that Paulus was compelled to issue a special order declaring them "dishonourable.")

The main Russian effort was directed against the western end of the 6th Army position, the "nose of Marinovka," where the defenders had little cover from either ground or buildings. On the second day the Russians sliced five miles off the perimeter. The 29th Motorised, which had led Guderian's Panzer group across White Russia in the summer of 1941, one of the finest units in the German Army, was finally destroyed. For forty-eight hours the 6th Army gave ground, until it was forced back onto the frozen gully of the Russochska. Then the Russian pressure slackened. Incredibly, Paulus had weathered the storm, but his army had expended the last reserves of energy. The whole system of reliefs had broken down. Units simply fought, and expired, where they stood. The cellar "hospitals" had to refuse to accept any more wounded, and many of the injured had to plead with their comrades to shoot them on the spot. Pitomnik airfield was overrun on the second day, and after that time supplies could be dropped only at night, although it was possible to land with a light airplane at the battered Gumrak strip, a few yards from Paulus' headquarters.

When the Russians resumed their attack on the 16th they made better progress, still applying the tactics of forcing the Germans back from three sides of the compass against the iron barrier of Chuikov's 62nd Army in the ruins of the city itself. On 23rd January they captured the Gumrak airstrip, and the last contact with the outside world was broken. For another week the battle dragged on—now back in its original seat, the battered buildings and underground warrens of the September fighting, the Krasny Oktyabr factory, the Matveyev-Kurgan. Then, on 30th January, the southern pocket collapsed and Paulus was captured. The rest of the 6th Army surrendered two days later.

During the night the Russians published a special communiqué, announcing the surrender and giving the names of all the senior officers (including Paulus) who had been captured. At midday on 1st February a special Führer conference was called at which this was discussed, and the record of proceedings at this meeting has been preserved.

Hitler is not at his best. He rambles and repeats himself. A certain air of fantasy[4] pervades the whole conference—one not dispelled by Zeitzler's extraordinary servility and his acquiescence in the unprec-

[4] Hitler refers three times to the story of a woman who committed suicide. His first version (recorded at the start of the conference) differs substantially from that which he told Jodl (after Zeitzler had departed) later that same afternoon:

"Such a beautiful woman she was, really—first-class. Just because of a small matter, insulted by a few words, she said, 'Then I can go, I'm not needed.' Her husband answered, 'Why don't you?' So the woman went, wrote farewell letters, and shot herself."

edented notion that officers of the General Staff should commit suicide rather than submit to capture. (However, the Führer's instincts have not deserted him so far as to prevent him from making an accurate forecast as to the future careers of Paulus and Seydlitz.)

HITLER. They have surrendered there formally and absolutely. Otherwise they would have closed ranks, formed a hedgehog, and shot themselves with their last bullet. When you consider that a woman has the pride to leave, to lock herself in, and to shoot herself right away just because she has heard a few insulting remarks, then I can't have any respect for a soldier who is afraid of that and prefers to go into captivity. I can only say: I can understand a case like that of General Giraud; we come in, he gets out of the car and is grabbed. But—

ZEITZLER. I can't understand it either. I'm still of the opinion that it might not be true; perhaps he [Paulus] is lying there, badly wounded.

HITLER. No, it is true—they'll be brought to Moscow, to the GPU right away, and they'll blurt out orders for the northern pocket to surrender too. That Schmidt[5] will sign anything. A man who doesn't have the courage in such a time to take the road that every man has to take some time doesn't have the strength to withstand that sort of thing. He will suffer torture in his soul. In Germany there has been too much emphasis on training the intellect and not enough on strength of character—

ZEITZLER. One can't understand this type of man.

HITLER. Don't say that. I saw a letter—it was addressed to Below. I can show it to you. An officer in Stalingrad wrote, "I have come to the following conclusions about these people—Paulus, question mark; Seydlitz, should be shot; Schmidt, should be shot—"

ZEITZLER. I have also heard bad reports about Seydlitz.

HITLER. —and under that, "Hube—The Man." Naturally, one would say that it would be better to leave Hube in there and bring out the others. But since the value of men is not immaterial, and since we need men in the entire war, I am definitely of the opinion that it was right to bring Hube out. In peacetime in Germany about 18,000 or 20,000 people a year chose to commit suicide, even without being in such a position. Here is a man who sees 50,000 or 60,000 of his soldiers die defending themselves bravely to the end. How can he surrender himself to the Bolshevists? Oh, that is—

ZEITZLER. That is something one can't understand at all.

HITLER. But I had my doubts before. That was the moment when I received the report that he was asking what he should do. How can he even ask about such a thing? From now on, every time a fortress is besieged and the commandant is called on to surrender, he is going to ask, "What shall I do now?"

[5] In fact, Schmidt was one of the very few senior officers captured in the Stalingrad pocket who remained loyal to Hitler throughout his captivity. And see p. 271.

ZEITZLER. There is no excuse. When his nerves threaten to break down, then he must kill himself.

HITLER. When the nerves break down, there is nothing left but to admit that one can't handle the situation, and to shoot oneself. One can also say that the man should have shot himself just as the old commanders who threw themselves on their swords when they saw that their cause was lost. That goes without saying. Even Varus gave his slave the order: "Now kill me."

ZEITZLER. I still think they may have done that and that the Russians are only claiming to have captured them all.

HITLER. No.

ENGEL.[6] The extraordinary thing is, if I may say so, that they have not announced whether Paulus was badly wounded when he was taken. Tomorrow they could say that he died of his wounds.

HITLER. Do you have exact information about his being wounded? The tragedy has happened now. Maybe it's a warning.

ENGEL. The names of the generals may not all be correct.

HITLER. In this war no more field marshals will be made. All that will be done only after the conclusion of the war. I won't go on counting my chickens before they are hatched.

ZEITZLER. We were so completely sure how it would end that granting him a final satisfaction—

HITLER. We had to assume that it would end heroically.

ZEITZLER. How could one imagine anything else?

HITLER. Together with such men in such surroundings, how could he have brought himself to act differently? If such things can happen, I really must say that any soldier who risks his life again and again is an idiot. Now if a private is overwhelmed, I can understand it.

ZEITZLER. It's much easier for the leader of an outfit. Everyone is looking at him. It's easy for him to shoot himself. It's difficult for the ordinary soldier.

HITLER. This hurts me so much because the heroism of so many soldiers is nullified by one single characterless weakling—and that is what the man is going to do now. You have to imagine, he'll be brought to Moscow, and imagine that rat trap there. There he will sign anything. He'll make confessions, make proclamations—you'll see. They will now walk down the slope of spiritual bankruptcy to its lowest depths. One can only say that a bad deed always produces new evils. With soldiers the fundamental thing is always character, and if we don't manage to instil that, if we just breed purely intellectual acrobats and spiritual athletes, we're never going to get a race that can stand up to the heavy blows of destiny. That is the decisive point.

[Zeitzler then recounts an anecdote whose purpose is to denigrate the importance of General Staff training.]

[6] Oberstleutnant Gerhard Engel, Army Adjutant to Hitler 1937-44.

HITLER. Yes, one has to take brave, daring people who are willing to sacrifice their lives, like every soldier. What is life? Life is the nation. The individual must die anyway. Beyond the life of the individual is the nation. But how anyone can be afraid of this moment of death, with which he can free himself from his misery, if his duty doesn't chain him to this Vale of Tears? Na!

[There is then some discussion of the official attitude to be taken concerning the surrender. Zeitzler leaves.

Enter Christian, Buhle, Jeschonnek, Jodl, and Keitel. Following the reading of situation reports from Africa and the Balkans, the subject of Stalingrad comes up.]

JODL. In regard to the Russian communiqué, we are checking that to see if there isn't some kind of error in it. Because a single mistake—for instance, a general who couldn't have been there—would prove that everything they published was taken from a list they captured somewhere.

HITLER. They say they have captured Paulus as well as Schmidt and Seydlitz.

JODL. I'm not sure about Seydlitz. It isn't quite clear. He may be in the northern pocket. We are ascertaining by radio which generals are in the northern pocket.

HITLER. Certainly he was with Paulus. I'll tell you something. I can't understand how a man like Paulus wouldn't rather go to his death. The heroism of so many tens of thousands of men, officers, and generals is nullified by such a man who lacks the character to do in a moment what a weak woman has done.

JODL. But I am not yet certain that that is correct—

HITLER. This man and his wife were together. Then the man fell sick and died. The woman wrote me a letter and asked me to take care of her children. She found it impossible to go on living, in spite of her children. Then she shot herself. That's what this woman did. She had the strength—and soldiers don't have the strength. You'll see, it won't be a week before Seydlitz and Schmidt and even Paulus are talking over the radio.[7] They are going to be put in the Lubianka, and there the rats will eat them. How can one be so cowardly? I don't understand it.

JODL. I still have doubts.

HITLER. Sorry, but I don't.

[A grumble follows about Paulus' promotion, similar to that expounded to Zeitzler.]

I don't understand that at all. So many people have to die, and then a man like that besmirches the heroism of so many others at the last minute. He

[7] The first broadcast was on 28th May, 1943. But Hitler misjudged Schmidt.

could have freed himself from all sorrow and ascended into eternity and national immortality, but he prefers to go to Moscow. What kind of choice is that? It just doesn't make sense—it is tragic that such heroism is so terribly besmirched at the last moment.

JESCHONNEK.[8] I consider that it is possible that the Russians have reported this on purpose. They are such clever devils.

HITLER. In a week they'll be on the radio.

JESCHONNEK. The Russians would even manage to let someone else speak for them.

HITLER. No, they themselves will speak on the radio. You'll hear it soon enough. They'll speak personally on the radio. They'll ask the people in the pocket to surrender, and they'll say the most disgusting things about the German Army—

Here the fragment ends. As so often in his recorded conversation Hitler seems coarse, superficial, and vindictive. But this, surely, is no more than smoke blowing from the chimney stacks; what of the inner furnaces of this satanic genius? What were Hitler's private convictions, which he turned over in his own mind, in the darkness of his bedroom, concerning the state of the war and the prospects of the Reich? The last chance of winning a total victory had gone. The "will," whose mystique he had invoked with mixed success in the past, and which he was to press with manic fervour when defeat loomed close, was now at a discount. It was a time for reason. Time was required to develop new weapons, diplomacy to exploit the stalemate which the new weapons could achieve. Now, for a brief period, as he saw the boundaries of his victory contracting—in Africa as in the East—Hitler was prepared to allow his generals to trade space for time. In a conversation with Jodl at this period he said, "Space is one of the most important military factors. You can conduct military operations only if you have space. . . . That was the misfortune of the French. In a single drive last year we occupied more territory than in our whole Western offensive. France was finished off in six weeks, but in this huge space one can hold on and on. If we had had a crisis like this last one, on the old German border along the Oder-Warthe curve, Germany would have been finished. Here in the East we were able to cushion the blow. We have a battlefield here that has room for strategical operations."

These "strategical operations" Hitler was now to submit, for several months, to the almost uninterrupted direction of his professional military

[8] Hans Jeschonnek, General of the Luftwaffe; Chief of Staff of the Luftwaffe 1942-43.

advisers. And they, starting well enough, were to bring the German Army, before the campaigning season was half over, to its third and most serious defeat.

On 6th February, Hitler's personal Kondor aircraft touched down at the Stalino airfield with a summons for Manstein to attend a conference at the *Wolfsschanze*. The army group headquarters had been established there only five days, and during that period Manstein had been adjusting his front with an almost reckless disregard for the old *Diktate* of rigid defence which were still (theoretically) mandatory. To cut through the indecision at OKH and Hitler's own habit of not replying to a request to which he did not care to accede, Manstein had evolved the formula of attaching a report that in default of a directive from OKH by a particular time or date (on whatever subject was at issue) the army group would act at its own discretion. This discretion, moreover, was resulting in a defence of remarkable elasticity. Russochska, Kantemirovka, Millerovo, had all been abandoned as the northern end of the German line was pulled back to the Donetz.

As if to accentuate the autocratic way in which he was handling his army group, Manstein had sent another memorandum to OKH "demanding" that a withdrawal to the Mius be authorized forthwith and attaching a number of secondary points. These ranged from lavish supply requirements and the drawing off of further reinforcement from the supine Kluge to barely veiled sarcasm regarding the prospects of the SS corps in a counteroffensive that OKW had planned for them.

Manstein would have been justified, therefore, in expecting a reception at the *Wolfsschanze* which might range from the chilly to the hysterically abusive. In fact, Hitler was at his most irresistible. He started by coming as close as he ever could to an apology. The responsibility for the tragic end of the 6th Army, he told Manstein, was exclusively his.

> . . . I had the impression [Manstein wrote] that he was deeply affected by this tragedy, not just because it amounted to a blatant failure of his own leadership, but also because he was deeply depressed in a purely personal sense by the fate of soldiers who, out of faith in him, had fought to the last with such courage and devotion to duty.[9]

[9] This impression contrasts with the stenographic report of Hitler's reaction on first hearing the news, although of course it is not incompatible with his suffering remorse later.

The two men then had a long discussion concerning the advisability of withdrawing from the eastern Donetz basin. Hitler, naturally, argued against it; but he seems to have done so calmly and rationally, switching with fluency from the economic and political plane to matters of detail—where he displayed "his quite astonishing knowledge of production figures and weapon potentials"—and back again. Throughout the interview Hitler was courteous and reasonable. He took no special umbrage at Manstein's assessment of the SS corps' abilities, agreed that the Luftwaffe field divisions had proved a "fiasco," and concluded by granting permission for withdrawal to the Mius.

Emboldened by this atmosphere, Manstein proceeded to raise that most delicate of subjects—the Supreme Command. Would this not be a good moment, he asked Hitler, to ensure "uniformity of leadership"; to appoint a Chief of Staff, *whom he must trust implicitly;* such a person to be vested with "the appropriate responsibility and authority"? It is some measure of Hitler's new mood that these suggestions, too, were received calmly. He had had "disappointments," Hitler explained. Blomberg, and after him Brauchitsch, had been found wanting at moments of crisis. There were some responsibilities that could not be delegated. Furthermore, he had already appointed Goering as his successor; Manstein would not feel, surely, that the Reichsmarschall was an appropriate person to fill the post of Chief of Staff? Yet it would clearly be unworkable if the Reichsmarschall were now to be subordinated to an appointment filled from the ranks of the professional, non-Party military.

Manstein could not but agree, and the two men seem to have parted on terms of mutual confidence. If the Führer's "intuition" in military affairs had led him into some difficulty, it could not be denied that he continued to show a masterly skill in his handling of subordinates.

With Manstein satisfied, Hitler now proceeded to the next stage of the reforms he was imposing on the German Army. He had decided on a radical overhaul of the tank arm, in terms both of its constitution and its equipment. And at the beginning of February, Schmundt, Hitler's personal adjutant, had made a preliminary approach to Guderian and asked him if he would undertake this task.

It is some measure of Guderian's political *savoir-faire* (of which other examples will be seen) that he set out a number of conditions of acceptance which would give him special powers—as it is of Hitler's own anxiety that he accepted them. But before discussing the start of the special relationship between these two men, which was to survive many vicissitudes before rupturing in the last hours of Germany's *extremis,* it is necessary to examine the state and the recent history of the Panzer forces.

By the start of 1943 the Panzer force had got into very poor shape. This decline was due as much to muddle and indecision on the quartermaster and industrial side as to operational mishandling. In terms of equipment the Germans were still relying entirely on the PzKw III and IV, of which the former was totally and the latter in many respects inferior to the Russian T 34.[10] As long ago as November 1941 a team of designers had visited the front to collect data on combat experience against the T 34, and to evolve a solution to Russian technical supremacy, but the whole of 1942 had passed with very little being done to implement their decisions, owing to a continuous stream of changes to the specifications, and directives to develop new designs and variations.

At the time the inquiry started it had already been decided to equip one battalion in every Panzer division with superheavy tanks of sixty tons, and specifications for this model (the Tiger) were already out for tender to Henschel and to Krupp, where Dr. Porsche was employed. No provision, however, had been made for a "general-purpose" tank, other than the upgunning of the two standard types. Most of the officers interviewed by the commission had recommended that the T 34 be copied, with only minor modifications to allow for radio installation and a powered turret traverse. Not unexpectedly, the natural vanity of the German designers led them to reject this idea, and several precious months slipped past while new plans were drawn and sent out for tender, this time to M.A.N. and Daimler-Benz. The Ordnance Office team had in fact worked out two separate designs, one for a forty-five ton general-purpose tank (Panther) and one for a light reconnaissance tank (Leopard). Leopard never got beyond the prototype stage, but its construction and trial absorbed much time during 1942.

[10] For the 1942 campaign the PzKw III had been regunned with a 50 mm., L 60 gun, which would penetrate the armour of the T 34 at short and medium ranges. The PzKw IV had been fitted with the 75-mm. L 46, which put it on equal terms with the 76-mm. L 42 of the Soviet tank, although it remained inferior in terms of mobility.

These improvements applied to new tanks, and to those sent back to Germany for refit, but there were many obsolete versions still in service at the end of 1942.

In tracing the development of the second generation of Panzers, one finds immediately the same pattern of conflicting personalities, departmental overlap, and haphazard co-ordination which characterises every facet of the Nazi war effort.

Foremost among the civilians is the shadowy figure of Dr. Porsche, an automobile designer of some renown, who had Hitler's ear. That Porsche was a genius of a kind cannot be denied. He had designed the S and SS series of Mercedes sports racers in the 1920's, the only vehicles (literally) of German prestige in that troubled era. When Hitler gave him a free hand to make an invincible Grand Prix car in 1933, Porsche came up with the six-litre Auto-union—the most powerful single-seater ever made, before or since, and a car which only three men could handle (and two of them subsequently died doing so). Porsche was an originator, not an analyst. He thought in concepts rather than evaluating detail,[11] and this made him an "engineer" after Hitler's own heart when the Führer was indulging in his expansive and unrealistic propensity to "table talk."

But as for designing weapons . . . that was another story. Indeed, Hitler would have done better to design them himself. Porsche's plans for a Tiger were quite impractical, and the Ordnance Office threw them out, even though they came up *via* the hallowed house of Krupp. But Porsche went to Hitler and persuaded the Führer to allocate to him resources for a superheavy tank three times the size of the Tiger, to weigh 180 tons. It was also decided to allow two other engineers, Grote and Hacker, to draw up plans for a "land monitor" of a thousand tons!

At the same time Hitler was under pressure from the artillery branch to press with the development of self-propelled "tank destroyers" (*Jagdpanzer*) and infantry-support guns (*Stürmgeschütze*). The origins of this lay in the obsolescence of the towed anti-tank gun (both the 37-mm. and the 50-mm. had been found valueless against the T 34) coupled with the very natural fear among artillerymen that its disappearance from service would result in a contracting of their own field of authority. In the spring and early summer of 1942 a number of Czech 38-T tank chassis had been fitted out with 75-mm. guns and used with success in anti-tank roles. At the time this had been conceived as a temporary measure, but the crews and drivers had been taken from the artillery, and when, like so many measures that are conceived as temporary, it

[11] The best example of this, and ironically, Porsche's most enduring memorial, is the Volkswagen car, which, although brilliantly original in conception, required ten years of detailed development before it became a commercial proposition.

attained permanent establishment its origins kept it within the artillery command structure.

Production of the *Jagdpanzer* was easier and quicker than that of a tank, and Hitler saw in this a way of rapidly raising the total armoured-vehicle strength figures. In this he was encouraged by the artillery, which also persuaded him that the development of the hollow-charge shell, with its superior powers of penetration, would lead to a decline in the ascendancy of the tank.

The result of Hitler's persuasion was twofold, and each was to have very serious effects on the battles in 1943. First, Dr. Porsche, who was quick to sense which way the wind was blowing, revised his design for a Tiger and "sold" it to Hitler. The new design (known subsequently as the "Ferdinand," or at the front itself as the "Elephant") had the appearance of a giant *Jagdpanzer* with a 100-mm. L 70 gun in a fixed mounting. It had, in fact, all the disadvantages of the *Jagdpanzer*—narrow field of fire, no secondary armament, restricted accommodation—and the complications and expense in construction of a tank, including 100-mm. belly armour. At all events, Krupp got the contract for them, and over ninety Ferdinands were produced. They were all committed to action on the same day, and few weapons in modern war were to have so inauspicious a beginning, or one that had such a disastrous effect on the main operation.

Meanwhile the Henschel Tiger was proceeding through its development stage, though suffering from the diversion of resources to its Krupp cousin. An experimental battalion of Henschels was put into battle on the Leningrad front in the autumn of 1942 and showed great promise in spite of unsuitable and swampy terrain. As a result of this operation the 88-mm. L 71 was standardised in both Henschel and Krupp versions, and so the Porsche design forfeited even the paper superiority of a larger gun.

The second result of Hitler's persuasion by the artillery school had been a gradual, but steepening, decline in the quantitative tank strength of the Panzer divisions. From a maximum of four tank battalions per division during the battle of France, the number had fallen to three at the start of *Barbarossa,* and was currently only two, with the third battalion being made up of the equivocal *Jagdpanzer*. Furthermore, the number of tanks in each all-tank company had fallen from a nominal twenty-two to seventeen, and in some cases to fourteen. This was partly due to the withdrawal from service of the PzKw II's, the "tin coffins"; partly to the fact that it was almost impossible to get new tanks allocated to old formations—they were used to build up "fresh" divisions; and partly to the reluctance of local commanders to allow damaged tanks to go back as far as the main repair workshops in Germany, and the bias toward attempting makeshift repairs in divisional garages, which in turn led to a high proportion of tanks being out of service.

The end result was that the Panzer divisions seldom had a strength of more than a hundred tanks at any given time, and a more usual figure was around seventy or eighty. In terms of sheer firepower these figures might not have been so bad if the *Jagdpanzer* battalions had really been up to strength, but here the division of authority between the armoured forces and the artillery produced the expected result, and the majority of the *Jagdpanzer* seldom found their way into the Panzer divisions proper, but were used to stiffen the motorised infantry and the Waffen SS.

When this is added to the confused picture the German armaments industry still presented in 1942—until the death of Dr. Todt and his succession by Speer, Daimler-Benz was still making civilian automobiles—it will be appreciated that the appointment of the Inspector General was long overdue.

Guderian had not seen Hitler since December 1941. The Führer, who had avoided a brush with Russian cavalry at Zaporozhe two days before[12] "seemed to have aged greatly. . . . His speech was hesitant; His left hand trembled." But Hitler was bent on ingratiating himself with Guderian just as he had with Manstein. The Colonel General noted with satisfaction that ". . . my books lay about on his desk," and that he was greeted with the same apologetic, almost suppliant demeanour which the commander of Army Group South had enjoyed at the time of his own visit. "I need you," Hitler told Guderian. "Since 1941 our ways have parted. There were numerous misunderstandings at that time which I must regret." Hitler went on to say that he "had reread my pre-war writings on armoured troops and had noticed that I had even then correctly prophesied the course of future developments."

Sweet as these words must have sounded, the deeds which accom-

[12] See p. 300.

panied them spoke even louder. For every power which Guderian had asked for when the subject was broached by Schmundt was to be granted him. The Inspectorate General, far from being a subordinate department of OKH—an *Amt*—was to be answerable neither to the Training Army (as orthodox military practice would have required) nor subordinate even to the Chief of the Army General Staff (as the accepted protocol between senior officers dictated), but directly responsible to the Führer. Guderian was given the powers and seniority of an army commander; control over all armoured and mobile troops in the Army; a direct line to the Army Ordnance Office and the Armaments Ministry; and most spectacular concession of all, equivalent powers over tank forces attached to or manned by the Waffen SS and the Luftwaffe.

The interview lasted little more than three quarters of an hour, and Guderian withdrew to special quarters which had been set aside for him at Vinnitsa, to enjoy the task of drawing up, as it were, his own letters patent. One more private empire was to be carved out of the disorderly agglomeration of private and departmental estates which made up the Nazi war machine.

There was another aspect to Guderian's appointment. Hitler had exploited the enthusiasm of a technical specialist to overcome the traditional scruples of a Prussian staff officer at so flagrantly bypassing the customary channels of command. The stiff-necked unanimity of the generals would be shaken and their executive power still further eroded by one of the Führer's own nominees. Hitler must have been confident that professional jealousy, operating through the OKH machine, would go some way toward balancing the special powers he was conferring on the Inspector General—but he can hardly have expected them to operate as soon as they did.

One reason for Guderian's having been given everything for which he asked—as it was for the Führer's aged and shaky appearance—was the continuing and (it must have seemed at Supreme Command headquarters) irreversible deterioration of the German southern front. The western wing of Army Group South was so broken up by the continuing Russian pressure that on 13th February, OKH had stepped in with some new command "boundaries." The old Army Group B, over which Manstein had been attempting to exercise control for the last month, was dissolved and its entire staff organisation withdrawn to Germany. Its strongest component, the 2nd Army, was transferred to Kluge, and the remnants were put into a new force of fresh troops which was assembling at Kharkov, Army Detachment Lanz.

General Lanz had three crack SS Panzer divisions in his command, *Leibstandarte, Das Reich,* and *Totenkopf*. But as practical reinforcement for Manstein this force was valueless. There was no direct signals link with Army Group South, nor, it seemed, was the matter of establishing one treated with any urgency by Lanz. It soon became clear that the detachment was in fact a special body appointed by Hitler to deny Kharkov to the Russians, without concern for the development of the battle to its left or right.

That same week two more deep Russian penetrations had threatened to lever open the position so lately and precariously established on the Mius. A very large force of Russian cavalry, upward of three divisions with some mechanised artillery, had pushed between the Fretter-Pico group and the 17th Corps. Moving at night, and making wide detours over the frozen ground to bypass the scattered pockets of German resistance, the horsemen had emerged at Debaltsevo on the main east-west railway some forty miles behind the "lines." Here they had intercepted two trains of reinforcements for the 17th Corps and put the occupants (literally) to the sword. Their presence meant that the awkward Taganrog-Mariupol railway was the only supply route left for the entire force defending the Mius River line.

Still more serious, a strong force of tanks, three mechanised brigades from Popov's 1st Guards Army, had pushed up the frozen valley of the Krivoi Toretš (which the commander of the 40th Panzer Corps, whose left flank the valley protected, had assured Manstein was "impassable") and established itself at Krasnoarmeiskoye, within tank-gun range of the main railway from Dnepropetrovsk, along which travelled the fuel and ammunition for the whole of the 1st Panzer Army, and the Hollidt and Fretter-Pico *Gruppen*. SS Viking was immediately committed against the Krasnoarmeiskoye break-through, but its efforts were unavailing. OKH apologetically explained that the division (which was composed of SS volunteers from the Baltic and "Nordic" countries) "had such severe losses that there were no longer enough officers available with command of the appropriate languages."

In the second week of February, then, Manstein's position was as follows. He had no effective contact with his left wing, the bulk of which, in Army Detachment Lanz, was tied to Kharkov, and the Russians had virtually complete freedom of action across a fifty-mile stretch of the Donetz on either side of Izyum. The effective strength of the army group in the south and east had been divided into three by the Russian penetrations at Krasnoarmeiskoye and Debaltsevo. On 15th February the SS Panzer corps withdrew from Kharkov—in spite of orders from Hitler, and from Lanz himself, that the city was to be held to the last—and the gap between Kluge's right and the first solid positions on Manstein's left flank had widened to over a hundred miles.

Hitler can be forgiven for seeing in the apparent disintegration of the whole southern wing the inevitable consequences of departure from his principle of holding fast regardless. On the day following the fall of Kharkov, Army Group South was warned to expect an "immediate" visit from the Führer, and he arrived at Zaporozhe, accompanied by Jodl and Zeitzler, on 17th February. This was the closest that Hitler ever approached to the fighting line throughout the war (until the fighting closed over his head in Berlin, in 1945), and for three days he remained there, with only a few anti-aircraft units and the H.Q. defence company between him and the outriders of Popov's group. Packets of Soviet cavalry had already filtered through as far as the north bank of the Dnieper, and on Hitler's last day at Zaporozhe some T 34's approached to within gun range of the airfield!

The real interest of the Zaporozhe conference was in the substance of the argument between Hitler and Manstein that took place over the three days. Hitler's great strength in disputation with his marshals was twofold. First, an inspired talent for strategy on a grand scale; the Norwegian campaign and the Ardennes plan of 1940 will stand to his credit in perpetuity in military textbooks. Second, a remarkable faculty for retaining figures and for tactical detail—the rate of fire of a new mortar, the ammunition requirement of a platoon, the rail capacity of a given network, even such minutiae as the best way to site an 88-mm. gun. His familiarity with the small change of tactical discussion allowed Hitler to talk on equal terms with most of his commanders, and to attain an ascendancy over them quickly even in their own expertise.

But although Hitler could plan a campaign and conduct a regimental action, his ability as commander in the middle scale, that of handling corps and armies—grand tactics—was less sure. His first real experience had come in the winter of 1941, when he dismissed Brauchitsch and plunged into the deep end, with the whole German Army in a state of febrile emergency. Then he had discovered—not by any rational process, but from the roots of his own conviction—the solution. Hold fast, yield not a foot, die where you stand. The coincidence of strategic requirement and despotic ordinance on this occasion has already been discussed. But when Hitler applied the same formula to the predicament of Paulus at Stalingrad, the result had been less happy. The Führer had been shaken, he had been filled with remorse, and he had allowed Manstein to adopt a different technique to deal with the new crisis on the Donetz. And what was happening? Ground, it seemed, and prestige with it, was being lost at an even greater speed. The front had lost all cohesion. The fighting was "fluid," certainly, and the current was against the Wehrmacht.

So when Hitler saw Manstein at Zaporozhe, he told him immediately that the retreat must stop—as from that day. Kharkov must be recaptured immediately. (The SS Panzer divisions must surely be ready by now?) Lanz was to be dismissed, and a frontal attack mounted. The penetrations by Popov and the cavalry group must be repelled at once. Of course Viking had been unable to do more than contain Popov. Was not one exhausted SS division grossly inadequate for such a vital assignment? Again, Manstein was always grumbling about his need for more reinforcements—but more than half his forces and three quarters of his armour were not even engaged! The 4th Panzer Army had been out of action for a fortnight. The 1st Panzer Army was leading with only one division.

It is not recorded what Zeitzler and Jodl said, if they were given the chance to say anything, but it seems likely that they were brought along only to support the Führer's strictures. Yet Manstein was unperturbed. Of course the situation was grave, he told Hitler, but in its very fluidity lay the germs of success. For the Russians could not simply drive due west through the Kharkov gap. For what objective would they be aiming—Kiev? Lvov? They had neither the supplies nor the mobility nor the strength in support to get there. They *must,* Manstein lectured his audience, be intending to wheel southward and attempt to force the army group back against the Black Sea. Yet Zhukov's chances of bringing off a second encirclement were quite different from November of the previous year. Then it had been the Russians who were attacking from their own starting lines, with their dumps accumulated over the weeks, and the Germans who were exhausted, fought to a standstill at the far end of uncertain communications. Now the Russians were worn out, their tanks had travelled hundreds of miles, the country behind them was ravaged and desolate, and within weeks would be rendered almost impassable by the thaw. When their spearhead lunged, as it must, toward the crossings of the upper Dnieper, then Hoth would be let loose again; then the three SS divisions could play their rightful role as avengers, and strike southeast to meet the 4th Panzer Army, catching the Russian armour in a noose. There would just be time to inflict this defeat before the thaw started, and in its wake Kharkov would fall "like a ripe apple." Thus during the thaw, the six-week respite when the better part was an impenetrable morass, the cohesion of the Wehrmacht could be restored.

sixteen | THE "CONSOLIDATION PERIOD"

In the last days of February the Russian offensive touched high water mark, and although the edge of the tide continued to creep forward in places the ebb force, as Manstein had predicted, had already begun. The front had moved over two hundred miles to the west in less than three months, and as they withdrew the Germans devastated the whole countryside and razed the towns.

> The Germans had burned villages down to the ground, laid low the orchard trees, trampled on the cultivated fields, effaced every sort of evidence of human occupation. In the farms they had taken the ploughs reapers, mowers, made a pile of them and then blown them up.

It was the Russians' first experience of an offensive war of movement on a large scale, and they were finding it very different from the fluid fighting of 1941 and 1942, when they had been falling back toward their dumps and railheads; when the forward troops had always been travelling to meet, instead of away from, supplies and reinforcements. The weather, the devastated communications, and their own inexperience in maintaining the traffic density required to support a deep penetration on a narrow front had combined to force a dangerous dispersal of effort on the Russian advance, which had broken down into four separate groups.

In the north, the two armies which had captured Kharkov, Rybalko's 3rd Tank Army and Kazakov's 69th, plodded westward, clashing intermittently with the Kempf detachment. South of Kharkov two more

Soviet armies, the 6th and the badly exhausted 1st Guards, were strung out down the long corridor they had opened between Izyum and Pavlograd, with their leading cavalry and a few stranded tanks reaching as far as the Dnieper near Zaporozhe. Still farther to the east the Popov group was in league around Krasnoarmeiskoye, with only fifty tanks left as runners out of four corps. Two other groups lay behind the German lines, the cavalry divisions at Debaltsevo and some infantry which had forced their way across the Mius at Matveyev.

At the beginning of the month, when both armies were almost "out on their feet" and the Germans seemed in the worse state, the risks of this dispersal were justified. If the Red Army could keep up the pressure until the thaw (went the *Stavka* calculation), then, in that quiet period, the Germans would straighten their line according to the normal dictates of military prudence and the many Soviet salients won in the last weeks would merge into a spectacular territorial gain. Manstein's remarkable coolness in thinning out his front to well past the accepted danger limit, in order to conserve his remaining armour, was something for which no allowance had been made.

For five days Hoth's two corps, the 48th and 57th Panzer, moved northwest to the line of the Krasnoarmeisk railway, while below Kharkov the SS Panzer corps completed its assembly. On 21st February, Hoth began to attack the Popov group and the left flank of the 1st Guards Army, and on the 23rd the SS began their drive southeast. Under this converging pressure the exhausted Russian armies began to disintegrate and retreat in an easterly direction in individual units. The appearance of the Tiger, with which SS *Das Reich* and *Gross Deutschland* had lately been equipped, was a further blow to Russian morale. Hitherto the only gun which could cope with the T 34 under all conditions had been the 88-mm.; but it suffered from vulnerability in a static role and while being manoeuvred into position. Now, mounted in an armoured turret, and on tracks so broad that it could travel where even the T 34 hesitated, it marked the end of the Russian tank's role as the undisputed queen of the battlefield.

Within a week the German pincers had closed with a meeting between the SS and Hoth's tanks. The Germans' shortage of infantry made it impossible for them to seal the pocket completely, and many of the Russian soldiers slipped away on foot or horseback to cross the still frozen Donetz at unguarded points. The Germans claim to have counted

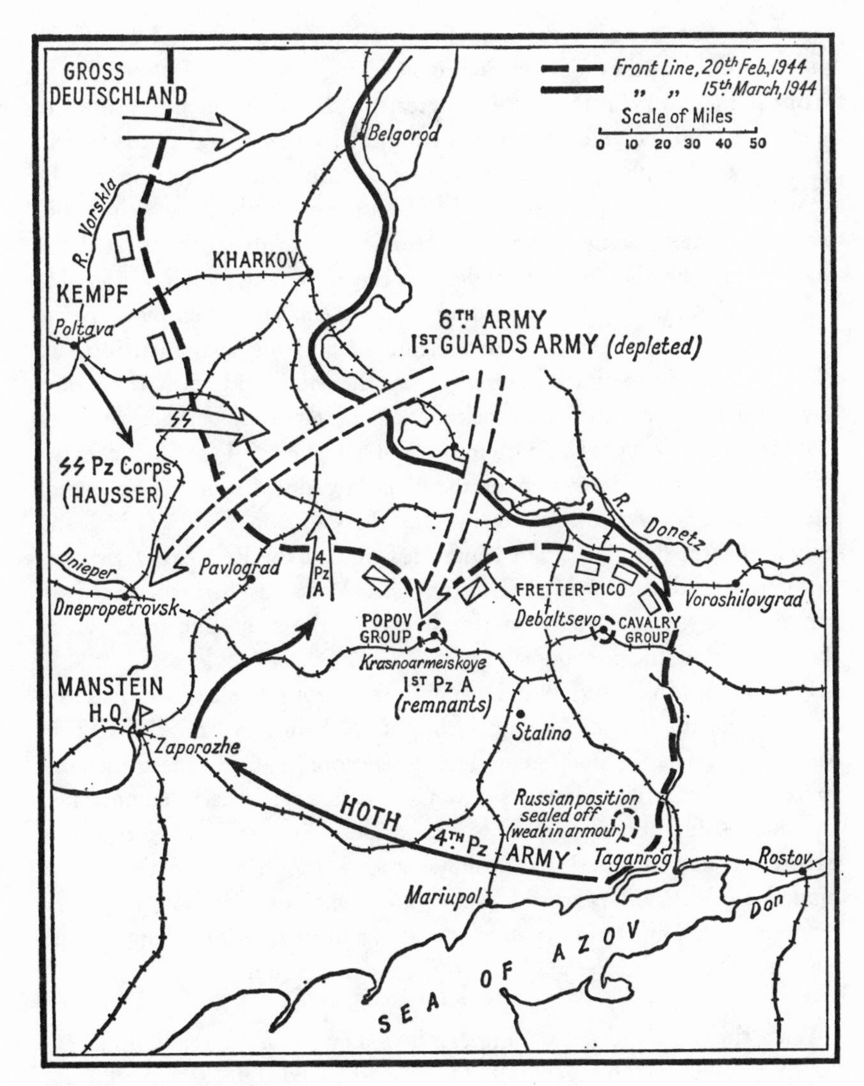

The "Miracle of the Donetz"

23,000 Russian dead on the battlefield and to have captured 615 tanks and 354 guns, but they took only 9,000 prisoners.

But if many of the Russians had escaped, they were in no position to block the continued advance of the combined Panzers and SS. Manstein at first toyed with the idea of forcing an immediate crossing of the Donetz and driving up the line of the Kupyansk railway, which would place his armour some eighty miles *east* of Kharkov, then rejected the plan in case he should be stranded by the thaw. Instead he ordered a closer pincer on the town, sending *Gross Deutschland* around to the north with a reinforced Kempf detachment and the combined force of Hoth and the SS to attack the town from the south and rear. After some days of hesitation the *Stavka* acknowledged (indeed, perhaps overestimated) the severity of Manstein's counterstroke, and its one concern became to avoid a repetition of the Kharkov disaster of the spring of 1942. Kharkov was evacuated by the Russians on 13th March, and Belgorod three days later. The southern fronts of the Red Army were reformed behind the melting Donetz, and Manstein had succeeded in bringing the Germans back to almost exactly the same line from which they had set out the previous summer.

Few periods in World War II show a more complete and dramatic reversal of fortune than the last fortnight in February and the first in March 1943. The German Army had done more, it seemed, than demonstrate once again its renowned powers of recovery; it had demonstrated an unassailable superiority, at a tactical level, to its most formidable enemy. It had repaired its front, shattered the hopes of the Allies, nipped the Russian spearhead. Above all, it had recovered its moral ascendancy. Already OKH began to consider, and the *Stavka* to contemplate with apprehension, a fresh German offensive in the summer.

With this in mind, and confident in a new, relaxed relationship between himself and his generals, Hitler had acceded to a request from Kluge that he pay a visit to the headquarters of Army Group Centre.

It was nearly two years since the Führer had visited the army group,[1] but the personnel at headquarters had changed little—save for the substitution of Kluge for Bock. In 1941 a few of these young officers were toying with a harebrained scheme for abducting Hitler. In March 1943 these same officers were still there, but older and wiser, they had

[1] See Ch. 4.

evolved a scheme altogether more systematic and more violent. The whole weight of the conspirators' machine, with all its ramifications in OKH departments and in the field, was to go into their attempt, whose failure must be reckoned one of the great tragedies of World War II.

All the conspirators were now agreed that Hitler must be assassinated. "Deposition," or some other such formula, was no longer practical, not only because of the physical difficulties it presented, but because no officer of the General Staff having active command would consider taking sides until "the living Führer," to whom he was bound by oath, had ceased to exist. Yet the only element in the state which had sufficient force to kill Hitler and deal with his bodyguard was the Army; and the only section of the Army where this was really feasible was Army Group Centre, at Smolensk. Another difficulty had vexed the conspirators, but they believed they had found the answer. This was the problem of taking over the central administration at Berlin in the period of confusion which would follow the Führer's death. For they now realised that even if a *coup* were successful, in the dark forests of Russia it would carry no more menace to the Nazi regime in Berlin than did the revolt of a Byzantine legion to the court of Imperial Rome. But by the beginning of 1943 plans had been prepared by Oster (now a general) and Olbricht, who was Chief of the *Heeresamt* on the staff of the Home Army, for a simultaneous take-over by the army in Berlin, Munich, and Vienna. All that was required, Olbricht told Tresckow in February, was the "Flash."

Tresckow had been working on the conscience of the nervous and malleable Kluge ever since the Field Marshal's appointment to the command of the army group. Kluge had inherited Schlabrendorff as his A.D.C. from Bock, and the young major, while browsing in his superior's files, had turned up a letter from Hitler referring to a large check, and a building permit, which the Führer had bestowed upon Kluge in October 1942. Schlabrendorff immediately told Tresckow, and the two joined in exerting spiritual blackmail on the miserable Field Marshal, which inflamed his troubled conscience and accentuated his habitual indecision. That winter they had even persuaded Kluge to meet Goerdeler, and the learned doctor travelled to the front (with papers provided by the Abwehr) and delivered a long and persuasive discourse on the aims of the conspirators and the necessity of eliminating Hitler.

That Kluge in fact invited Hitler to Smolensk with the intention of assassinating him seems very unlikely. More probable is the traditionally ambivalent attitude of the General Staff: If the Führer went unscathed, a "normal" visit would do Kluge and his interests no harm (Kluge cannot have been relishing the unusual attention which had lately been lavished on Manstein); if the coup were successful, then not only would he emerge as Commander in Chief, but he would arrive there with his honour unsullied. Indeed, there is some doubt whether Kluge even took the risk of personally inviting Hitler at all. Tresckow claims that he suggested it to Schmundt, who persuaded Hitler.

Once the Führer's plans were known, the first stage of the "Flash" operation began. Admiral Canaris, with his (as it seems) customary disregard for what was probable or prudent, summoned a "Conference of Intelligence Officers" at Army Group Centre. Kluge appears to have put a brave enough face on this, but he cannot have felt too happy to see Canaris' aircraft disgorge a veritable posse of anti-Party figures on Smolensk airfield less than a week before the date fixed for Hitler's arrival. One of them, General Erwin Lahousen, had brought with him a batch of special British time bombs with three different types of fuses, in case direct means of "execution" should fail. And fail, of course, they did—because of Kluge's last-minute change of heart.

The original plan had been that Hitler was to be shot by Lieutenant Colonel Freiherr von Boeselager and his fellow officers of the 24th Cavalry Regiment, a crack unit billeted at Smolensk at that time. Boeselager claimed that all the men of this unit were reliable and that they would be able to overpower the SS bodyguard. But no sooner had Canaris and his colleagues left, with the plans for the Berlin end of the conspiracy perfected, as they thought, to dovetail with the projected *attentat* at Smolensk, than Kluge lost his nerve. He could not authorise Boeselager's action, Kluge told the shattered Tresckow,

> *at this time,* because neither the German people nor the German soldier would understand such an act . . . we ought to wait until unfavourable military developments made the elimination of Hitler an evident necessity.

But Tresckow and Schlabrendorff were determined to kill Hitler, and Lahousen had brought them the means. For two days they "practiced" on a disused Red Army firing range, and learned to master the particular type of explosive they intended to use. It was composed of compact slabs of nitrotetramethanium, which came in sections, like a child's building bricks. The sections could be "built up" to construct a bomb of whatever force was required—they had been developed by the British for parachuting to saboteurs of the Resistance—and were actuated by three types of fuses: a ten-minute one (subsequently used by Stauffenberg in the *attentat* of 20th July), a half-hour, and a two-hour setting. Tresckow first inclined to depositing a ten-minute bomb in the conference itself, but both he and Schlabrendorff agreed that this would be a needlessly extravagant way of achieving their purpose, and one which carried the disadvantage that a number of Regular Army officers who were intended to assume executive positions in the new regime (including Kluger Hans) might be killed. No, the most satisfactory solution, they agreed, would be an air accident; this would leave, for a matter of hours at least, an element of doubt in the situation, which the Berlin section would be able to exploit; the uncertainty over the details of Hitler's end would be a comfort to many who shrank from the crude reality of an assassination.

After the usual postponements Hitler and his entourage appeared at Smolensk on 13th March. There were two aircraft, a personal bodyguard of twenty-five SS men, Dr. Morell, Fräulein Manziali (the vegetarian cook), and a mass of different categories of adjutants and couriers. During the afternoon Schlabrendorff had an experience which still more sharply emphasized the care with which Hitler protected his own life and the magnitude of the task which faced the conspirators. While the Führer and Kluge were standing at the wall map, Schlabrendorff wandered over to the table where Hitler had placed his cap on entering the hut; idly picking it up, the young A.D.C. was staggered to find it "heavy as a cannon ball." It had been lined with four pounds of hardened tungsten steel, including the peak, as a protection against snipers in upper windows!

When the time came for the Führer's party to leave, Tresckow approached one of Hitler's staff officers, Colonel Brandt, and asked him a favour. Could the Colonel take back with him to East Prussia a couple of bottles of brandy which Tresckow had promised to a friend of his at Rastenburg, General Helmuth Stieff? Brandt raised no objection, and said that he would. Kluge and Tresckow travelled in the same car as Hitler to the airfield; behind came Schlabrendorff with the "brandy bottles." There were two Kondor aircraft waiting, but to their relief, the conspirators saw Brandt making for the Führer's own, while most of the SS bodyguard travelled in the other. At the last moment Schlabrendorff actuated the fuse (it was at a half-hour setting) and handed the parcel up to Brandt. The door closed and the aircraft trundled down the runway. The two officers watched it become airborne and disappear into the grey cloud banks toward East Prussia.

The minutes ticked past. Tresckow had telephoned Berlin, spoken to Captain Gehre, who was in turn to repeat the password to Oster and Olbricht. On a half-hour fuse they calculated that the explosion would occur just as Hitler's aircraft approached Minsk, and would certainly be announced over the radio by one of the fighter escort. But Minsk receded, and Vilna, and Kaunas (Kovno), and after two hours a routine teleprint announced that the Führer's plane had landed safely.

> A great hope fell.
> You heard no noise,
> The ruin was within.

What a heart-stopping moment for Schlabrendorff! Truly the imagination quails at the prospect which opened before him. Had the bomb been discovered, perhaps accidentally, and neutralized? Or had it simply failed, and if that was the case what would happen to the parcel? Heaven knew, there were enough reasons why Brandt, an ardent Nazi and quite outside the plot, might unwrap the "bottles"—either the better to present them to Stieff or, perhaps, to sample one with a few friends. With these thoughts racing in his mind, Schlabrendorff's first concern was nonetheless for his colleagues in Berlin, and he immediately telephoned Berlin to tell Gehre that the operation had miscarried. Next, he had to decide how to cover his own traces. Schlabrendorff could be forgiven if, sensing already the Gestapo's torture instruments on his flesh,[2] he had been concerned solely with his own safety. It might

[2] After his arrest (in 1944) Schlabrendorff was taken to the Gestapo prison in the Prinz Albrechtstrasse.

> One night I was taken out of my cell . . . in the room to which I was taken there were four people, Commissioner Habecker, his woman secretary, a sergeant of the SD in uniform and an assistant in plain clothes . . .
>
> The torture was applied in stages. First my hands were tied behind my back; then a contrivance was applied to both hands, which gripped all ten fingers separately. On the inner side of this instrument were spikes, which pressed against my finger tips. The turning of a screw caused the machinery to contract so that the spikes penetrated into the fingers.

just be possible, with help from Canaris' Abwehr, to slip out of the country either to Sweden or Switzerland before the hue and cry went up. But no such thought seems to have crossed his mind. Instead he put through a telephone call to Brandt at Rastenburg. As he was put through, Schlabrendorff, not trusting his own voice, handed the receiver to Tresckow. The parcel? No he hadn't delivered it yet, said Brandt, it was lying about the office somewhere. Should he . . . ? No, replied Tresckow, it would be best if he kept the parcel in his possession, there had been a mistake, Major von Schlabrendorff would be travelling to headquarters tomorrow and would bring the right one . . .

The next day Schlabrendorff took the regular courier plane to Rastenburg with two real bottles of brandy, which, he could not but notice, weighed more than, and looked different from, the parcel in Brandt's possession. He immediately called on Brandt.

> . . . I can still recall my concern when [he], unaware of what he held, smilingly gave me the parcel and gave it a jerk that made me fear a belated action. Feigning a composure that I did not at all feel, I took the parcel and drove to the neighbouring railway junction of Korschen.
>
> From Korschen a sleeper train left for Berlin in the evening. I got into a reserved compartment, locked the door and with a razor blade opened the packet. Having stripped the cover, I could see that both explosive charges were unaltered. . . . I dismantled the bomb and took out the detonator. When I examined it I found to my great surprise what had happened. The fuse had worked; the glass globule had broken; the corrosive fluid had consumed the retainer wire; the striker had operated; but —the detonator cap had not reacted.

The Devil's hand had protected Hitler.

To persons not of the Roman Catholic faith, especially to those such as scientists, technicians (and even military historians), who are concerned with fact and reality, the purported existence of cosmic forces can seem an irritating abstraction. Yet there are occasions when the eternal struggle between good and evil seems more than a convenient adjunct to a code of behaviour evolved by the priesthood for disciplining the lower classes and assumes a disquieting magnitude, which towers over the puny "self-determination" of mortal man.

The second torture was applied in the following way: I was strapped face down on a frame resembling a bedstead and my head covered with a blanket. Then instruments like stovepipes were shoved up over my bare legs with nails fixed on the inner side. Once again a screwing mechanism contracted these tubes, so that the nails bored into my thighs and shins.

With the onset of the muddy season in the middle of March, OKW had time to contemplate its projected strategy for 1943. For the first time in twenty years Hitler was silent. He had no ideas. Looking back on Hitler's conduct in this period, we can see 1943 as a plateau of reason and orthodoxy standing between the extravagant ambitions of the post-Munich period and the nihilist defensive with which the war ended. Even in his private circle Hitler spoke little of grand strategy, but discoursed at length on the new weapons which would restore the military ascendancy of the Reich. He had conceived no grandiose tasks for the Army, save to preserve what it had already won, while the war machine of the Reich was injected with the enormous human and mineral wealth of the conquered East and "the new Europe."

This attitude suited the generals very well. It meant that they could plan a battle without having to feel that its course and direction lay at the whim of some vague political or strategic concept of which they had been told nothing. Manstein's first plan, which he had broached to Hitler as early as February, was to wait for the Russians' summer offensive to develop and hit them "on the backhand." He envisaged giving up the whole Donetz basin and luring the Russians as far as the lower Dnieper. The whole weight of the Panzer force would then strike southeast from Kharkov and pin the attackers against the Sea of Azov. On that occasion Hitler was already being compelled to swallow a lot of strong medicine, and this plan (expounded before Kharkov had even been recaptured) was too much for him to stomach. He rejected it on vague political grounds—the effect on Turkey and Rumania—and although the plan cropped up at intervals during the spring it gradually fell out of favour before an alternative "forehand" (i.e., anticipating a Russian offensive) stroke farther north.

Manstein's "backhand" plan was brilliant in conception, and might well have resulted in one of the most classically perfect battles of *riposte* ever fought. Its chief risk was not the political one, but the danger that the Russians might strike too far north—at Belgorod, for example, rather than at Izyum—and force an armoured battle before the Germans were ready for it. With this in mind, a majority of professional opinion began to build up for a more limited (and obvious) action against the "Kursk salient," a huge and menacing-looking bulge in the German line, more than a hundred miles square, which separated the fronts of Manstein and Kluge.

The first essential, before these plans could be given any body, was that the Panzer force be invigorated. For although the Tiger had shown itself to be a magnificent tank and the Panther promised well, every officer who had taken part in the Donetz battle recognised that it had been a special case. New tanks, harder training, and improved tactics were going to be needed to cut through the Russians' numerical superiority, just as they had in 1941, for the production of the Reich could never match that of Russian factories, now settling down after the upheavals of the 1942 evacuation.[3]

The man who carried the responsibility for the new policies in the development of the Panzers was Colonel General Guderian. Hitler had granted him exceptional powers, but it was not long before professional jealousy, aroused, it cannot be doubted, as much by the nature of the powers as by the manner of their conferment, started to thwart Guderian's scheme. The opening paragraph of the Assignment of Duties that Guderian had drawn up for Hitler's signature read:

> *The Inspector-General of Armoured Troops is responsible to me for the future development of armoured troops along lines that will make that arm of the Service into a decisive weapon for winning the war.*

A footnote then defined "armoured troops" as

> tank troops, rifle components of *panzer* divisions, motorised infantry, armoured reconnaissance troops, anti-tank troops and assault gun units.

Together with Guderian this definition had been agreed upon by Jodl, Zeitzler, and Schmundt, all of whom were consulted for the first draft, which was drawn up at Hitler's headquarters at Vinnitsa, on 21st February. On the 22nd, Guderian had flown to Rastenburg, where the majority of the OKW headquarters staff were still located, and shown the draft to Keitel and Fromm. Their reaction has not been recorded, but between 22nd February and the 28th, when the formal copies went back from Rastenburg to Vinnitsa for Hitler's signature, someone at Rastenburg had altered the footnote so that it now read:

[3] In 1942, German tank production had been 4,280. In 1943, 6,000. In 1944, when the reforms inaugurated by Speer and Guderian had taken full effect (and in the teeth of heavy bombing by the West), production rose to 9,161. The Russian figure for 1943 was 11,000; and for 1944, 17,000. But the latter figure may have included self-propelled guns, and if these are included in the German totals the discrepancy between the two rates is not as marked as is often claimed. Production rate of *Stürmgeschütze* (assault guns) and *Jagdpanzer* (self-propelled anti-tank guns) was 778 in 1942, 3,406 in 1943, and 8,682 in 1944. (Ogorkiewicz 217-18.)

> . . . anti-tank troops and *heavy* assault guns.

Thus at one stroke the Inspector General was confined to the self-propelled 88-mm. on the Panther chassis, and the mass of 75-mm. on Mark IV and 38-T chassis—which made up 90 percent of assault gun production—was kept outside his authority.

It was a foretaste of (in Guderian's words)

> difficulties and lack of co-operation from certain quarters [i.e., the General Staff] which occurred over and over again.

The protocol of the German Army dictated that the departments of OKH be subordinated to the Chief of Staff, whose permission had to be obtained for each and every visit to other units. Department heads exercised no influence over the Training Army and the schools, nor were they allowed to publish any written material. Guderian's task was not made any easier by the fact that his own views as to the conduct of operations were in opposition to those of practically every other senior officer in the Army.

In an effort to win support, Guderian asked Hitler for a conference at which he could expound his views to the Führer and a few senior commanders, and his notes for this conference open with a categorical assertion:

> The task for 1943 is to provide a certain number of *panzer* divisions with complete combat efficiency capable of making limited objective attacks.
>
> For 1944 *we must* prepare to launch large-scale attacks.

Thus he could expect opposition, not only from all the specialists who felt their own departmental authority threatened, and from the General Staff, which resented his extraordinary powers, but his whole audience could unite in protest before the Führer at the pusillanimity of allowing a whole year to pass with nothing more heroic than "limited objective attacks."

Furthermore, Guderian had, most imprudently, sent a précis of his notes to the office of Hitler's adjutant in advance of his arrival. The result was that on his arrival at Vinnitsa, on 9th March, he had to face not a small and select audience but a large and disgruntled General Staff clique, including the senior officers of infantry and artillery (who were particularly concerned to retain their authority over the assault guns), and many other artillery officers. Panzer commanders were conspicuous by their absence.

Undaunted, the Inspector General proceeded with his address, which included reading aloud an article by Liddell Hart on "Armour for the Attack"![4] "A Panzer division," he told his listeners, "only has complete combat efficiency when the number of its tanks is in correct proportion to its other weapons and vehicles." The original composition of a Panzer division, Guderian reminded them, was four tank battalions, and it was at this target, of four hundred tanks in each division, that he intended to aim.

> It is better to have a few strong divisions than many partially equipped ones. The latter type need a large quantity of wheeled vehicles, fuel and personnel, which is quite disproportionate to their effectiveness; they are a burden, both to command and supply, and they block the roads.

Guderian had already complained to Hitler about the alteration of the definitive footnote in his Assignment of Duties. He now tried to convince his audience that in order to attain the required figure of armoured strength in individual Panzer divisions, it would be necessary to draw in the self-propelled guns from non-Panzer formations, and to direct all future production into the Panzer divisions proper. But at this suggestion

> . . . the whole conference became incensed. All those present, with the single exception of Speer, disapproved, in particular of course, the gunners; Hitler's chief adjutant also spoke up against me, remarking that the assault artillery was the only weapon which nowadays enabled gunners to win the Knight's Cross.
>
> Hitler gazed at me with an expression of pity on his face, and finally said, "You see, they're all against you. So I can't approve either."

In this way the development of the Panzer force was stunted at the outset. But in spite of this the tactical reforms inaugurated by Guderian and Speer's overhaul of the production machine combined faster even than Hitler expected to build up a very formidable striking force. By early summer all the Panzer divisions in the south had been taken out of the line and rested, and in battalions which were being re-equipped with the Tiger and the Panther the crews had been sent back to Ger-

[4] The article had first appeared in the *Daily Mail* of 21st December, 1942. Guderian's notes for this conference offer a salutary corrective to the view that the German Army was "finished" after Stalingrad.

many for training and familiarisation courses at the factory. The result was a marked revival of morale and confidence—although it helped to defeat Guderian's larger purpose, as it was grist to the mill of those who could not wait before launching another major operation. But before examining the background to these decisions (which Guderian himself consistently opposed from the start) it is necessary to look at another aspect of the German "consolidation program" of 1943.

After the first period of haphazard oppression and brutality the German attitude to the occupied East became more systematic. As the probable duration of the war lengthened into the distance, so the attractions of the East as a vast reservoir of slave labour had grown, and in March 1942 a special office of labour "allocation," the *Generalbevollmächter für den Arbeitseinsatz,* henceforth referred to as the GBA, was established. The GBA started off as a subsidiary of the Four-Year-Plan, but soon grew to be a sovereign power in its own right,[5] competing with the rival authorities of SS, *Ostministerium,* and military government to multiply the administrative chaos which prevailed in the whole territory, and whose overworked relief valve was the suffering of the Russian people. Head of the GBA was Fritz Sauckel, a second-string Nazi who had missed out on the first round of appointments and was intent on making up for his late start. Sauckel had originally been recommended by Rosenberg for that plum of the "golden pheasants," the Ukraine, but had been defeated by Koch. Once installed at the GBA, he lost no time before ordering his erstwhile protector about:

> I must ask you to exhaust all possibilities for speedily shipping the maximum number of men to the Reich; recruitment quotas are to be trebled immediately.

To a Gauleiter conference Sauckel declared:

> The unparalleled strain of this war compels us, in the name of the Führer, to mobilise many millions of foreigners for labour in the German total war economy, and *to make them work at maximum capacity* . . .
> You may rest assured that in my measures . . . I am guided by neither sentimentality nor romanticism . . .

Familiarity with the utterances of Nazi executives allows us to recognise that translucent veneer which precedes and accompanies measures of

[5] The politics of Nazi administration in the early days of the war are discussed in Ch. 3.

the most horrific brutality—and the actions of the GBA are no exception. Masses of Russian civilians were swept up at random and press-ganged into unheated freight cars with neither food nor sanitation. They were given no time to collect their belongings or even to tell their families, and many, need it be said, died on the journey

> after being exposed to arbitrary abuse by the accompanying German personnel.

The result was that still more of the population went underground and joined the Partisans, and passive hostility to the Germans became universal. The administration retaliated by treating "evasion" of the labour draft on a par with Partisan activity. An OKW report dated 13th July, 1943, speaks of

> an intensification of countermeasures: among others, confiscation of grain and property; burning down of houses; forcible concentration; tying down and mishandling of those assembled; forcible abortion of pregnant women.

"The population reacts particularly strongly," it was woodenly recorded, ". . . against the forcible separation of mothers from their babies, and of school children from their families." In the Ukraine, province of the sadistic Koch, scenes of not so much mediaeval as of pre-Roman barbarism were a daily occurrence as columns of "labour volunteers" were driven through the streets to the marshalling yards by guards with stock whips.

> Relatives of departing workers were not allowed to hand them food and clothes, the crying women being ruthlessly pushed into the muddy streets with rifle butts.

The requirements of the GBA began almost immediately to clash with the activities of the SS, and each organization blamed the other for "deficiencies"; their complaints echoing higher and higher into the Nazi stratosphere, where they further poisoned and obscured the personal relations of the Diadochi. SS General Stahlecker had to submit a report on the fact that only 42,000 Jews had been killed out of a total of 170,000 registered in the area of operations allotted to *Einsatzgruppe A*. Apologetically he offered a double excuse:

> . . . the Jews form an extraordinarily high proportion of specialists who cannot be spared because of the absence of other reserves. Furthermore, Einsatzgruppe "A" only took over the area after the severe frosts set in, which made mass executions more difficult to carry out . . .

With the advent of warmer weather, though, it seems that this particular problem was eliminated, for Kube was able to report, "In the past ten weeks we have liquidated about 50,000 Jews . . . In the rural areas of Minsk, Jewry has been eradicated without jeopardising the labour situation."

Nonetheless, the "labour situation" was deteriorating, owing to the increasing failure of the GBA to fulfil its quotas through the local commissariat machinery. To Sauckel's complaints the Reich commissars excused themselves by blaming the SS, whose "reckless" anti-Partisan activity was alienating the population. By Directive No. 46, of August 1942, Hitler had formally taken the responsibility for "order" in the zones of the civil administration out of the province of OKW and granted it to Himmler. Himmler had delegated the task to SS General von dem Bach-Zelewski,[6] an energetic thug for whom the Führer had a particular admiration. "Von dem Bach is so clever," Hitler would say, "he can do anything, get around anything." (It should be pointed out that what we understand by "clever" Hitler would describe as "artistic"; he used to say that Goebbels and Speer were "artistic." The nearest translation which we can get for Hitler's "clever" is "unscrupulous.") After the disastrous winter of 1942-43, when the Partisans were in control of huge areas of central Russia, Bach had instituted a series of "drives" in which tracts of "suspected" territory were systematically laid waste, the villages burned down, and the inhabitants killed on the spot.

Even Kube was put out by this. How could he fulfil his labour obligations, he asked, under such conditions?

> The political effect of this enterprise on the peaceful population is disastrous, because of the shooting of so many women and children.

The pacific Lohse was even more outspoken:

> Should this take place without regard for age and sex and economic interest, for instance the Wehrmacht's need for specialists in armament plants?

The SS was not going to stand still while a few weak-willed civilians attempted to saddle it with the results of their own incompetence. By the transmission of orders directly to GBA officials, instead of passing

[6] And see p. 391, *et seq.* Readers will be interested to learn that Bach was not brought to trial until 1951. A German court in Munich then gave him a ten-year "suspended sentence."

them back to Himmler and across to Sauckel, the SS was trying to confirm its supreme authority with one more intruder in the occupied area. At Christmas 1942 "Gestapo Müller" had drafted an order, the opening sentence of which still chills the blood:

> For reasons of military importance, which need not be elaborated further, the RFSS and Chief of the German Police has ordered that by the end of January 1943 at the latest, at least 35,000 able-bodied persons be transferred to concentration camps. In order to attain this figure, beginning immediately . . . workers who have tried to escape or violated their contracts [sic] are to be delivered as fast as possible to the nearest concentration camp.

The "reasons of military importance" were the desire for more subjects for experiment. These experiments were medical and sexual, of the crudest kind, and it may be thought that 35,000 more than exhausted the various permutations that science could devise for their suffering. However, by this time the labour situation was so critical that the GBA was extremely reluctant to release this number, and Sauckel was coming under fire from another quarter.

To the practical men of OKH the indiscriminate oppression of the civilians in the rear areas was a constant source of irritation. Furthermore

> . . . if labour recruitment continues, the danger could arise that Army and local war economy demands will no longer be met. It is also necessary to leave certain labour reserves in the Army Group Area for a possible building of fortifications.

Kleist even went so far as to draft a special memorandum ordering that recruitment be solely on a voluntary basis—and took steps to see that "voluntary" meant what it said. This drew an immediate howl of protest from Koch, who made Sauckel send Hitler a telegram:

> Unfortunately several Commanding Generals in the East have forbidden the labour conscription of men and women in the conquered Soviet territories—for political reasons, as Gauleiter Koch informs me. [This was a lie.] My Führer! I ask you to countermand these orders so as to enable me to carry out my assignment.

No sympathy need be wasted on Sauckel, a typical representative of the Nazi breed of administrator, of whom too few shared his ultimate fate—judicial hanging. But it must be conceded that his job was not a happy one. Under pressure from four sides; from Speer in the Armaments

Ministry, from Goering under the Four-Year Plan, from the SS and the army commanders, he soon found himself falling from favour at the Nazi court. In his diary for 24th April, 1943, Goebbels recorded:

> Speer came in late in the afternoon. He remained until the evening, which gave me a chance to discuss the general situation fully with him. He has had a long talk with Goering, and what he reported to me about it was all to the good. Goering has so far adhered to our line and intends to do so in future. Speer, however, also reported that he still gives the impression of a rather tired man.
>
> Speer told me about a so-called manifesto that Sauckel addressed to his organisation within the Reich and the occupied areas. [This document appears to have been a more or less typical combination of bombast and special pleading concerning the GBA's achievements and failure to meet all demands put upon it.] . . . This manifesto is written in a pompous, terribly overladen, baroque style. Sauckel is suffering from paranoia. When he signs off with the words "Written on the Führer's birthday in an aeroplane over Russia," it smells . . . It is high time that his wings were clipped.

Superimposed on the chaotic interplay of administrative rivalries were the crackpot (but tenaciously held) views of racial propriety. With so large an influx of *Untermenschen* into the homeland while the Army was abroad, the danger of "polution" could not be discounted. Workers were being imported in about equal proportion of men and women. Girls between the ages of fifteen and twenty-five were (when they survived "mishandling" on the journey to their destination) put into sexual thrall, either in army and SS brothels, or at "rest centres" in the Reich. Others were sent to concentration camps, where the proclivities of the guards determined whether they suffered normal sexual abuse or "experimentation" ending in death. Still others were sent *au pair* (an early, and stringent, example of this arrangement) to German homes, where their fate depended entirely on the whim of the householder. Women over twenty-five were drafted directly into industry, where they were expected to fulfil the dual role of workers and doxies to the alien male population.

Nonetheless, a risk remained that there might be some contamination of German women. The *Ostministerium* had early directed that

> Prisoners of Asiatic features should in no case be brought as labour to the Reich.

And sexual intercourse with German women was punishable by death. However, as the normal scale of punishment for foreign workers acknowledged no graduation between whipping and deprivation of rations at one end of the scale and death at the other, it is hardly surprising that this deterrent was not always effective. From time to time reports of "scandalous" cases are to be found in the German newspapers.

An element of tragedy pervades the German defeat in *Zitadelle*. Perilously close to annihilation at Moscow, dreadfully mauled at Stalingrad, this magnificent army had twice recovered. Now, once more invigorated, overhauled, equipped with new and formidable weapons, it was to throw away the prospect of victory in a series of trivial errors and miscalculations whose sum was disastrous. If "tragedy" is too strong a word, no observer can avoid a sense of frustration at the persistent abuse of this wonderful machine. And so it is all the more important to remember that just as the Nazi state rested on a basis of total brutality and corruption, so the parts of the army machine, the actual weapons with which the soldiers fought, Tigers, Panthers, Nebelwerfers, Solothurns, Schmeissers, came from the darkened sheds of Krupp and Daimler-Benz; where slave labour toiled eighteen hours a day; cowering under the lash, sleeping six to a "dog kennel" eight feet square, starving or freezing to death at the whim of their guards.

seventeen | THE GREATEST TANK BATTLE IN HISTORY

Of all the operations in World War II none is so evocative of 1914-18 as the German attack on the Kursk salient, the ill-fated *Fall Zitadelle* in high summer of 1943. Rightly acclaimed as the greatest of all tank battles —at its height there were close to three thousand tanks on the move at the same time—it was from first to last a colossal battle of attrition, a slugging match which swayed to and fro across a narrow belt of territory, seldom more than fifteen miles deep, in which mines, firepower, and weight of explosives (rather than mobility and leadership) were the decisive factors. There is another feature of the offensive, which broke the Panzer force and irrevocably handed the strategic initiative to the Russians, which evokes the Great War, and this is the procrastination and argument which preceded its launching. The plan can be seen acquiring a momentum of its own, which ends by sweeping along all its participants, some protesting, some intoxicated, to a doom whose inevitability they have all come to recognise.

Manstein had originally intended to strike against the Kursk salient immediately after his victory at Kharkov in March, but with the imminence of the thaw and the difficulty of getting Kluge to exert corresponding pressure from the north, the project was shelved. It came up again in April at a Chiefs of Staff conference which Zeitzler called at OKH headquarters at Lötzen. By this time Manstein was more inclined toward the "backhand" stroke, which involved giving up the whole Donetz basin and staging a major Panzer offensive southeast from Kharkov, but Zeitzler judged that an attack at Kursk would be less risky, would entail no preliminary sacrifice of ground, and "would not make such heavy demands on the reserves." A memorandum suggesting a convergent attack by Kluge (with Model's 9th Army) and by Manstein (with Hoth's 4th Panzer Army) was submitted to Hitler on 11th April. The Führer, however, was unable to make up his mind. Zeitzler's memorandum had suggested that between ten and twelve Panzer divisions, with supporting infantry, would do the trick. Hitler thought that this was not enough, and when Zeitzler argued that only five had been needed to recapture Kharkov the Führer replied that victory there had been due to the employment of the Tiger, "of which one battalion was worth a normal Panzer division." For the spring offensive Hitler was determined to include the Panther as well.

The argument dragged on for some weeks, with the Führer, it seems, genuinely undecided and the production of the Panther—which was experiencing a variety of teething troubles—trickling along at a bare twelve per week. During April the ripples of argument spread outward and affected the whole of the Nazi high command. Jodl, Chief of Staff at OKW, was against *Zitadelle* because he believed that it was dangerous to empty the strategic reserve when so many new crises threatened to develop in the Mediterranean. Zeitzler countered with the paradoxical argument that the Wehrmacht was now *so* weak in the East that it could not stand still and "wait to be hit" but must do something to draw Russian fire. There was also, inevitably, the personal element. Warlimont has described how

> Zeitzler was not interested in these far-off problems, and the fact of his being excluded from them, as Chief-of-Staff at OKH, was a constant source of anger to him. He urged all the more the execution of "his" offensive, and complained to Hitler of Jodl's intrusion into his sphere of responsibility.

And in fact the OKW staff was still further excluded from matters relating to the Eastern front. From being an over-all advisory body, it was now little more than a second army operations branch for theatres other than the Russian. Only one man, Hitler, possessed complete insight into the whole strategic picture, and persons who advised him, whether on military, economic, or political questions, did so on the basis of their restricted and departmentalised knowledge. One result of this was that in the case of *Zitadelle* the majority of those in favour were generals with active commands on the Russian front, and those against it (with the exception of Guderian) did not have access to the detailed figures which its protagonists could juggle to support their case.

If a personal element underlay the dispute between Zeitzler and Jodl, between Kluge and Guderian it was open and notorious. The two men barely spoke on even the most ceremonial occasions, and in May, Kluge wrote to Hitler, asking his permission to challenge the Inspector General to a duel. Kluge, as C. in C. of Army Group Centre, was passionately in favour of *Zitadelle*. His apparent triumph over Guderian in December 1941 had turned sour, for while Kluge's command had remained static and inglorious for over a year, Guderian had returned from the shadows with immense power and influence. It was not difficult for Kluge, and others who were jealous of the Inspector General, to represent his opposition to *Zitadelle* as originating in a fear that the operation might lead to a reduction in his power of control over the Panzer force, and thereby to drum up support for the operation simply because Guderian was against it.

In the meantime Hitler was continuing to sound opinion among the field commanders, which he did through Schmundt and his adjutant's staff. These showed one surprising exception to the unanimity which Zeitzler alleged. For Model, who was to command the 9th Army under Kluge, reported that he was highly dubious about its prospects. Aerial reconnaissance and patrol activity showed that the Russians, at any rate, had no doubt as to where and how the Germans would strike, and were preparing energetically to meet them.[1] To this the Zeitzler school retorted with a change of rationale. If the Russians were in fact going to give battle there, was it not an admission that the area chosen was of vital importance, and would result in a substantial part of the Russian armour being brought to battle?

Meanwhile the weeks were slipping by, and with the accumulating Russian strength the original concept of *Zitadelle* was inexorably changing from a short, sharp blow that was to throw Soviet offensive plans out of gear to a head-on trial of strength which would settle the whole course of the summer campaign. At the beginning of May, Hitler was still undecided as to whether to issue the directive or not, and a con-

[1] There is some conflict of evidence as to what Model's final recommendation actually was. Under interrogation by Liddell Hart, Manstein asserted that Model told Hitler the offensive was feasible if he was given adequate strength, and that this lay behind Hitler's decision to wait for the Panthers. But when he came to write his memoirs, Model is not mentioned at all. Guderian (admittedly not an unprejudiced source) says that Model was against it under any conditions.

ference of army and army group commanders was called at Munich on the 3rd for a discussion of prospects. At this conference, which lasted two days, only Guderian spoke forcibly against the offensive in any form (although he was supported by Speer in an address on the weapons and production aspect). Zeitzler and Kluge were very enthusiastic, and Manstein, "as often when face-to-face with Hitler, was not at his best." The commander of Army Group South could only say that the chances of success "would have been excellent in April" but that now he found it hard to form an opinion. Indeed, the best arguments against the operation seem to have come from Hitler himself, who had opened the discussion with a concise résumé of Model's report and concluded by saying, "Model drew the correct deduction . . . namely that the enemy was counting on our launching this attack and that in order to achieve success we must adopt a fresh tactical approach."

However, Hitler still would not make up his mind definitely, and in a characteristic device to postpone doing so, he returned to the question of the Panthers. Inquiry showed that only about 130 of these tanks had actually been completed and of these less than 100 had been delivered. The original production schedule stipulated that 250 were to be ready by the end of May. Speer explained that the earlier difficulties had now been overcome, that the target of 250 could now be comfortably exceeded, and that 324 would be available by 31st May. This meant that if the Panthers were to be employed in quantity the offensive would have to be postponed until June. A putative date, 13th June, was fixed around which the contingency planning could be based, pending a final decision.

These figures had been produced at a separate tank-production conference held at the Chancellery a week after the Munich discussion, on 10th May. And it was at the close of this conference that Guderian had approached Hitler and their celebrated exchange, in which Hitler admitted that the thought of the *Zitadelle* operation "made [his] stomach turn over," took place. In a state of considerable agitation, Guderian had asked Hitler why he wanted to attack *at all* in the East in 1943. Keitel then interrupted and said, "We must attack for political reasons," to which Guderian replied, "How many people do you think even know where Kursk is? It's a matter of profound indifference to the world whether we hold Kursk or not . . ." Thereupon Hitler, after confessing his own misgivings, said that he had "as yet by no means committed [himself]."

If the generals who favoured *Zitadelle* had known the truth about the Russian preparations, they could hardly have been so enthusiastic. The first appreciation of the German plan had been drawn up by Vatutin as early as April, and forecast with remarkable prescience the final shape of the operation. For the next two months the Russians were shoring up the flanks of the salient with guns and armour at a considerably faster pace than the German concentration opposite them.

To co-ordinate the three "fronts" involved, and to evolve a counter-offensive plan which was to come into operation as soon as the German impetus slackened, the *Stavka* had sent down Zhukov and Vasilievski to Kursk at the end of April. They calculated that the main shock of the assault would fall on Vatutin's "Voronezh front," opposite Belgorod, and there were placed two veteran armies of the Stalingrad fighting, the 21st and 64th (now designated the 6th and 7th Guards armies) and a very strong tank force, the 1st Armoured Army. The bulk of the salient, including its northern corner, opposite Model, was under Rokossovski's "central front," and this was steadily reinforced with artillery until at the end of June it contained more artillery than infantry regiments, with the fantastic figure of over 20,000 pieces, of which 6,000 were 76.2-mm. anti-tank guns and 920 *Katyusha* multiple rocket throwers. Anti-tank and anti-personnel mines were laid in a density of over 4,000 per mile. All the defending units were subjected to an intensive and repetitive course of training on their reaction to the expected German attack. A Red Army captain has described how his brigade

> anticipated five possible places where they [the Germans] may strike and at each of them we know alongside whom we shall be fighting, our replacements and command posts. The brigade is stationed in the rear, but our trenches and shelters are ready up in front, and the routes by which we are to get there are marked out. The ground, of which we have made a topographical survey, has been provided with guide marks. The depths of fords, the maximum loads of bridges are known to us. Liaisons with division have been doubled, codes and signals are arranged. Often alerted by day or night, our men are familiairised with their task in any eventuality . . .

This confidence, and certainty as to their role and objectives, was as definite in the "front" commands as it was in the small field units. Once the salient had been strengthened, reinforcement turned to the "western" and "Bryansk" fronts of Popov and Sokolovski, who were to hold themselves ready to strike against Model's left flank in the Orel bulge when Zhukov judged the moment ripe. A new "reserve front" called the "steppe front" was created and put under Koniev, as a reservoir of fresh units which could be directed on critical points once the battle had started, and for the exploitation—which Zhukov and Vasilievski knew would surely follow once the German force was spent. Most of the Russian tank strength was integrated with the defence for close support, but one very strong force, the 5th Tank Army, was held back in Koniev's front for mobile action against the Panzers, should any of them break through. Indeed, few major operations since the disastrous "Nivelle offensive" of April 1917[2] have been so long and so carefully anticipated as the German attack against Kursk in 1943. The new atmosphere in the Red Army was summarised by a tank captain when he declared, "At the beginning of the war everything was done in a hurry, and time was always lacking. Now we go calmly into action."

Paradoxically, while Russian preparations were going ahead with such energy and resource, the Germans were suffering continual delays and rumours of alteration and cancelling. June came, and the Panther target had duly been reached. But now the reports of Russian preparations were so alarming that it was decided to wait for another three weeks' production, which would allow two extra battalions of Panthers to go to Model's divisions in the north, in addition to those with the 11th Panzer, *Gross Deutschland,* and the three divisions of Hausser's SS Panzer corps. This meant that D-Day was again postponed, from the second week in June to the beginning of July. Manstein now came out in the open and protested that the operation was no longer feasible and must be abandoned, but it was too late. The united stand of orthodox General Staff opinion, Keitel, Zeitzler, Kluge, had persuaded the Führer, whose mind, once made up, was never altered. H-Hour was set for

[2] At the beginning of 1917 the new Commander in Chief of the French Army, General Nivelle, began to draw up plans for assaulting a vulnerable sector of the German line, held by nine German divisions, with forty-four French divisions. Plans were circulated down to N.C.O. rank, and some fell into German hands in February. In March the Germans withdrew from their vulnerable salient and stepped behind the Hindenburg Line, which they reinforced with an additional thirty-four divisions. Nivelle, in spite of misgivings in the French Cabinet and among his own colleagues, insisted on mounting the offensive directly against the new German position, but with practically no alteration to the tactical planning. The result was a complete disaster. A recent account can be found in *Dare Call It Treason,* by Richard M. Watt.

the 4th of July, "Independence Day for America," gloomily observed the Chief of Staff of the 48th Panzer Corps, "the beginning of the end for Germany."

In the light of his earlier misgivings it must be assumed that Hitler, in agreeing to authorise *Zitadelle,* was not guided by purely tactical considerations, but as to what these considerations were, there is no direct evidence. Some observers, Warlimont in particular, believe that the Führer's pathological aversion to withdrawal was exploited by Zeitzler, who maintained that this was the only alternative. Jodl's arguments concerning the dangers that threatened in the Mediterranean theatre and the rumours of Mussolini's insecurity in Italy combined to make it vital to check any Russian approach, however distant, on the Balkans. But there was also the mesmeric effect which the prospect of a great battle can have on all associated with its planning. The generals argued, and with reason, that they had always managed to penetrate the Russian positions at the first blow; it was later, when the armoured impetus ran out into the endless space of the plains, that difficulty had begun. This time they had limited their aim—a mere seventy miles, less than forty miles for each arm of the pincer! This, surely, was within the capabilities of troops who had taken hundreds of miles at a single thrust forward in the previous campaigns. The firepower and mobility of the attacking forces would be greater than in 1941 or 1942, their degree of concentration much higher, their objectives incomparably less ambitious. Was it not the case that no power on earth could stand up to the first shock of the German Army in a major offensive?

Certainly, by any standard other than that of the Soviet formations opposing them, the German order of battle, as it finally took shape in the last days of June 1943, looked very formidable. The number of Panzer divisions had been raised from the ten originally allocated to seventeen by ruthlessly stripping the rest of the front of its armoured protection. In the 9th Army, Model had no fewer than three Panzer corps, and two army corps of supporting infantry. The southern pincer, Hoth's 4th Panzer Army, was the strongest force ever put under a single commander in the German Army. His attack frontage, flanked by three army infantry corps, was from west to east—the 3rd Panzer, *Gross Deutschland,* the 11th Panzer, SS *Leibstandarte,* SS *Das Reich,* SS *Totenkopf,* the 6th Panzer, 19th Panzer, 7th Panzer—nine of the finest divisions in the German Army, shoulder to shoulder along less than thirty miles of front!

In the last days before the attack a strange feeling, not so much of confidence as of fatalism, pervaded the German tank forces—if this strength, this enormous agglomeration that surrounded them on every side, could not break the Russians, then nothing would. In dutiful obedience to security regulations long since nullified by Russian intelligence, the Panzer officers discarded their black uniforms before visiting the forward positions to make their final reconnaissance. As they peered across no man's land at the enemy they saw

> a far-flung plain, broken by numerous valleys, small copses, irregularly laid out villages with thatched rooves, and some rivers and brooks; of these the Pena ran with a swift current between steep banks. The ground rose slightly to the north, thus favouring the defender. Large cornfields covered the landscape, making visibility difficult.

This was the ground on which was to be fought the last major offensive by the German army in the East; the greatest armoured battle in history, and one of the most important and most bitterly contested engagements of World War II.

On 2nd July the *Stavka* had issued front commanders a warning that the attack could be expected at any time between the 3rd and the 6th, and on the night of the 3rd-4th a Czech deserter from an engineer battalion of the 52nd Army Corps recounted that all units had been issued a special issue of schnapps and rations for five days. Judging that the attack was imminent, Vatutin ordered a bombardment of German forward positions and assembly points, and this was carried out at an intense rate by Russian medium artillery for four hours, although the anti-tank artillery was under strict orders to stay silent. While they were suffering this destructive, and ominous, attention from the Soviet artillery, the waiting German soldiers received a personal message from the Führer:

> Soldiers of the Reich!
>
> This day you are to take part in an offensive of such importance that the whole future of the war may depend on its outcome. More than anything else, your victory will show the whole world that resistance to the power of the German Army is hopeless.

That curious lack of imagination and adaptability which is a recurrent characteristic of the German military attitude, and whose leaden influence upon the tactical planning of *Zitadelle* has already been the subject of remark, soon became uncomfortably apparent in its execution. Once again the old *Blitzkrieg* formula—Stukas, short, intense artillery bombardment, massed tanks and infantry in close contact—was fed into the computer, with little regard for the changed conditions save a simple arithmetical increase in the strength of the respective components.

At two o'clock in the afternoon the German tanks, about two thousand strong in the first wave, clambered out from the sunken lanes and dried-up *balkas* where they had been lying and moved slowly forward, hatches closed, across the billowing yellow-green corn of the upper Donetz valley.

> As we advanced the Russian artillery ploughed the earth around us [wrote the radio operator in a Tiger]. Ivan, with his usual cunning, had held his fire in the weeks before, and even that morning when our own guns were pounding him. But now the whole front was a girdle of flashes. It seemed as if we were driving into a ring of flame. Four times our valiant "Rosinante" shuddered under a direct hit, and we thanked the fates for the strength of our good Krupp steel.

The Germans started with virtual parity in armour to their opponents (although no German account will admit this), and a definite qualitative superiority in the Tigers and Panthers, but Russian artillery was incomparably stronger in weight, numbers, and direction. Manstein's guns had been unable either to saturate the forward Russian defence zone or to achieve much in the way of clearing avenues through the mine fields. The result was that many of the tanks were disabled by mines in the first half mile and were soon overtaken by the supporting infantry. The Panzer crews had received strict orders that

> . . . *in no circumstances* will tanks be stopped to render assistance to those which have been disabled. Recovery is the responsibility of engineer units *only*. Tank commanders are to press on to their objective as long as they retain mobility. Where a tank is rendered immobile but the gun is in working order (e.g., from mechanical failure or track damage), the crew will continue to give fire support from a static position.

This was virtually a death sentence on the crews of disabled tanks, for the Russian guns were so thick on the ground that they could pick off the crippled Panzers within minutes of their striking a mine. There were also special tank-destruction squads of infantry, which had been positioned in slit trenches in the middle of the mine fields and which, as will be seen, operated with particular and gruesome success in the north, against Model.

German tactics were to advance in a succession of armoured wedges—

known as the *Panzerkeile*—with the Tigers bunched at the tip of the wedge and the Panthers and PzKw IV's fanned out behind. Lightly armed infantry with carbines and grenades moved close behind the tanks, and heavier forces with mortars followed at the base of the wedge in tracked personnel carriers. This tactic amounted to a rejection of the traditional principle of the Panzer army, as a sword, to be used in a deep, narrow thrust to the enemy's rear, and substituted an axe which was to break down the opposing front along a considerable length. It had been forced on the Germans by the tenacity of the Red Army in holding close to the sides of the breach, and the multiplication of their firepower in the last year, which made independent action by the Panzers too dangerous—at least in the early stages of the battle.

Both Model and Manstein were in fact using the same tactics Montgomery had employed at Alamein, opening the battle with the armour in an infantry-support role, in the hope that there would still be enough battleworthy tanks left for exploitation once a breach had been made. But here, in contrast to Alamein, the defenders' strength was at parity with the attackers', and the way in which they had prepared for the battle allowed them to keep back a large proportion of their armour until a late stage in the fighting.

The Russians had developed a method of fire control, known to the German troops as a *Pakfront,* based on the use of groups of up to ten anti-tank guns under a single commander which would concentrate on a single target at a time in broadsides. The mine fields had been laid so as to channel attacking tanks into the field of fire of clusters of these groups, which were sited in depth over about five miles. As each broadside could be expected to take out one tank, it can be seen that the approaching Panzers would have to endure very heavy losses before they could close with the gunners and attain their initial objectives.

In the operational orders it was made clear that the tanks were not expected to deal with all the Russian gun positions on their own, but to leave this to the accompanying infantry, although they were, of course, to engage guns which were actually holding them up. The difficulty in this plan was first, that the number and depth of the *Pakfront* had been greatly underestimated; second, that the Russian guns were themselves protected by machine-gun and mortar nests, with strict orders not to fire except at German infantry, and then only in support of their own battery.

In the result it was not until the late evening that the extent of the

Russian defences became apparent, and by that time the *Panzerkeile* had become badly distorted. The great majority of the Russian anti-tank guns were the standard 76.2-mm. L 30, a weapon which was no longer capable of penetrating the frontal armour of the Tiger at other than point-blank range, and many of the monster tanks thus managed to pass through the first belt of defences with only superficial damage. In their rear, though, the PzKw IV's had suffered severely, and many of the Panthers had broken down or been disabled in the mine fields. Darkness fell on the battlefield with the mass of the attacking strength still entangled in the gun positions and slit trenches of the first belt, but with a few isolated packets of Tigers deep in the main Russian defence zone. During the night groups of *Panzergrenadiere* struggled across the battlefield in attempts to reach the stranded tanks and give them some protection; and the short hours of darkness were savaged by the glare of star shells and Very's, of dense white currents of tracer, and the hot orange stab of portable *Flammenwerfer* as they clashed with the Russian tank-killing squads and infantry patrols.

At four o'clock on the morning of 5th July the sun rose on a classic tableau of positional warfare, almost as if the Great War had been running continuously since 1917, with only the influx of more sophisticated equipment to mark the passage of time. Clouds of brown smoke from burning corn and the thatched roofs of villages rolled across the battlefield under a gentle westerly breeze, streaked at intervals with the black, oily discharge of a flaming tank; the continuous rattle and chatter of small-arms fire was overlaid by a steady cannonade from the Russian 76-mm.'s and the scream of *Katyusha* rockets; periodically the high-pitched slap of the 88-mm.'s told of the Tigers' defending themselves some three to four miles away. Tremendous exertions by the German infantry during the night had rewarded them with mastery of the first belt of Russian defences, to the extent at least of silencing most of the anti-tank guns there, although many sharpshooters remained and picked off engineers as they tried to clear the mine fields. It was still proving almost impossible to make any headway against the second, and strongest, belt, whose gunfire, pre-ranged onto the points the Germans were now occupying, raked them mercilessly. The Russians had moved up many of their own tanks during the night and parked them in hull-down positions which had been prepared in the weeks preceding the battle. This meant that on the second day their firepower was almost as strong as it had been on the first, in spite of the elimination of their outpost line. The Germans, in contrast, were already badly strung out between the leading Tiger companies, whose situation reports and calls for help crackled incessantly across the short-wave radio, and the truncated stumps of the *Panzerkeile,* struggling to re-form for a second attack.

Of the two arms, which were meant to converge across the sixty-mile base of the salient, Model's had fared worse. For it was here in the northern sector, spearheading the attack of the 47th Panzer Corps, that the ninety Porsche Ferdinands were being used. Like the Henschel Tigers with the SS at Belgorod, they managed to break into the Russian defence system with little difficulty because of their heavy armour plating. But within hours of their appearance the Russian infantry had discovered that these monsters had no secondary armament. Devastatingly effective against the T 34, formidable against static gun positions, they were useless against infantry in slit trenches. They soon became separated from the lighter tanks that escorted them and had given some protection with their own machine guns, and fell victim one by one to groups of Soviet infantry, which would board them while they were on the move and squirt flame throwers over the engine ventilation slats. Guderian's own judgment was:

> They were incapable of close-range fighting since they lacked sufficient ammunition [i.e., high-explosive as well as armour-piercing] for their guns and this defect was aggravated by the fact that they possessed no machine-gun. Once [the Ferdinands] had broken into the enemy's infantry zone they literally had to go quail shooting with canon. They did not manage to neutralise, let alone destroy, the enemy rifles and machine-guns, so that our own infantry was unable to follow up behind them. By the time they reached the Russian artillery they were on their own. Despite showing extreme bravery and suffering unheard of casualties, the infantry of Weidling's division did not manage to exploit the tanks' success . . .

Nor was the advent of the Panthers as devastating as had been hoped. The Chief of Staff of the 48th Panzer Corps reported that they

> did not come up to expectations. They were easily set ablaze, the oil and petrol systems were inadequately protected and the crews were insufficiently trained.

After twenty-four hours of fighting, the Russian front had only been dented in one place—the left centre of Manstein's attack, by the combined strength of the 48th Panzer Corps and the SS. Here the Russians had been driven back across a defence zone about two and a half miles wide to a line of four small hamlets which bordered a stream, Sawidowka-Rakowo, Alexeyevka, Luchanino and Ssyrzew. During the night the *Panzergrenadiere* had succeeded in clearing the houses on the south side, and Hoth decided to force the 3rd Panzer and *Gross Deutschland* across the stream at first light on 5th July. But during the night a cloudburst raised the level of the stream by several feet and effectively increased its breadth by turning the fields on either side into a glutinous swamp. Under cover of darkness the Russians had moved tanks and guns forward into the buildings and ruins on the opposite side, and the two Panzer divisions were severely punished by direct fire while they assembled in dense formation during the morning. Throughout the day the engineers struggled under cover of an erratic smoke screen to lay their bridging equipment. Over their heads a vicious artillery duel raged between the Russian guns and the massed tanks of the 3rd Panzer and *Gross Deutschland,* with hourly raids by the Stukas as feeble compensation for Hoth's lack of heavy artillery. By nightfall the Germans had suffered heavy casualties and not gained a yard. During the night of 5th-6th July both divisions were pulled back and re-formed. A Russian counterattack had won back part of Sawidowka, and the 3rd Panzer was moved back to the left wing to recapture the village and get across the stream at its confluence with the Pena; the 11th Panzer was to come in on the right of *Gross Deutschland,* in another effort to make a breach between Luchanino and Ssyrzew —a task which it was hoped would be facilitated by the progress of SS *Leibstandarte* and SS *Das Reich,* both of which had ground their way forward to a depth of about four miles on the right of the 48th Panzer Corps.

By 7th July, the fourth day of the attack, the ground had dried out enough for German armour to get across the stream, and *Gross Deutschland* captured Ssyrzew. At the same time the attacks of the 3rd Panzer were slowly prising the defenders away from their positions on the Pena. During the evening Russian fire slackened, and the 48th Panzer Corps was able to cross the line of the stream at will. Hoth was now almost halfway through the Russians' defence zone, and right up to their main "gun line." On the right of the 48th Panzer Corps, Hausser's three SS divisions had penetrated rather deeper, but unlike Knobelsdorff, the SS commander had not succeeded in taking the Russian front back along a

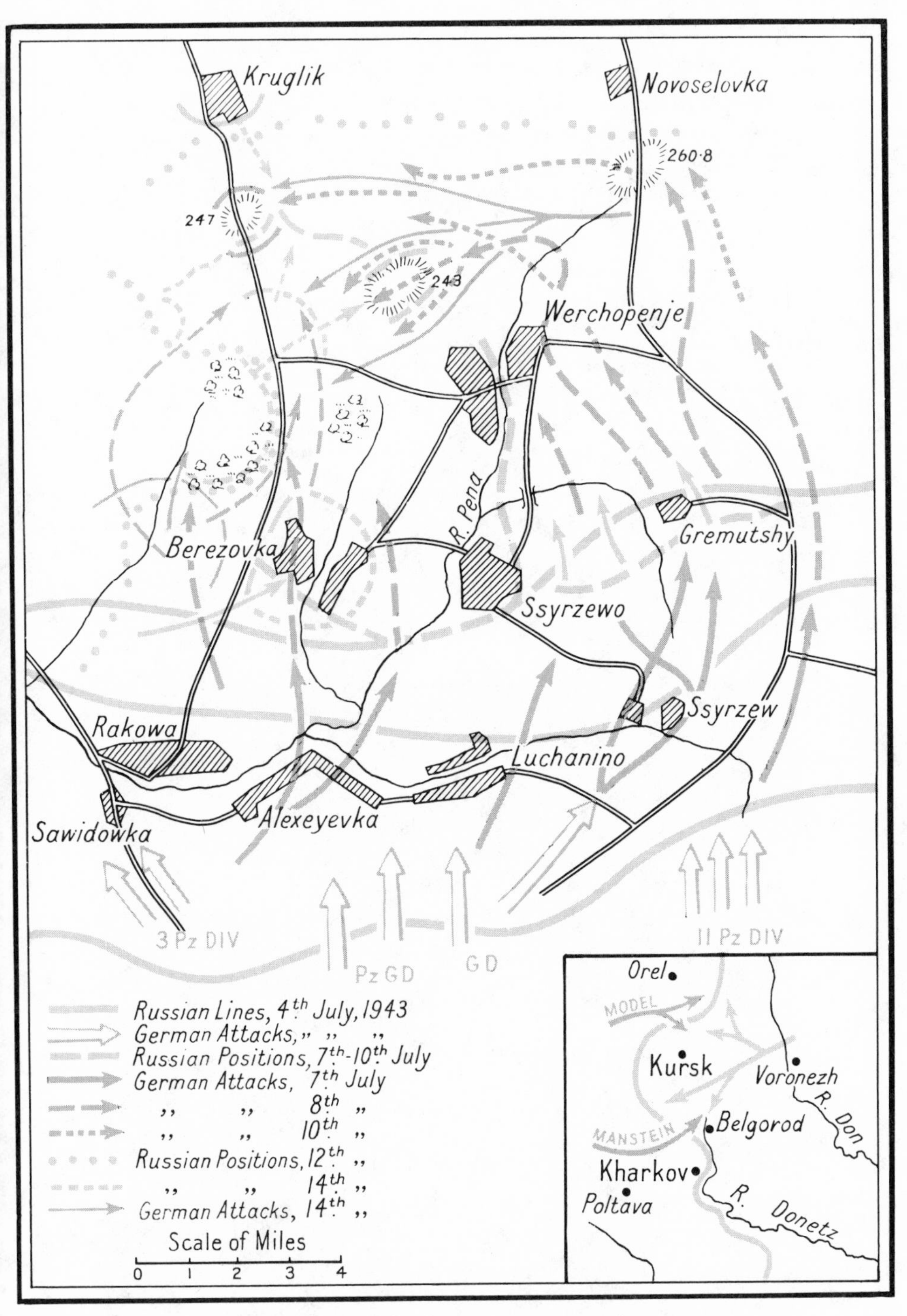

The Death Ride of the Fourth Panzer Army

continuous stretch; instead each division had punched a hole of its own, through which at great cost it had struggled northward, under a continuous enfilade fire from its flanks. On 9th July it was plain to Hoth that the crisis of the battle was approaching, for the majority of his troops had been in action without respite for five days. The rations and ammunition with which they had started the battle were running low, and the intensity of Russian fire was making it very difficult to operate breakdown and refuelling services for the armour. The only gleam of light seemed to be in the centre of Knobelsdorff's front, where *Gross Deutschland* had succeeded in forcing a battle group through the village of Gremutshy, which lay across the main Russian defence zone. During the afternoon and evening of 9th July, General Walter Hoernlein, the commander, managed to follow the battle group with the *Panzergrenadier* regiment and about forty tanks, and began to swing them westward, behind the Russian gun line, with the intention of loosening the defences that were holding up the left flank of the Panzer corps. This was rewarded that night when the Russians withdrew from Rakowo, where they had been blocking the path of the 3rd Panzer, and also from Knobelsdorff's right flank, which allowed the 11th Panzer to establish some contact with SS *Leibstandarte*.

Hoth had been in consultation with Manstein that same night, and on the morning of the 10th he told Hausser and Knobelsdorff that they were to clean up the breach with assault guns and *Panzergrenadier* and collect all their available tank "runners" for what he hoped would be the breakthrough thrust against the last Russian line between Kruglik and Novoselovka. For two days the infantry of the 3rd Panzer and *Gross Deutschland* struggled to force the Russian gate back on its hinge. In continuous and savage fighting they cleared the group of villages that straggled down the vale of the Pena and by the evening of the 11th had forced the surviving Russians back into the woodlands above Berezovka. A rectangular salient, about nine miles deep and fifteen across, had been driven into Vatutin's position—a poor result after the effort and casualties of a week's fighting, but enough, at least, to allow the tank commanders to reassemble out of range of the Soviet artillery. That night *Gross Deutschland* pulled out of the front and handed over to the 3rd Panzer.

On Hausser's front even this limited goal remained out of reach. The SS infantry was so heavily engaged protecting the divisional flanks that individual commanders had great difficulty in extricating tanks from the tips of the separate spearheads. On the 11th, *Das Reich* and *Leibstandarte* managed to effect a junction, but *Totenkopf* was still on its own, and suffering from the Russian propensity to kill all prisoners wearing its insignia.

This was the last, and fiercest, fight of the purely Germanic SS. After *Zitadelle,* Himmler opened the ranks of his army to an increasing flood of recruits from the occupied countries, and to general criminal riffraff from the Reich civil prisons. But the men who fought at Kursk had all passed through the brutalising school at Bad Tölz, where the regime was such that the recruit had been turned out

> as if he had just been unpacked to hang on a Christmas tree, incredibly pink, fresh and teutonic, his well-flattened pockets containing only a modest supply of paper currency which did not bulge, his paybook, his handkerchief creased according to regulation, and one prophylactic.

Having taken the SS oath, the recruit had to pass through the armoured warfare school, where ". . . he might be called on to dig himself into the ground, knowing that within a prescribed time the tanks would drive over his head, whether the hole was completed or not. If he was an officer cadet he might be required to pull the pin out of a grenade, balance it on his helmet, and stand to attention while it exploded."

Now these men were face to face with the *Untermensch* and finding to their dismay that he was as well armed, as cunning, and as brave as themselves.

In spite of the erosion of their position in the south, Zhukov and Vasilievski cannot have been anything but satisfied with the position on the night of 11th July. The blocking of Model's attack had left them free to face Hoth, and the whole of their mobile armoured reserve, the 5th Armoured Army, was still uncommitted. Realising that a final trial of strength with the Panzers was less than twenty-four hours away, Zhukov placed this force under Vatutin's orders, and during the night of the 11th-12th July it began to move forward to meet the expected eruption of the 48th Panzer Corps and Hausser's SS. At the same time Sokolovski was ordered to lead off the succession of counterattacks which the *Stavka* had planned against the Orel bulge, on Model's denuded left flank, with Popov to follow him after a further forty-eight hours.

All this was, of course, unknown to Hoth, but reports from the 3rd

Panzer indicated that the Russians' defences between Kruglik and Novoselovka were hardening hourly, and their activity against his left flank was increasing. Regardless of the state of Hausser and Knobelsdorff, the commander of the 4th Panzer Army was determined to force his armour through into open country before a defensive scab could form over the thin membrane exposed in the remaining Russian defences. On 12th July the whole mobile strength scraped together from the three forces of Kempf, Hausser, and Knobelsdorff, about six hundred tanks in all, started on their death ride. Before noon they were in a head-on collision with the armour of the Soviet 5th Army, and under a gigantic dust cloud and in stifling heat an eight-hour battle was joined. The Russians were fresh, their machines were unworn, and they had a full complement of ammunition. In addition two of their brigades had been equipped with the new SU 85, a self-propelled 85-mm. gun which had been mounted on the T 34 chassis as a mobile answer to the Tigers and the new L 70 gun of the Panther. The Germans, in contrast, had in many cases just come from fierce fighting in support of the close infantry actions of the preceding days. Many of their tanks had been patched up by engineers in the field, and were soon, particularly in the case of the Panthers, to break down again. In addition

> We had been warned to expect resistance from *pak* [fixed anti-tank guns] and some tanks in static positions, also the possibility of a few independent brigades of the slower KV type. In fact we found ourselves taking on a seemingly inexhaustible mass of enemy armour—never have I received such an overwhelming impression of Russian strength and numbers as on that day. The clouds of dust made it difficult to get help from the *Luftwaffe,* and soon many of the T 34's had broken past our screen and were streaming like rats all over the old battle-field . . .

By the evening the Russians were in possession of the battlefield, with its valuable lumber of disabled hulks and wounded crews. A sharp counterattack on Knobelsdorff's left had recaptured Berezovka, and the exhausted *Gross Deutschland* had to go straight back into action to prevent the 3rd Panzer from being cut off. The following day Hitler sent for Manstein and Kluge and told them that the operation should be cancelled forthwith. The Allies had landed in Sicily and there was a danger of Italy's being knocked out of the war. Kluge agreed that it was impossible to continue, although Manstein, with a most uncharacteristic lack of judgment, asserted ". . . on no account should we let go of the enemy until the mobile reserves which he had committed were decisively beaten."

Hitler, however, overruled him (in an exceptional reversal of their customary roles), and that evening the Germans began slowly to withdraw to their starting line. Guderian, who had seen his cherished Panzer arm shattered in ten short days, retired to his bed with dysentery. Only the indefatigable Tresckow persisted in his activity. Seeing in this disastrous defeat fertile soil in which to sow his intrigues among the generals, he had approached Kluge and suggested to the Field Marshal that he and Guderian resolve their differences and work together "to reduce the powers of Hitler as Supreme Commander," an exceedingly loosely worded objective, which covered everything from assassination to mild constitutional reform. Kluge, characteristically, agreed, provided that Guderian would "take the first step." Tresckow therefore approached Guderian, who was by that time in a hospital, awaiting an operation on his intentines. However, the Inspector General declined to have anything to do with the idea as

> My very exact knowledge of Field-Marshal von Kluge's unstable character prevented me from accepting . . .

So it was that the private jealousies and suspicions of the generals played their part in obstructing internal "reform," just as they had vitiated the possibility of success by force of arms.

In the short life of the "Thousand-Year Reich" eight more months had passed by.

eighteen | THE AFTERMATH

One member of the Nazi hierarchy, at all events, was not deluded. Heinrich Himmler saw that the failure of the *Zitadelle* offensive meant that the war was lost. The question which now exercised him was how to moderate defeat and save his own skin, and as on the two previous occasions which we have already recorded, his mind began to toy with the idea of a palace revolution.

Although he seems to have relished the terror he could inspire in his fellow Germans, Himmler never fully comprehended the hatred in which he was held abroad. To foreigners and neutrals the head of the SS liked to posture as an executive, a senior bureaucrat characterised by his steadfast opposition to Communism—one, in short, who was ideally suited by his position to preserve order in the country during any period of "difficulty" and, were an international agreement to be formulated, to lead Germany punctiliously back into "the family of nations." In his domestic circle, where, as a family man Himmler found it easier to strike such an attitude, he had an old acquaintance, a Dr. Carl Langbehn, whose respectable record made him an ideal piece to introduce onto the board at this stage.

Langbehn was a close neighbour of the Himmlers' on the Walchensee, and their daughters went to school together. Once, in the past, Langbehn had approached Himmler on behalf of his old law tutor, a Professor Fritz Pringsheim, when the latter had been thrown into a concentration camp (because of his Jewish ancestry). Himmler had obliged, and not only ordered Pringsheim's release but authorised a permit with which the old man was able to leave the country. It cannot be imagined that the relations of the two men were very close, for Langbehn, who practiced as a constitutional lawyer, would indulge periodically in ostentatious gestures of opposition to the regime. At the time of the Reichstag Fire Trial he had offered to defend the Communist leader, Ernst Togler. Later Langbehn undertook the defence of Dr. Günther Gereke, the Labour Minister under the Papen government, whom the Nazis were trying on a variety of fabricated charges. Emboldened, perhaps, by the friendship and protection of his powerful neighbour, Langbehn assumed prominence in the circle of "the conspirators," and consorted with Goerdeler and Johannes Popitz, to whom he propounded the importance of engaging Himmler in the plot.

If the National Leader felt any embarrassment at his neighbour's treasonable activities he bore it manfully. On at least one previous occasion Himmler had gone so far as to communicate his own doubts about the future to Langbehn, and fortified by this recollection, the lawyer now began to raise, as tactfully as possible, the notion of a meeting with Popitz.

No agenda for this meeting was ever formalised, which is hardly surprising, but its purpose was twofold. First, to test the reaction of the National Leader to the idea of a palace revolution; second, if Himmler was willing, to discuss in general terms the basis of an approach to the Western Allies for ending the war. It would seem that both sides started with the intention of double-crossing each other. The conspirators hoped to use the SS to get rid of Hitler, then turn on it with the full weight of the Army. Himmler simply intended to use Popitz and Langbehn as a respectable "front" for the opening of negotiations. If the suggestion (of ending the war in the West) should be favourably received, there would be no difficulty over Himmler's assuming the executive authority to implement it, for *Treuer Heinrich*—the cognomen had been bestowed on him by Hitler personally in recognition of his perfect loyalty—had at his command the most perfect machinery for a *coup d'état* that has ever existed, in the police, the Gestapo, the SD, and the SS.

Coups—always a sensitive, indeed a traumatic subject in totalitarian circles—were occupying much of the business at the Führer's headquarters at this time. While the first cautious approaches were being revived between Himmler and the conspirators in Berlin, Hitler's own entourage had been thrown into alarm by events in Rome. On 25th July, Mussolini had been deposed and Marshal Badoglio had taken over the government.

For several days the collapse of his ally dominated Hitler's thoughts, to the exclusion of regular military business and in particular to the detriment of the close direction of *Zitadelle,* now grinding wearily to a halt on its old starting line.

Several fragments of the daily conference transcripts over this period have been preserved, and extracts from them will show the state of excitement and confusion which prevailed:

HITLER. [after a long, speculative back and forth with Keitel and Jodl about alternative political developments in Italy] . . . Although that bugger [Badoglio] declared immediately that the war would be continued, that won't make any difference. They have to say that, but it remains treason. But we'll play the same game while preparing everything to take over the whole crew with one stroke, to capture all that riffraff. Tomorrow I'll send a man down there with orders for the commander of the 3rd *Panzergrenadier* Division to the effect the he must drive into Rome with a special detachment and arrest the whole government, the King and the whole bunch, right away. First of all to arrest the Crown Prince and to take over the whole gang, especially Badoglio and that entire crew. Then watch them cave in, and in two or three days there'll be another *coup.*

How far are they [presumably the 3rd *Panzergrenadier* Division] from Rome?

JODL. About a hundred kilometres.

HITLER. A hundred? More like sixty. That's all they'll need. If he drives in with motorised troops he'll get in there and arrest the whole works right away.

KEITEL. Two hours.

JODL. Fifty to sixty kilometres.

HITLER. That's no distance.

WAIZENEGGER. [an Oberstleutnant on the OKW staff] The division has forty-two assault guns, my Führer.

HITLER. Are they down there with the division?

WAIZENEGGER. Yes, with the division.

HITLER. Jodl, work that out right away.

JODL. Six battalions.

KEITEL. Ready for action. Five only partially ready.

HITLER. Jodl, work out the orders for the 3rd *Panzergrenadier* Division to be sent down, telling them to drive into Rome with their assault guns without letting anyone know about it, and to arrest the government, the King and the whole lot. I want the Crown Prince above all.

KEITEL. He is more important than the old man.

BODENSCHATZ. [General of the Luftwaffe, and liaison officer between Goering and the Führer's headquarters] This has to be organised so that they can be packed into a plane and flown away.

HITLER. Straight into a plane, and off with them.

BODENSCHATZ. [laughs] Don't let the *Bambino* [i.e., the Crown Prince] get lost at the airfield.

HITLER. In eight days the thing will be reversed again. Now I want to talk to the Reichsmarschall.

BODENSCHATZ. I will inform him immediately.

It is apparent from the repetitive and at times almost incoherent quality of Hitler's speech (particularly noticeable in the piece with which this fragment opens) that he was in a highly nervous condition. As so often in times of stress, he wants to lean on Goering, from whose placid and cynical outlook he would draw strength. And, as was becoming more and more frequent, Goering was not there, and when located, seems to have made only moderate sense. Finally the Reichsmarschall was reached by telephone (it was still only 10 P.M.), and there is a record of Hitler's part—though not, naturally of Goering's—in the ensuing conversation:

"Hello, Goering? . . . I don't know. . . . Did you get the news? . . . Well, there is no direct confirmation yet, but there can't be any doubt that the Duce has resigned and that Badoglio has taken his place. . . . In Rome it is not a question of possibilities, but of facts. . . . That's the truth, Goering, there's no doubt about it. . . . What? . . . I don't know, we are trying to find out. . . . Of course that's nonsense. He'll keep going, but don't ask me how. . . . But now they'll see how *we* keep going. . . . Well, I just wanted to tell you. Anyway, I think you ought to get over here right away. . . . What? . . . I don't know. I'll tell you about that when you arrive. But try to adjust yourself to the fact that it's true."

Goering then rang off. Hitler said to the room at large, "We had a mess like this once before. That was on the day the *coup d'état* took place *here*." (The assumption is that he was pointing on the map to Belgrade and referring to the *coup* by King Peter and General Simović in March 1941.) "We changed things there, too."

Two and a half hours later there was still no sign of Goering, and it is evident that Hitler's temper had not improved. Walter Hewel, a representative of the Foreign Ministry, suggested that some sort of guarantee of the Vatican should be offered.

HITLER. That doesn't make any difference. I'll go right into the Vatican. Do you think the Vatican embarrasses me? We'll take that over right away.

For one thing, the entire diplomatic corps are in there. It's all the same to me. That rabble is in there. We'll get that bunch of swine out of there. Later we can make apologies. That doesn't make any difference.

HEWEL. [changing his tack, it may be suggested, to the Führer's mood] We will find documents in there.

HITLER. There, yes, we'll get documents, all right. The treason will come to light. A pity that Ribbentrop isn't here. How long will it take him to draft the directive for Mackensen? [Mackensen was the German Ambassador in Rome.]

HEWEL. It may be finished.

HITLER. All right.

HEWEL. I'll check on it right away.

HITLER. Will that be a journalistic essay of twelve pages? I'm always afraid of that with you people. It can be done in two or three lines.

Now, Jodl. I've been thinking of something else. If our people in the East want to attack tomorrow or the day after—I don't know if the units have been assembled yet—I would recommend letting them do that. Because then *Leibstandardte* can still take part. For if they have to wait for the stock anyway— [SS *Leibstandardte* was under orders to move to Italy, but was stranded in Russia by shortage of rolling stock.]

KEITEL. The rolling stock.

JODL. There's something in that. It would be better if *Leibstandardte* consolidates its position before leaving.

HITLER. Yes, that would be good. Then this one division, the *Leibstandardte,* can be taken away. They must be moved first, but can leave their stuff there. They don't have to take their tanks along. They can leave them over there, and get them replaced here. By getting Panthers here they will be perfectly well equipped. That's obvious. By the time the division is here it will have its tanks.

HEWEL. May I interrupt? It's about the Prince of Hesse. He stands around all the time. Shall I say that we don't need him?

HITLER. All right. I'll send for him and say a few words to him.

HEWEL. He bothers everybody, and wants to know everything.

HITLER. I would start by giving him all the proclamations we have collected here. They have been made public anyway. Philip can read these, all right, they aren't dangerous. But be sure to give the order not to let him have the wrong thing.

So much for the scene at Hitler's headquarters. The total time devoted to the campaign in the East was about forty seconds (the exchange with Keitel and Jodl quoted above). But there, on the southern front, a situation was building up every bit as menacing as that which prevailed in Italy.

This situation arose out of a persistent faulty appreciation by the German commanders on the spot, and in particular, by Manstein. Although they recognised that the attempt to squeeze out the Kursk salient had been a failure, they still clung to the view that the fighting there had sucked in most of the Russian armour and that the rest of the summer could be devoted to a series of minor tactical "solutions" which would straighten out and consolidate the front before the onset of winter. The Germans would not accept the fact that, as the weaker army, they would have to confront a Russian initiative in both summer and winter instead, as had been the rule until then, of alternating their offensive and defensive with the seasons.

Manstein is perfectly frank about this. "We hoped to have given the enemy so much punishment in the course of *Zitadelle* that we could now count on a breathing space in this part of the front," he wrote, and that "Southern Army Group [i.e., himself] decided to withdraw a substantial weight of armour from that wing for the time being, in order to iron things out in the Donetz area."

The result was that practically all the Panzers were withdrawn from the old front around the Kursk salient. The majority of units were seriously understrength, and still further depleted by a rigorous inspection program which took out all vehicles in need of repair and sent them back to the maintenance depots at Kharkov and Bogodukhov. These establishments became so choked with work that after 1st August tanks and assault guns had to be sent as far back as Kiev, even for minor repairs, to the running-gear and gunnery-control equipment. By the time the different Panzer divisions engaged in *Zitadelle* had sorted themselves out, it was painfully clear that "a substantial weight of armour" was going to be very hard to find. Manstein managed to scrape up a few oddments to stiffen the headquarters of the 3rd Panzer Corps and sent these south, together with the whole of the 3rd Panzer Division, one of the units which had suffered least in the previous battle. However, as it was intended that this force, the rump of the German armoured strategic reserve, into whose creation Guderian had put so much energy less than six months earlier, was now to strike two blows in succession, on the Donetz and on the Mius, still further reinforcement was needed.

The only force available consisted of the SS Panzer corps which had been withdrawn from the Kursk battle at an early stage, and was thus in better shape than the regular Panzer divisions. In addition the usual priority which the SS enjoyed in the allocation of equipment had allowed it to replace its losses more rapidly. From Manstein's point of view the drawback to using the SS lay in the fact that he had to get permission from OKW, and thus from Hitler in person. This would have been a delicate enough affair at the best of times, but during the last week of July, Hitler (as has been seen) was preoccupied with the Italian revolution and had himself decided to transfer the SS Panzer corps to Italy. A hectic exchange of messages took place between Army Group South and OKW, with intermittent direct intervention by the Führer. First, all of the SS was ordered to Italy; then only *Leibstandardte;* then it was ordered to Kharkov, except *Leibstandardte,* which was first to open the 3rd Panzer Corps attack, and then entrain for Italy; then, and finally, the whole corps was ordered south for the counterattack, subject to the withdrawal and entrainment of *Leibstandardte* when rolling stock became available—*vide* the conversation between Hitler and Jodl recorded above.

But now Hitler imposed a new condition. Presumably from pure intuition—it certainly cannot have been from intelligence reports, as none of the army commanders shared this opinion—the Führer thought that a fresh Russian offensive between Rylsk and Belgorod was more than likely, and he forbade Manstein to use the SS in the Donetz operation, but ordered him to proceed immediately against the Russian bridgeheads across the Mius.

In the meantime Russian tank strength in the Kursk salient was gradually returning to pre-*Zitadelle* figures. The figure of thirty-five armoured divisions, with which Rokossovski and Vatutin had started the battle, hardly shrank at all, although naturally the effective strength of many were diminished. Having been left masters of the field, the Russians were able to pull in and repair the majority of lightly damaged tanks by the end of July. Moreover, their spares problem was greatly simplified by the fact that they had fought the battle with only one type of tank—the T 34—while the Germans had engaged five separate types of tanks[1] and two assault guns. The number of Russian tanks operational on 5th July in the salient was about 3,800; by 13th July this had fallen to under 1,500; yet by 3rd August it had risen again to 2,750. It seems unlikely that any new equipment was sent to Rokossovski (al-

[1] This refers to basic, or chassis, variants; viz. PzKw III and IV, Panther, Porsche, and Henschel Tiger. Differences in armament compounded the varieties. Assault guns were mounted on both PzKw IV and Skoda chassis.

though three fresh regiments of self-propelled artillery went to Vatutin at the end of July), for both Sokolovski, to the north, and Koniev and Malinovsky, below the Donetz, were charged with the undertaking of supporting offensives at this time. Consequently this recovery must be attributed to skill and energy in the field workshops—and would probably have been even more spectacular if it had not been for the number of crews who had been made casualties in the early stages of the battle and for whom there were no trained replacements.

In this brief lull both sides proceeded with the urgent groundwork to their own plans. The Germans, if anything, were the more leisurely. Their notion was for "a short sharp punch to straighten out First Panzer Army's position south of the Donetz," (opposite Koniev) then "to use the whole of our armour to wipe out the big enemy bridgehead in Sixth Army's sector and to restore the Mius front." The *Stavka,* on the other hand, was pressing on with its own plans for breaking up the entire German front in South Russia. This plan was typical of all of the Russians' major operations in the East (with the one brilliant exception of Stalingrad), being unimaginative, lacking in finesse, and dictated by the size of their forces and the limitations of the subordinate commanders.

After worsting the Germans in the head-on *Materialschlacht* at Kursk, the Russians planned three separate secondary offensives which had the purpose of keeping the German reserves dispersed, together with the usual opportunistic and ill-defined aim of picking up any ground that was going if a weak spot should be discovered. Of these Sokolovski's, at Orel, was purely diversionary. But that launched under Koniev on the upper Donetz was intended as the northern arm of a gigantic pincer aimed at Kharkov, against which the supporting thrust was to come from Malinovski in the southern sector.

Thus it can be seen that the despatch of all the battleworthy armour to the extreme south of the front, to settle the score with Malinovsky, was the most dangerous course the Germans could follow. Paradoxically, Hitler, who sensed an impending attack on Kharkov and was trying to hold the SS back to deal with it, aggravated the danger. For had Manstein been allowed to attack first on the Donetz, against Koniev, the strength of Russian resistance would have warned Army Group South that something was afoot. As it was, the 3rd Panzer Corps and two of the SS Panzer divisions started their attack against Malinovsky on 30th

July. The Russians were very short of armour (German intelligence identified only one tank brigade in the entire Mius bridgehead), and within a few days their infantry was being forced back over the river, having suffered over seventeen hundred casualties in prisoners alone. The 3rd Panzer Corps also claimed the capture of four hundred anti-tank guns and two hundred field guns.

This was one of the last German tactical successes on the Eastern front, and it was to have immediate, and serious, strategic consequences. For even as the 3rd Panzer Corps was counting its spoils, Rokossovski's twenty-four armoured divisions were moving into position, 350 miles to the northwest.

While the military situation deteriorated, both in Russia and Italy, the days moved inexorably forward to that appointed for the meeting between Himmler and Popitz. Nothing had occurred, since the subject had first been broached by Langbehn, to moderate the Reichsführer's conviction that world events would soon make some form of personal "reinsurance" desirable. Yet he cannot have contemplated this direct confrontation with one of the leading conspirators against the National Socialist regime and the Führer's life with a completely easy mind.

The final details of the meeting were arranged by Himmler's personal Chief of Staff, SS Obergruppenführer Karl Wolf, for 26th August. The four men—Himmler, Wolf, Popitz, and Langbehn—met in a room at the Reich Ministry of the Interior. After some opening pleasantries (and there can have been few occasions in history when "pleasantries" sounded so brittle) Popitz got the ball rolling. He was very discreet.

"He approached the subject from the angle of the critical military and political situation which had arisen for Germany. Was it not possible that things had perhaps got a little beyond the *Führer's* control? Should he not be relieved of some of the many burdens which he bore? Should he not be, perhaps, reduced somewhat in cares and in power, with, of course, a resulting devolution of authority upon some strong personality—Himmler himself, perhaps; who better?—who should take action to save the Reich?"

Not surprisingly, there is no record of what Himmler said. All the participants are now dead, and we shall never know, but the likelihood is that he did little more than mumble noncommittally and leave the questioning to Wolf. At all events, what he did *not* do was to clap both

Popitz and Langbehn into prison for their gross sedition, and Popitz later told Goerdeler that Himmler was "not averse in principle" to the suggestions. Within forty-eight hours Langbehn received a travel permit for Switzerland, where he was to "feel out . . . the reactions of the Allies to a change of régime."

So far, so good. But the catch in the situation, from the conspirators' viewpoint, remained. They had chosen to confide in one of the most astute and ruthless operators in a jungle world where they themselves were mere novices, dilettante amateurs, gentlemen of principle. Himmler's vanity may have been touched by Popitz's suggestions. But he was far too astute not to recognise the essential enmity between himself, and everything he represented, and the conspirators. The naïveté of Popitz's suggestions emphasised rather than concealed the fact that the elimination of the Führer was but a preliminary stage, to be followed by a purge of the whole Nazi apparatus. And he who thought that he could beat Himmler to the draw, particularly after the events of June 1934, was a bold man indeed.

Whereas, though Himmler may have been fairly confident of his ability to deal with the conspirators at a moment of his own choosing, he was also uncomfortably aware of other, personal, enemies in less vulnerable positions. The constant power struggle in the higher echelons of the Nazi Party has already been the subject of discussion,[2] and it was sharpening and becoming more obsessive as the decline in Germany's fortunes steepened and the possibility of a successor to the Führer came to be openly discussed. It was sudden intervention from this quarter which put paid to the "Langbehn mission," and which ensured for the unfortunate doctor a more gruesome demise than he might otherwise have suffered.

Langbehn left Berlin with his wife at the end of August and travelled to Berne. Here he sought out British and American intelligence officers. Whether out of impatience or disappointment, he soon extended the scope of his contact to include "Allied" and even neutral missions, and to them he spoke less guardedly. It may be that he feared for his own skin if he did not return to Himmler with at least some form of positive encouragement. At all events, one of these lesser agencies (probably the Free French) sent a telegram to London asserting, . . . HIMMLER'S LAWYER CONFIRMS THE HOPELESSNESS OF GERMANY'S MILITARY AND POLITICAL SITUATION AND HAS ARRIVED TO PUT OUT PEACE-FEELERS.

[2] See pp. 59-69.

The telegram was sent in a cipher which the Germans had already broken, and it was decoded (independently of each other) by the Abwehr and the SD. Canaris at once warned Popitz that there was trouble ahead, but both imagined that it would be moderated by Himmler's protection. They had reckoned without the forces of personal rivalry which seethed around that focus of power. Schellenberg, the head of the SD, and himself no stranger to the world of cloak and dagger,[3] acted fast. He had known for a long time that there was something up between Himmler and Langbehn; now he saw the opportunity to put his chief in a cast-iron frame.[4]

When Langbehn and his wife crossed the border into Germany, they were immediately arrested on Schellenberg's orders. Thinking at first that this might be some sort of device to cover their tracks, and still confident in the patronage of the Reichsführer SS, the Langbehns consented meekly enough, and made no immediate attempt to contact Himmler. Schellenberg, meanwhile, had taken the text of the decoded telegram to "Gestapo Müller," Himmler's deputy, and a man who made no secret of his ambition to succeed the National Leader. Müller knew exactly what to do with such an incriminating document. He showed it immediately to Bormann. That same afternoon it was on Hitler's desk.

Matters had now gone so far, so fast, that Himmler must have been considerably taken aback when confronted with the facts concerning an emissary whom he still fondly imagined to be "conferring" in Berne. Regrettably, the trail of evidence peters out here, and we have no knowledge of the last act. Himmler must have talked his way out of it, as the whole affair blew over. Langbehn remained in jail, and was later transferred to Mauthausen concentration camp. Himmler managed to postpone his "friend's" trial on one pretext and another, so Schellenberg and Müller were unable to get anything on Popitz. The latter, with considerable courage, twice tried to approach Himmler and secure Langbehn's release, but the Reichsführer, not unnaturally, refused to see him.

[3] At the age of twenty-nine he had played a brilliant part in the kidnapping of two British intelligence agents, Captain Payne-Best and Major R. H. Stevens, at Venlo, in Holland. For this he had been decorated personally with the Iron Cross by Hitler.

[4] Or did he? Professor Trevor-Roper, the leading authority on the relations between the high Nazis, believes that Schellenberg's loyalty to Himmler rules this out. The Langbehn episode still has many unexplained features. But a clue to Schellenberg's attitude can perhaps be found in his behaviour in the affair of the Jewish trains. (See Ch. 22.)

Then, after the 20th July *attentat,* Popitz was arrested[5] and Langbehn hauled out and sentenced. Anything the wretched doctor may have said at that time went unheard in the discordant chorus of the Freisler trials. Even so, the Reichsführer took precautions. A letter from Kaltenbrunner to Lammers is a good example of the SS tail wagging the constitutional dog:

> I understand that the trial of the former minister Popitz and the lawyer Langbehn is to take place shortly, before the People's Court. In view of the facts known to you, namely the conference of the *Reichsführer*-SS with Popitz, I ask you to see to it that as a practical matter the public be excluded from the trial.
>
> I assume your agreement and I shall dispatch about ten of my collaborators to make up an audience. As to any others present, I request the right to pass on their admittance.

The SS at last had its fun with Langbehn (after he had been sentenced to death), and tortured him "in the most barbarous and horrible manner," finishing by tearing off his genitals.

One curious side issue is that Himmler does not seem to have borne any ill will against Schellenberg for his part in the affair. It may be that the Reichsführer never realised who had betrayed his scheme, or that he blamed only Müller. Two years later, indeed, Himmler was to make use of Schellenberg himself, casting the head of the SD in the same role as that which Langbehn had played at Berne—but that is to anticipate the events of April 1945.

While Dr. Langbehn had been exploring the possibilities of a separate peace in Berne, Army Group South was slowly disintegrating under a sequence of hammer blows that alternated along the length of its front. On 2nd August the 4th Panzer Army was monitoring such heavy traffic with its radio intercepts that it reported to Manstein that a fresh Russian offensive there was "inevitable within two or three days." Air

[5] Some idea of the state of mind of the wretched Popitz at this time can be gathered from Himmler's own account (*Vierteljahreshefte fuer Zietgeschicte* Vol. IV, 575):

> . . . since that time . . . Herr Popitz looks like a cheese. When you watch him he is as white as a wall; I should call him the living image of a guilty conscience. He sends me telegrams, he telephones me, he asks me what is the matter with Dr. X, what has happened to him; and I give him sphinx-like replies so that he does not know whether I had anything to do with what happened or not.

reconnaissance also showed that still more Russian armour was being moved south into Vatutin's command from Sokolovski, at Orel. Manstein immediately ordered the return of the 3rd Panzer Corps, with its two additional SS divisions, from the Mius—but the ink was barely dry on the teleprint when Vatutin's attack broke, at dawn on 3rd August.

The Russians had struck at the junction of the 4th Panzer Army and Army Detachment Kempf, west of Belgorod. Both these formations were very short of tanks, and were forced to give ground immediately. The majority of their vehicles were still under repair or being serviced, and were overrun in the field workshops. Thrusting down the interarmy boundary, the Russian armour fanned out to the west and south, opening a gap between the two German commands which by 8th August had widened to over thirty miles. The 3rd Panzer Army was still lumbering along the railway network from Zaporozhe, diverted from Kharkov to Poltava, then to Kremenchug. "It was clear, however, [Manstein wrote] that no action by these forces, nor indeed by those of the Army Group as a whole, could provide any long-term answer to the problem. Our divisional casualties were already alarmingly high, and two divisions had broken down completely as a result of the continous overstrain. . . . It emerged beyond any possible shadow of doubt that the enemy was now resolved to force an issue against the German southern wing."

Just before the Russian offensive started, Manstein had asked OKH for two Panzer divisions to be transferred to the northern wing of his front from the adjoining command of Kluge's Army Group Centre. The subject came up at a Führer conference for which an almost complete transcript exists. Besides the incidental light which is shed on other matters, this record is of unique interest in showing how the war in the East was run at the highest level.

HITLER. [who has been holding forth about the situation in Italy] . . . I don't know where the Duce himself is. As soon as I find out, I'll have him brought out by parachutists. In my opinion, that whole government, like that in Belgrade, is a typical *putsch* government, and one day it will collapse, provided that we act immediately. I can't take action unless I move additional units from the East to the West. In case your offensive[6] can't be carried through, we must make plans for reorganising your line. Are these your maps?

[6] The "offensive" referred to was that of the 3rd Panzer Corps against the Mius bridgehead of Malinovsky.

ZEITZLER. They are. Marked according to report.

HITLER. Will you please explain your over-all position to me. The point is I can't just take out units from anywhere. I have to take politically reliable units. That means, first of all, the 3rd SS Panzer Division, which I can only take from the Army Group South. That means that you will have to send other units down, and one can free those units only by liquidating this whole business, by giving up this whole bulge. Perhaps the front should also be shortened at other, minor, places.

KLUGE. Well, my Führer, the present situation is that a certain pressure from strong elements is being felt here. However, this has not had the full effect, because they[7] had trouble crossing the Oka River. Unfortunately, they were able to score a fairly deep penetration yesterday, in the area of the 34th Army.[8] However, this is being compensated by counterblow, although our own forces there are relatively weak. Here was the break-through, in the area of the 297th Division, which could be compensated somewhat by the withdrawal of the whole line.

HITLER. Are you on *that* line?

KLUGE. No. On *this* line.

ZEITZLER. The other map shows the exact position today. Up there is the withdrawal.

HITLER. Please show it to me on this map.

KLUGE. Well, the present situation is that yesterday there was a very strong attack here, although it was not as strong as we had expected. Rather, it was weaker and narrower, and although it resulted in a certain penetration, it could be stopped. Principally, there were large tank attacks—here there were 150 tanks, of which 50 were knocked out. The plan is now to go into this so-called Oka position, to cross over here, to shorten the Bolkhov bulge—tonight. I would like to take permission along right away to move the line away from Bolkhov and to shorten the whole business here. In general, our intention is to retreat here again, and then to move into this line. That's the immediate plan. After this minor withdrawal has been carried out, the general withdrawal should take place. In preparation for this movement—which has to take place in a very constricted area, especially here in the north—the *Gross Deutschland* has advanced with its reconnaissance forces; has thrown back the enemy here; although it struck fairly heavy resistance here. I don't know how that is going to develop today. At any rate, they are to reach the edge of this land, which is marked as swamp. Actually, that is no swamp; unfortunately, it is terrain which can be crossed safely.

ZEITZLER. This morning the enemy made stronger attacks.

KLUGE. Has he attacked there?

ZEITZLER. Yes, and also an armored brigade.

[7] He means the Russians; very frequently in this conference, "they" or "he" means the Russians.

[8] There is a mistake in the transcript; no German army numbered higher than 25 existed; probably he meant the 4th Army, which was fighting in this area.

KLUGE. We knew about that already yesterday. In this area the enemy has two infantry divisions—that is, two good ones—plus one armored brigade, and another brigade is being moved up.

HITLER. Tell me, where are those hundred Panthers?

KLUGE. They aren't there yet. They are just being assembled after having been unloaded.

ZEITZLER. The last trains were all there on the 26th.

KLUGE. They are there. Their crews are, too, though not all of them.

HITLER. And where are they?

KLUGE. In Berdyansk. Well, there is rather strong pressure here, which is not limited to this place but unfortunately also extends to this very weak salient, which, in my opinion, is the most dangerous place. It is held by some jumbled miscellaneous troops that first tried to hold this line but have now been pressed back. The following development could become very unpleasant if the enemy could take this road to the station at Reseta. We are still using that for moving from south to north. For that reason I have requested that the 113th Division be committed up here with the 4th Army, next to the Orel rail line and next to the highway . . .

ZEITZLER. The Führer has already given his permission.

KLUGE. . . . and to compensate for that, I would like to pull out two divisions; first one division which was supposed to be sent *there* right away, and then another which we really also wanted to put into this locality, in order to strengthen this wing; because *here* we ought not to retreat one step more. That would be a very unpleasant development.

Here there are strong forces on the advance, which are far superior to our own—even tanks, but comparatively few of them; the mass of their tanks is pressing down in this direction, toward the *Gross Deutschland,* and of course *here,* too.

HITLER. They must be gradually losing their tanks, too.

KLUGE. Certainly, that is clear. We have knocked out quite a number of them. Just the same, he is still attacking with strong tank units so that at present we have our hands full in coping with this crowd. That is the present situation.

Now we want to withdraw into this shortened Oka line, and on this basis the evacuation of Orel and everything that belongs to it is supposed to take place, and then—

ZEITZLER. Then the next point is the *Gross Deutschland,* sir.

KLUGE. My Führer, I still wanted to add that in order to create a sound foundation for further movements, Model[9] and I both feel that the attack of *Gross Deutschland,* which is in progress, and another attack are necessary and would establish a solid line.

HITLER. I don't think that this will work any more. Will the *Gross Deutschland* have to go through the woods?

[9] Generaloberst Model was then commander of the 9th Army, belonging to Kluge's Army Group Centre.

KLUGE. Certainly not. That would have been forbidden. But the attack of the 253rd Division—

HITLER. I want to review the over-all situation again. The problem is to pull out a fair number of units in a very short time. This group includes, first of all, the 3rd Panzer Division, which I must take from the Army Group South, which itself has to cover a very broad front. . . . In other words, it is a very difficult decision, but I have no choice. Down there I can only accomplish something with crack formations that are politically close to Fascism. If it weren't for that, I could take a couple of army Panzer divisions. But as it is, I need a magnet to gather the people together. I don't want to give up the Fascist backbone, because in a short time we will rebuild so many things. I am not afraid that we can't manage that, if we can hold northern Italy.

KLUGE. My Führer, I want to call attention to the fact that nothing can be pulled out of the line now.[10] That is completely out of the question at the present moment.

HITLER. Just the same, it must be possible—

KLUGE. We can only withdraw troops when we reach the Hagen Line.

ZEITZLER. Let the *Gross Deutschland* get to this point, then draw them out, keep them here for a while, and the 7th Panzer must leave soon—

KLUGE. We could not anticipate these over-all political developments. We couldn't guess that this would happen. Now a new decision must be made, first of all that Orel must be evacuated after we have moved out our own vital material.

HITLER. Absolutely.

KLUGE. Then there is another question: This rear line, the so-called "Hagen Line," is still under construction?

HITLER. Yes, unfortunately.

KLUGE. There is nothing that can be done about it. We have a huge number of construction battalions and God knows what all else. We have been having cloudbursts every day, of a style you couldn't even visualise here. All of the construction battalions had to keep the roads in shape; they are supposed to have been back in the Hagen Line long ago in order to finish it, but I needed them up front to straighten things out.

HITLER. Perhaps the rain will stop soon.

KLUGE. I certainly hope so. It was a little better today.

HITLER. But you have to admit, Marshal, that the moment your troops

[10] The following discussion refers to the tempo of the German retreat from Orel to winter positions along the Dnieper, the so-called "Hagen Line." Originally this was supposed to be a step-by-step retreat, from Orel to Karachev, from Karachev to the Desna, and from the Desna to the Dnieper, extending from the end of August well into the fall of 1943. In order to free troops for action in Italy, Hitler wanted to force Kluge to start this retreat much earlier and to withdraw at greater speed than had been originally planned. The German line was back on the Dnieper at the end of September. The German retreat has been characterised as methodical, the daily withdrawals from 5th August to 22nd September averaging from one and a half to three and a half miles.

reach approximately that line, quite a number of your divisions can be moved out.

KLUGE. My Führer, I want to call your attention to the fact that four divisions—

HITLER. —are very weak.

KLUGE. I have four divisions which are completely exhausted.

HITLER. I'll grant you that. But how many of the enemy's divisions are smashed?

KLUGE. Well, in spite of that. Now we come to the question of the so-called Karachev position, my Führer. If I move into that position, which isn't ready, and if I am attacked again with tanks and everything else, they will break through with the tanks, and then when they have broken through with their tanks, the moment has come. I am just mentioning that again because this is a good opportunity, because we might get into a very difficult situation. I would like to suggest again that it might be more practical to move all the way back behind the Desna River while we are at it. We must have the Karachev position anyway as a skeleton, as it is now and as it will be after two more weeks of work, in order to give the troops support on the retreat. Therefore my suggestion is that it would be more practical to move right behind the Desna now.

HITLER. Here you are safe and *here* you are not.

KLUGE. Bryansk—this part of the line is good, but this other piece is not fully constructed yet.

HITLER. That part is not better than this one. If you put these two pieces at Bryansk together, then they make up as much as—

KLUGE. But then I have to have time to construct them. I can't do that—

HITLER. You would have to construct the other one anyway.

KLUGE. Yes, I would here. But not over here at the Desna.

HITLER. Not here.

KLUGE. I have to build from here to here, and over there I wouldn't have to build anything.

HITLER. But that is practically the same length—

KLUGE. But this one is better because nothing can happen to me on this whole line.

HITLER. They won't attack here. They'll come down this way.

KLUGE. That's the decisive point. But then, my Führer, I won't be able to draw back as early. First I have to construct the Hagen Line; I must have that in order, I can't just go back in a mad rush.

HITLER. Nobody said anything about a mad rush.

KLUGE. But at any rate, not much faster than was planned.

HITLER. What was your timetable?

KLUGE. Timetable as follows: In about five days—

HITLER. Altogether, when will you be back on that line?

KLUGE. We had not intended to be back in there before the beginning of September.

HITLER. That's impossible, Marshal, completely impossible.

KLUGE. Naturally, under these circumstances everything has changed a little. But it will take at least four weeks before the position is even usable.

ZEITZLER. Do it in two moves. Perhaps you can stay here until the line is ready.

KLUGE. That won't work for the following reasons: maybe for a short time, but not in the long run; the rail capacity to Orel is fifty trains, but the moment we lose Orel it decreases to eighteen trains per day, which would be a very unpleasant situation.

ZEITZLER. You won't need very many trains if you are in this position.

[It is noticeable that, although paying lip service to Kluge's seniority with an occasional "sir," Zeitzler takes Hitler's part throughout the discussion, to the extent almost of being insubordinate.]

KLUGE. No, that won't work. I don't even have facilities to unload them.

ZEITZLER. If your troops are here, this strip of rail line is of no value.

KLUGE. No, not any more. I just wanted to emphasise that if I give up Orel, I have to retreat in one move; but the important thing is that I have my positions prepared behind me.

ZEITZLER. If you can hold here for six or seven days, then you gain that much time and a few units here will be freed.

KLUGE. But the calculations must always be based on the situation in the rear. I must have at least a moderately strong position, or else they'll overrun me, and then I'll be in the hole again and won't be able to spare any troops.

ZEITZLER. Sir, on this line you will gain six or seven days.

KLUGE. You mean here? Oh no, the enemy will reach that in two or three days.

ZEITZLER. If you could hold this for six or seven days, then you could move the line down from here to here, so that in ten days you would be here.

KLUGE. You mean now?

ZEITZLER. Yes.

KLUGE. That would mean a headlong retreat in this whole area, which, in my opinion—

ZEITZLER. Perhaps the army group can make new calculations.

HITLER. Just the same, Marshal, we are not masters of our own decisions here; in war, decisions are frequently necessary—

KLUGE. My Führer, if you order me to do it quickly—but then I would like to direct your attention to the fact that this plan is contradictory to that of the Hagen Line, which isn't finished.

HITLER. The other one isn't finished either, at least not at that point, and anyway, the Russians won't attack where the position is finished.

KLUGE. For instance, I could do the following, my Führer: I could move back into this position, the construction of which is more advanced up here and also here, although over here practically nothing has been done. In that case I would have to allow for giving way a bit here, but then this has to be built.

HITLER. Certainly, that is supposed to be built, as a precaution; but I don't want a withdrawal at this point now, because that will have to be done anyway in the winter, when the Russians attack. Model has built that up very solidly. It ought to be possible to build some sort of a position in that time. At the time of the advance we managed to build a position anywhere we had to stop, and to hold it. Those bastards over there can dig a position in two days, and we can't push them out of it.

KLUGE. My Führer, actually the question is that of tanks. That is the main point. He batters so hard with his artillery and tanks that he gets through after all.

ZEITZLER. Sir, in my opinion, moving back into *this* line would free half of the divisions, which could then be pulled back here, and you could have them dig for six days. Then that position would be ready.

KLUGE. No, that doesn't solve the problem. In my opinion, the earliest time for occupying the Hagen Line would be in—let's see, today is the 26th—in about four weeks, if we cut it fine maybe three or four weeks, but that's absolutely the earliest.

HITLER. Well, we just can't wait that long. We must free some troops before that. It's no use.

KLUGE. Sauckel won't be able to get his workers out before that.

HITLER. He has to. Look how fast the Russians can evacuate.

KLUGE. But, my Führer, that is an enormous crowd. He'll jam up all my bridges over the Desna.

HITLER. How many people are in here, anyway?

KLUGE. Several hundred thousand.

ZEITZLER. Two hundred and fifty thousand men, I was told—

HITLER. What are 250,000 men? That's nothing at all.

KLUGE. My Führer, I need my forces for fighting now. I can't use them for all kinds of other things.

HITLER. On the contrary, I would herd these people out of there immediately and put them to work on the position here.

KLUGE. We've already tried that. At the moment they are all harvesting. The rye has just been mowed. They have no idea of what is coming. If we move them back for construction work they'll run away in the night. They'll run to the front just to mow their rye. All these are difficulties. Nothing has been organised.

HITLER. What is going to be done with the harvested rye? Is it going to be burned?

KLUGE. Certainly, we'll have to. Probably we will burn it, but I don't know whether we'll have time. We'll have to destroy it somehow. Especially the valuable cattle we have here, and back here I have lots of guerillas who are not finished off. On the contrary, they are making themselves felt again. They were suddenly reinforced by a huge parachute operation, here. And then there was this famous cutting of the rail line at four hundred points.

HITLER. All that may be perfectly true, but it doesn't alter the fact that this has to be done. I think that the Army Group South is in a much worse position. Look at the kind of sectors it has. One of its divisions, the 335th, has a front of forty-five kilometres.

KLUGE. But, my Führer, I don't know how the impression originated that we didn't have long sectors too. That's where the 56th Division was, they had more than fifty kilometres; and the 34th had forty-eight kilometres. That calculation is not correct.

HITLER. That's true, you did have such sectors when we started out.

KLUGE. At the time we started out—

HITLER. On the whole, the Army Group Centre had an entirely different type of division sector.

KLUGE. Our sectors became narrower through the mass attack, but we still have thirty kilometres and more apiece. Our front is already thinned out to that extent.

HITLER. That's no comparison.

KLUGE. Up there in the sector of the 3rd Panzer it is very thin too.

HITLER. How is the situation here?

ZEITZLER. They haven't attacked here. The latest seems to be that they have pulled out their motorised corps here, and have replaced them with rifle corps of the Guard. They may be resting these corps in order to use them over here. I'm a little worried about that place, because they moved that parachute army up there too. I'm not sure what he plans to do with that. Their railroad traffic is a little heavier, so that I am of the opinion that they are pulling troops out by rail. Or else they are bringing them up there. So we have to watch that. They evidently had too heavy casualties, so they've stopped trying to do that with motorised units. They are pulling out here. I have spoken to Manstein about this business, this as well as that. He called up again today. Now that the *Leibstandardte* has left, he wants to reconsider whether to attack[11] at all. I think it would be sensible to wait. This small matter doesn't need to be cleaned up, since the pressure is not too great.

HITLER. How soon can the *Leibstandardte* leave?

ZEITZLER. The first train leaves tomorrow night. We are counting on 12 trains per day. After four or five days, 20 trains. The whole movement, consisting of 120 trains, will take from six to eight days.

HITLER. Only 120 trains?

ZEITZLER. Yes.

HITLER. Now, now, Zeitzler.

ZEITZLER. It might even be 130 trains.

HITLER. I'm afraid it will be 150.

ZEITZLER. It doesn't make much difference if there are ten or fifteen more or less.

[11] This attack was the second of the two operations planned by Manstein for the end of July, against Koniev's bridgehead across the Donetz at Izyum. As has been seen, it never materialised.

HITLER. What are they leaving behind? Are they going to leave the Mark IV's here or are they going to take them along?

ZEITZLER. Last night the order came through to leave them behind, because they would get new ones. I counted on that—but we should put pressure on the *Leibstandardte* to leave them there. As far as I know Sepp,[12] he'll take them along unless we send someone down there to make sure. The best thing is for them to leave them.

HITLER. The two divisions that remain behind are weak anyway. It would be better to give them additional tanks, and one must see whether one can't give the Tigers to one of them. He[13] is going to get two Tiger companies anyway.

ZEITZLER. I agree that they shouldn't.

HITLER. That will be enough for the *Leibstandardte*. How many Tigers is that—two companies?

ZEITZLER. He is going to get two more companies—twenty-two Tigers.

HITLER. On top of that he's going to get a hundred Panthers, his whole battalion. And then he must get replacements for his Mark IV's in the rear.

ZEITZLER. What should have gone there as replacements I'll hold back, and he can have them, too.

HITLER. Maybe assault guns, too, so that he can leave his behind. That would strengthen the two remaining divisions. Then the next one to leave would be the *Reich*. The *Reich* can also have part of its equipment to these or other units, and can get its hundred Panthers replaced in transit.

ZEITZLER. That way we are saving a lot of equipment.

HITLER. We are saving a lot of equipment, and the units out here are getting stronger. And then Manstein has to get some more supplies for his divisions. For instance, the 16th *Panzergrenadier* Division must also get something.

ZEITZLER. When the other Panzer divisions arrive, they can take over some of it—

KLUGE. Well, my Führer, then we are confronted by a new situation.

ZEITZLER. Perhaps the army group can work out a plan for what is the earliest possibility, and what the risks involved would be.

KLUGE. We'll sit right down. I brought my G-3 along. We will go over it once more. But everything still depends on the construction of the Hagen Line. I don't want to slide back into a position that is practically nonexistent.

HITLER. This is how I really feel about it: If there weren't this pressing danger down here, I would have committed the two divisions you are getting right away, instead of the 113th.

KLUGE. Yes, my Führer. Now there will be no forward commitment of

[12] Sepp Dietrich, SS Oberstgruppenführer, commander of the SS Panzer Division *Leibstandardte*, later of the 5th and 6th Panzer armies.
[13] Evidently Dietrich.

these two divisions. There will be no attacks at all here; that would be useless; that would be senseless. That was all planned under the condition—

HITLER. Just secure the rail line so that it can be used.

KLUGE. According to the original plan, we would have had plenty of time to do this.

HITLER. Wouldn't it be possible to detach some units for the purpose of building up the position?

KLUGE. You mean take them away from Model? Though the others follow up every day—and we with the miserable remains of the 11th, 212th, 108th, with the 209th—

ZEITZLER. Those are the ones that were smashed up.

HITLER. All right then, draw out the broken divisions, fill them up, and build with *them.*

KLUGE. All right then, I'll have to free some units somehow. Unfortunately, I also need troops to secure the roads along which these caravans are moving, or else they'll all be knocked off behind the forest of Bryansk, because everything is swarming with guerillas who have been reorganized.

HITLER. I, too, have to make difficult decisions, very difficult decisions.

KLUGE. I can well believe that.

HITLER. But there is nothing else to do.

KLUGE. But I absolutely cannot spare any units until this operation has been finished. We'll see how we can manage things afterward.

HITLER. You must see to it that you finish it as soon as possible. I can tell you this much: *Gross Deutschland* will be taken away in the near future, and secondly, you will have to give up a few units for that position down there. You will have to give up a few Panzer—and a few infantry divisions—

KLUGE. Not Panzer! I have—

HITLER. Yes, we'll pull them out and they'll be refitted in the West.

KLUGE. But I can't do anything without Panzer divisions!

HITLER. But certainly you don't care about that "junk." You can easily spare that.

KLUGE. What junk?

HITLER. You yourself said, "That's just junk."

KLUGE. I did not say that!

HITLER. Yes, it slipped out. That's why we're going to take them away from you.

KLUGE. No, my Führer, I didn't mean that. I have so little left, just a little bit. What I wanted to indicate was that the situation is hardly tenable any more.

HITLER. Yes, you have no Panzers. That is why I say they can be taken away and refitted in the West. We can always get them back again. Meanwhile they can be filled up. And finally, the men have deserved it. It would be all wrong to do it the other way. I can have these divisions reorganized in the West, and the Western units can be moved up here. The most important thing is to get the 9th and 10th SS divisions ready quickly.[14] Today I got an

[14] SS Divisions 9 and 10 were organised in France during the winter of 1942-43.

opinion how the "Goering Division" stands up in combat. The English write that the very youngest, the sixteen-year-olds, just out of the Hitler Youth, have fought fanatically to the last man. The English couldn't take any prisoners. Therefore I am convinced that these few divisions composed of boys, which are already trained, will fight fantastically well, because they have a splendid idealistic spirit. I am completely convinced that they will fight fantastically well.

ZEITZLER. Well, the Field Marshal and I will sit down to it later.

KLUGE. I'll have to think it over once more, my Führer. Now that I know what the over-all purpose is, I will act accordingly.

HITLER. As I said, the most important thing is that I get the SS corps out. Manstein needs something down there as replacement. I don't know yet what I'm going to give him. Perhaps the 7th Panzer Division, which could be pulled down here if he could close this up over here. But he has to have replacements or else he won't be able to hold any of this business. And he needs a couple of infantry divisions. He can't hold this mess here. Of course, if the worst comes to the worst, we can't do anything except shorten the lines down here, too. But we have to realise that that would be a desperate situation. It's certainly not pleasant. These are very difficult decisions, decisions that bring us to a critical point. But I'm considering all the alternatives. Difficult to do anything up there at Leningrad because of the Finns. I also considered whether we could give up this down there—

ZEITZLER. If we decide on something up there, we have to do something down there, too.

HITLER. There wouldn't be much profit in this.

ZEITZLER. Yes, we would gain something. The enemy isn't doing anything big now.

HITLER. If the worst comes to the worst, we may even have to give *that* up.

ZEITZLER. That's easier to do than this—

HITLER. How many do you think we can get out of here? We have to be strong here or else they will start landing operations at Novorossisk again. First, everybody says give it up, but then I hear Kleist,[15] or whoever else is down there, yelling, "That's impossible." With such limited forces in this position, it is impossible to counteract and the enemy will only start attacking in this area. If that happens, we won't be able to bring any more ships in. We are still getting our ships through, but that would finish it. What is already committed is all right, but I can't bring any more in.

ZEITZLER. We could try to form a small bridgehead. Then we could hold on there for a while.

HITLER. I'm afraid we can't hold it, but we can try. We'll have to think about that.

[15] Generalfeldmarschall von Kleist commanded Army Group A, which, after retreating from the Caucasus, held the southernmost part of the German Eastern front. Novorossisk, a port on the eastern side of the Kerch Strait, was still in German hands at this time.

KLUGE. On our extreme northern wing we can fall back on our prepared positions at Velikiye Luki, as I suggested. We can strengthen the line there, too.

ZEITZLER. That's been planned, sir, but that won't free any of your troops.

KLUGE. No, that wouldn't free any troops. We can't give up anything else, except that salient. Then we'll have to swallow Kirov and leave everything else the way it is, although I would have liked to improve things here a little, but unfortunately, it isn't possible.

HITLER. We can retreat here, too.

KLUGE. Perhaps we could free a division here, but that is a complicated story because the position there is already—

ZEITZLER. The positions there are particularly good.

KLUGE. The positions are good. They were built with terrific effort.

HITLER. But you'd rather have Kirov?

KLUGE. Yes, I would like to retake that. That is always a base for the enemy.

ZEITZLER. That would be more costly.

KLUGE. Under the present circumstances it is completely impossible—

ZEITZLER. You can free something up there only after you have withdrawn here.

HITLER. Will I see you again?

KLUGE. No, I intend to return immediately. Heil, my Führer.

[Field Marshal von Kluge withdraws.]

HITLER. If only the SS corps were out of there already.

ZEITZLER. The *Leibstandardte* leaves tomorrow at the earliest, at the rate of twelve trains per day.

HITLER. The SS corps equals twenty Italian divisions.

ZEITZLER. He[16] must get *Gross Deutschland* and the 7th Panzer down there. If Kluge stays on this line for a week and moves half of the freed divisions over there, it ought to work. If a division can dig itself in for six days in a sector, it has got something. He is still mentally adjusted to slow movements, and can't get away from this idea. Perhaps it will come to him. In my opinion, everything would be all right then.

On 8th August, Zeitzler flew out to see Manstein at Army Group South headquarters. The Field Marshal "told him quite plainly that from now on we could no longer confine ourselves to such isolated problems as whether such and such a division could be spared, or whether the Kuban bridgehead should be evacuated or not." There were only two possibilities, Manstein went on. Either the entire Donetz region was evacuated immediately, or he was drafted a further ten divisions from

[16] Manstein.

the other sectors of the Eastern front. (This would have brought Manstein's army group to a strength equivalent to the total of all the other German forces engaged in Russia.)

Naturally, Zeitzler did not commit himself, and after he had left repeated enquiries to OKH produced nothing more than formal acknowledgements. In the course of the following week the Russian pressure against Kharkov intensified to such an extent that Manstein was faced with the alternative of "locking up" Army Group Kempf in the city to play the role of a minor Paulus or abandoning the town altogether. Hitler sent him a special message, ordering that the town be held at all costs and pointing out that its fall would produce "an unfavourable effect" on the attitudes of Bulgaria and Turkey. "However true that might be," was Manstein's acid comment, ". . . the Army Group had no intention of sacrificing an army for Kharkov." The town was evacuated on 22nd August. But the unfortunate General Kempf was sacrificed. His command was redesignated "8th Army," and he was succeeded by General Wöhler. Manstein's own part in this affair is equivocal, to say the least. "Although I had got on well with General Kempf, I did not oppose the change," and the man appointed in Kempf's place had been Manstein's Chief of Staff of the 11th Army. It seems likely that both Manstein and Hitler concurred, if tacitly, in this choice of a scapegoat.

By now Himmler was not the only leading Nazi to have taken alarm at the way events were developing. Henriette von Schirach (who always claimed that Hitler tried to kiss her when she was twelve years old) has left an account of a contrived meeting between her husband and Goering. The meeting began in an innocent atmosphere of *Gemütlichkeit* in a "secluded, velvet-lined room" of a well-known Vienna restaurant. (Baldur von Schirach, founder of the *Hitlerjugend,* was at this time the Gauleiter of Vienna.)

"A famous composer played the piano, followed by Goering, who played improvisations from *Der Freischütz.* The Goerings were in a happy frame of mind. Hermann had just bought a new leather brief-case in the blue Luftwaffe colours, which he proudly showed us, also a bottle of Jean Desprez perfume, which he could obtain only in Vienna."

However, Schirach had set the stage with the express purpose of getting the Reichsmarschall into an amenable frame of mind, and judging that this had been attained at the conclusion of the impromptu piano solo, he approached him. Schirach's argument was that of self-interest, with the familiar disguise of concept of duty to the Reich, appeal to the subject's vanity, and a formal lip service to the "strain" on the Führer. He urged Goering to "speak to Hitler privately," but as this was coupled with more robust talk of action—"I and my Hitler Youth are with you, the Luftwaffe is strong, and there are plenty of men who are prepared to act . . . We must make this a common cause . . . As Reichsmarschall it is expected of you!"—what he was really saying must have been perfectly plain to those who were listening.

At the conclusion of this address

"Goering looked at him [Schirach] without batting an eyelid. He looked a little sad, as if he had heard this kind of complaint before. Then he picked up one of his exquisite imported cigarettes, fingered it for a while, and lit it very slowly. He sank deep into the red chair and looked at us.

" 'To speak to Hitler alone, what an idea! I never see him alone these days. Bormann is with him all the time. If I could, by God, I would have gone to see Churchill, a long time ago. Do you think I am enjoying this damned business!' "

At this point Emmy Goering, who had been long enough in the immediate vicinity of the Führer to know what conversations were, and were not, dynamite, pressed her white hand against Hermann's mouth and said:

" 'Let's not talk about it any more, all will be well in the end.' "

At this stage we must return to that most shadowy and perplexing of subjects—the state of Hitler's mind. For this was the mainspring of the German campaign in the East; the Führer's daemonic genius, in defeat as in victory, exerted influence at each shift in the fortunes of battle. In the late summer of 1943 three distinct and separate factors brought themselves to bear on Hitler's mind, warped it, and left its flexibility permanently impaired—a change whose subsequent course on the war will soon be illustrated.

The first, and most obvious, factor was the failure of *Zitadelle*. This had been an exclusively professional affair, conceived, prepared, and directed by the officer corps. They had chosen the ground, the weapons, the timing. Hitler's only intervention had been at a strategic level (and when the issue of the battle had already been decided). Hitler had been uneasy about the operation from the start. And two of the generals he most trusted, Guderian and Model, had shared his misgivings. Yet the weight of professional opinion, Keitel, Zeitzler, Manstein, had been against them. The result? Total defeat; disintegration of the Panzer reserve; retreat to the Dnieper and beyond.

Hitler's distrust of professional army officers had received spectacular, and almost simultaneous, confirmation by their behaviour in Italy. He regarded the Badoglio *coup* as a classic example of behavior by the military *clique,* which will not hesitate to overthrow the Party once it loses confidence in the outcome of the war. (In his radio address at midnight of 20th July, 1944, Hitler was to revert to this when he referred to the "attempted stab in the back . . . as in Italy," and this comparative formula was adopted in all the loyal Orders of the Day that were drummed up from outlying commanders in the days following the *attentat.*) Their loyalty Hitler had always mistrusted; their obedience in the field could be ensured only by the closest supervision, it seemed, for Hitler had not forgotten 1941; their professional competence, even when they were left to themselves, seemed doubtful.

In whom, then, could Hitler trust? He was far too shrewd a politician, too perceptive an observer of human frailties, not to be aware of the defeatism and intrigue to be found in his immediate entourage. Langbehn and Himmler; Schirach and Goering; Guderian and Goebbels. The three men closest to the Führer had been approached. They had taken no action, it was true, but neither had they arrested the appellants for treason. Between August and December 1943 there were five separate attempts on Hitler's life. Chance of circumstance inhibited them, and of their existence Hitler was never made specifically aware. But his instincts told him that he was in danger; his excuse for starting the war, made almost in jest—"At any moment I can be eliminated by a criminal or a madman"—had now taken on the force of an omen. The explosion of the Stauffenberg bomb was less than a year distant.

So, feeling himself increasingly isolated, and uneasily conscious of the enormity of the forces he had roused outside Germany in addition to the apprehension and discontent within her frontiers, Hitler stood completely alone. Alone in the sense that he was separated—whether out of suspicion or distaste—from the company and influence (however transitory) of rational minds. The resultant vacuum was not filled, but contaminated by influences at the same time feverish and malignant. Bormann, Fegelein, Dr. Wulf the astrologer, Dr. Morell . . . there were many waiting in the wings, ready to exploit for their own ends the Führer's isolation and disillusionment. And we can see that from this time on Hitler's mentality alters direction; his descent into a Faustian world of true insanity begins imperceptibly to steepen as it approaches the precipice of the 20th July *attentat*.

Indeed, by the late summer of 1943 the morale of the whole Wehrmacht, from top to bottom, had suffered permanent change. Its courage and discipline were unimpaired. But hope was tainted, and humanity, where vestiges of it remained, was extinguished.

August came in stifling heat; then September, the days crisper but with an evening fog. The old battlefields of 1941 rose and receded—Bryansk, Konotop, Poltava. To the rattle of machine guns, as a few last scores were settled with the local population, and the thud of demolition charges, the German Army retreated across European Russia, leaving a trail of smoke, of abandoned vehicles and loose-covered shallow graves.

BOOK IV | Nemesis

Fight on with us against hated Bolshevism, bloody Stalin, and his Jewish clique; for freedom of the individual, for freedom of religion and conscience, for the abolition of slave labour, for property and possession, for a free peasantry on its own land, for your own homestead and freedom of labour, for social justice, for a happy future for your children, for their right to advancement and education without regard to origin, for state protection of the aged and infirm . . .

Goebbels, January 1945
(Doc. Occ. E, 18-19)

Who compels us to keep the promises we make?

Himmler to d'Alquen

COMMANDERS AND DISPOSITIONS OF OPPOSING FORCES IN SPRING 1944

German	*Sector*	*Russian*
Army Group North		*"Fronts"*
(Küchler)	Baltic-Lake Ilmen	Leningrad (Govorov)
18th Army (Lindemann)		
16th Army (Hansen)	Lovat-Dvina	Volkhov (Meretskov)
		1st Baltic (Bagramyan)
Army Group Centre		
(Busch)		
3rd Panzer (Reinhardt)	In reserve to south of Polotsk	3rd Belorussian (Chernyakovski)
4th Army (Heinrici)	Vitebsk-Orsha	2nd Belorussian (Zakharov)
9th Army (Model)	Rogachev-Kalinkovichi	1st Belorussian (Rokossovski)
2nd Army (Weiss)	Along Pripet	
Army Group South		
(Model, after Manstein)		1st Ukrainian (Vatutin)
4th Panzer (Rauss)	Fluid; see sector maps	
8th Army (Wöhler)		2nd Ukrainian (Koniev)
1st Panzer (Hube)		
Army Group A		
(Schörner, after Kleist)	See sector maps	3rd Ukrainian (Malinovsky)
6th Army (Hollidt)		4th Ukrainian (Tolbukhin)
3rd Rumanian (Dumitrescu)		
17th Army (Jaenecke)	Crimea	Caucasus (Yeremenko)

nineteen | "THE FLOODGATES ARE CREAKING"

Now it was late October. At evening those same banks of inert and freezing clouds which had first roused foreboding in the invader two autumns before returned to the battlefield. Each night the temperature fell below zero, to petrify the soft ridges of glutinous mud which bordered every road across the steppe. During the day a weak sun would thaw the surface, but with the shortening of its parabola and the increase in the hours of darkness, the ground turned hard as concrete.

As the winter of 1943 approached, a feeling of gloom and despair permeated the German Army, a dull conviction that the war was lost—yet without sight of its end. They were still deep inside Russia. Unlike the winter of 1944-45, when they were aroused to heroic frenzy in the defence of their homeland, they found themselves slowly retreating across a bleak and hostile landscape, always outnumbered, perpetually short of fuel and ammunition, constantly having to exert themselves and their machinery beyond the danger point. And behind them lay bitter memories of what midwinter was like. Major Gustav Kreutz, an artillery officer with the 182nd Division wrote:

> Towards the end of the month we at last got some replacements, new assault guns [these probably were self-propelled 75-mm. on Skoda chassis] up to battalion strength. They were mostly young chaps from the training barracks with a few officers and n.c.o.'s who had seen action in Italy. In no time they were complaining about the cold. They kept fires going during the day as well as at night, and were breaking up a lot of wooden outhouses for fuel which would have been valuable later. I had occasion to speak sharply to them about this and one of them answered that on

> that day the thermometer had fallen to ten below, and was this not abnormal? I told him that soon he would count himself lucky when the thermometer was not ten but twenty-five degrees below, and that in January it would fall to forty below. At this the poor fellow broke down and sobbed.

Kreutz adds a characteristic comment. "I found out that Lieutenant P—— had a good combat record in Sicily, and so took no disciplinary action. Later, he was killed fighting in the gallant defensive battles around Zaporozhe."

Throughout the last half of 1943 the German Army in the East was in a steady decline. In the three months immediately following the suspension of *Zitadelle,* Manstein's army group received only 33,000 replacements, although it suffered 133,000 casualties. Nominal strength fell lower even than these figures suggest because all the satellite troops had been taken out of the line. The remnants of the Italian force had returned to their home country, and the Hungarians and Rumanians, who were now more interested in fighting each other, had to be kept on anti-Partisan duties in the rear and as far apart as possible.

The equipment situation continued to deteriorate, especially in the Panzer units, for with Guderian's enforced sick leave the principles he had tried so hard to enforce came to be neglected. First, there was a reversion to the bad old practice of proliferating new divisions. Besides the dangerously false impression of strength which this practice gave to the order of battle in the map rooms, it also had the effect of starving the experienced divisions at the front of equipment, as most of the material from the factories went into the new formations. In the absence of the Inspector General, various subordinate departments began to stake out their claims on industrial production. The SS tried to corner the production of Panthers for its own divisions; the artillery succeeded in diverting a large proportion of the assault-gun output, and even went so far as to get a ruling that the PzKw IV was to be stopped altogether and the production of assault guns increased. The tail end of the Mk IV turret production was to go to the Todt organisation for installation in concrete fortifications in the West. Production of a quadruple 20-mm.-gun AA tank was stopped just as it was about to get started, and a new design for one fitted with twin 37-mm. was put under way.

The Russians, in contrast, continued to produce an enormous volume of armour with a minimum of variations. T 34 chassis were coming out of the factories at the rate of nearly two thousand per month, and they were about evenly divided between "normal" types, T 34/85 and SU self-propelled guns. Soviet ordnance had developed two new anti-tank guns, a long-barrelled 100-mm. and a 122-mm., and the "100" was now being installed in the SU in place of the 85-mm. Neither of these guns had the refinement, in terms of muzzle velocity or quality of projectile, of the German "86" and "long 75" variations, but sheer weight of shot achieved the same effect if they scored a direct hit. Such heavy ammunition restricted the amount that could be carried and made the crews' quarters unpleasantly cramped, but the Russians' numerical superiority, their familiarity with extreme discomfort, and their enthusiasm for new machinery more than compensated for this.

The Germans had been pressing the development of the Panther vigorously, and by the winter of 1943-44 most of the troubles which had plagued them during the battle of Kursk had been eliminated. Tank for tank, the Panther was superior to the T 34/85—though certainly not to an extent which made up for its scarcity. However, the Russians were far advanced in their development of the Stalin, which, although it weighed only forty-seven tons, was intended to carry the new 122-mm. gun. The Stalin was based on the old KV chassis, but with an improved hull front and larger turret. Although it was not quite a match for the later models of Tigers, its mobility and relatively low weight allowed it to keep up with the mass of advancing armour, a feat which the heavy German machines often found impossible and which forced them to operate independently. Furthermore, by concentrating on gun *design* and chassis *development* the Russians were still managing to confine their production to only two basic chassis, the KV and T 34—a happy state of affairs, which continued until the end of the war.

By this time (the late autumn of 1943), too, the Red Army was starting to draw dividends on the enormous volume of aid which was flowing from the West, and particularly from the United States. After the first rather haphazard deliveries of early 1942, the aid program began to take on a rational shape and to play a very significant part in maintaining the Soviet war economy. The Russians preferred to make their own weapons, which were in almost every case superior to those which the Allies were offering them,[1] but would never have been able to concentrate on weapons production to the extent that they did without the American con-

[1] Exceptions: a number of aircraft types, notably the Mustang, by North American, and the Mitchell medium bomber. A larger number of Dakota transports were also delivered, and the Russians made several variants of these under licence.

tribution of "soft" goods. These included all kinds of material, from sheet steel to shoe leather; clothing, blankets, tents, radio sets; enormous quantities of canned food, iron rations (even fruit juice!) and first-aid packs. Most important of all, perhaps, were the trucks—particularly the White half-track—which began to put the Red Army infantry on wheels for the first time in its history.

The Germans were thus faced with a dangerous and worsening prospect. While their own numbers and firepower remained static or declined, their enemy was gradually raising his strength in both these respects. He was also improving his mobility, which was allowing a deeper thrust and a longer momentum to his attacks. But the real danger from the German side was the absence of coherent direction at a strategic level. The separation between OKW and OKH, between Zeitzler and Jodl, which had first made its mark on the planning of *Zitadelle,* was now irreparable. The only effective liaison between the two departments was through Hitler himself. There was no over-all plan to which the army group commanders could refer and within whose framework they could operate at their discretion. Even the flow of directives, which had been such a feature of 1941 and 1942, had slackened off, and their place had to be taken by a succession of personal "conferences," usually quite inconclusive, between Hitler and individual army group commanders.

For example, before he could get unqualified permission to retreat to the line of the Dnieper, Manstein had to "confer" with Hitler no fewer than seven times.[2] During the period that Hitler was making up his mind, very little could be done about fortifying the line, for neither the OKH Works Department nor Koch, who controlled all the local resources, would allocate materials or civilian labour for the purpose. Then, in the last fortnight of the withdrawal, following Manstein's visit to Rastenburg on 15th September, a perfect orgy of bloodshed and confusion broke out as all the various Reich agencies started to evacuate and to save what they could from the wreck.

OKH had ordered that a strip fifteen miles wide along the east bank of the Dnieper was to be made a "waste zone" and completely "sterilised" of buildings, water, utilities, and of course, people. Meanwhile

[2] The most important conferences were those on 27th August at Vinnitsa, on 2nd September at Rastenburg (Kluge also attended this conference), on 8th September (with Kleist) at Vinnitsa, and on 15th September at Rastenburg.

Goering, through the Four-Year Plan officials, Sauckel through the GBA, and Bormann through Koch and others of his nominees in the *Reichskommissariat* were all trying, more or less independently of one another, to plunder what they could, under a general edict that machinery, public buildings, cattle, horses, and men of military age were not to be left to the enemy "in a usable condition."

Yet when the Germans did get safely to the right bank of the river they did not find themselves safe for long. The total length of Manstein's front was nearly 450 miles. To protect this he had only thirty-seven infantry divisions and seventeen Panzer and *Panzergrenadier* divisions, many of which had been reduced to little more than regimental strength. The balance of his front was further upset by Hitler's determination to hold bridgeheads at Zaporozhe, Dnepropetrovsk, Kremenchug, and Kiev. The closing months of 1943 passed, like the autumn, in a sequence of bitter local battles, whose cumulative effect was to sap German strength almost beyond repair.

In November, faced by a simultaneous threat to Kiev and/or eruption from the Russian bridgehead on the lower Dnieper below Kremenchug, Manstein finally managed to extract some reinforcements from Hitler. These consisted of an infantry division and two refitted Panzer divisions (the 14th and the 24th), two Panzer divisions from OKH reserve (the 1st and *Leibstandarte*), and one new Panzer division (the 25th). All these formations were committed before Christmas 1943, but Russian pressure and their delays in arrival, owing to the loss of the vital railway junctions west of Kiev, prevented their ever being used as a mass. Minor tactical successes were gained in the large bend of the Dnieper, but it proved impossible to hold Kiev, and as the New Year came the German line in southern Russia presented a perilously lopsided appearance.

The persistent lack of strategic appreciation which discolours German deployment in the East for nearly eighteen months after *Zitadelle* must be Hitler's responsibility. Yet Hitler cannot but have known what he was doing or, rather, what he was intending to do. The Führer was no Hapsburg buffoon who moved corps and armies about according to the dictates of his digestion. Records of his conversations when individual tactical problems were being discussed (records which in every case were kept, and subsequently quoted, by those whose interest it is to show Hitler in the worst possible light) show him to be shrewd and rational. But in his over-all conduct of the campaign—or retreat, as it came to be—

Hitler seems to have been fighting alone, against the whole of professional army opinion. Partly, of course, this was because of his contempt for the General Staff. "No General will ever pronounce himself ready to attack; and no commander will ever fight a defensive battle without looking over his shoulder to a 'shorter' line," Hitler complained at one of his conferences. Also, it does seem as if Hitler preferred to regard the experiences of the 1941 winter as typical—as showing that the Russians could be held and slowly worn down provided that sufficient "will" was exerted—and the experiences of 1942 as being due (as many Germans still claim) to "exceptional" circumstances, like the positioning of the Rumanians on the flanks. Hitler was also obsessed with the importance of space, although he seldom allowed his commanders to make proper use of it in defence, and there is no doubt that he deluded himself into believing that all was well by brooding over the OKW wall maps and the immense Eastern territories they showed remaining between the Red Army and the Reich frontiers, just as he deluded himself by counting divisional "numbers" and ignored the new quality of the Red Army by making comparisons with 1941, when its *nominal* superiority in numbers and tank power had been almost as great.

Hitler also tended to rely too much on the military equation: Space equals time. Privately he was already convinced that he was fighting a defensive war. In December 1943, nearly a year before it became Goebbels' *leitmotif,* he told Manstein that the coalition would break up as a result of its own internal tensions. And once it is realised that Hitler's strategic purpose was to create conditions under which the coalition would lose heart at the apparent impossibility of its task and the incompatibility of its separate partners, then his determination to make it fight for every inch—even where this was in contravention of strictly military principle—is easier to understand. However, even in this aim, Hitler was not entirely consistent. For just as in 1944 he estimated the Americans as the weakest party to the Grand Alliance and denuded the Eastern front to punish them in the Ardennes offensive, so in 1943 he was prepared to thin out the East in order to accumulate behind the West wall a strong enough force to throw them into the sea. Yet when Guderian, who agreed with this strategy in principle, tried to persuade Hitler that it could be followed with security only if the Eastern front was first deliberately reduced in length as well as in strength, Hitler would not listen to him.

This is all the more surprising when one remembers that of all the generals, Guderian was the one whom Hitler respected most, and in whom his trust endured longest.[3] Guderian has described how the two of them had breakfast together early in January 1944:

> . . . at a small round table in a rather dark room. We were alone . . . only his Alsatian bitch, Biondi, was there. Hitler fed her from time to time with pieces of dry bread. Linge, the servant who waited on us, came and went silently. The rare occasion had arisen on which it would be possible to tackle and perhaps to solve thorny problems . . .

But when Guderian tried to persuade Hitler to authorise an immediate start on a line of heavy fortifications deep in Poland, he found that he had "stirred up a hornets' nest." Hitler, after claiming that he was "the greatest builder of fortifications of all time," objected that the railway system was inadequate to transport the materials which such a system would require, on top of the needs of the front, "and, as usual, bluffed by reeling off exact statistics which his listener was not for the moment in a position to contradict." Guderian, with his customary frankness, explained that he was thinking of establishing a line as far back as the Bug and the Niemen, and as the rail bottleneck began only after Brest-Litovsk this was not a valid objection.

As this involved a phased withdrawal of between two and three hundred miles, it is not surprising that Hitler refused to listen. Next, the Inspector General doggedly pressed on to a subject which every senior officer who could catch Hitler alone used to raise, and which was even less welcome to the Führer than that of a major withdrawal—namely, the appointment of a *generalissimo* who would have "supreme responsibility" for the East. Hitler countered with the usual arguments, that it would be a slight to Goering, that he could not get rid of Keitel, and so forth. Neither man could come out in the open with his real motives—Guderian that he believed Hitler's leadership to be disastrously incompetent, Hitler that he did not trust the Army sufficiently to allow it independence.

The result was that the two men parted with nothing achieved and their relations, if affected at all, slightly poorer—though not so much poorer, we may think, as they would have been had Hitler known that Guderian had already tried to raise the "leadership question" with two of the principals at the Führer's own court.

[3] That is, of the "independent" generals. Keitel and Jodl, Hitler regarded as "office boys."

Goebbels, whom Guderian had tackled immediately after the collapse of *Zitadelle,* "had pronounced the problem a thorny one, but had nevertheless promised to do what he could at an appropriate time." (Actually, he never did anything.) Himmler, too, was approached, but he presented an impression of "impenetrable obliquity." (In fact, as has already been shown, and will later be further demonstrated, Himmler agreed with Guderian, but preferred to go about things in his own way.)

Jodl had been the most crushing of all. After listening without comment he simply said, "Do you know of a better supreme commander than Adolf Hitler?"

After a short breathing space in the middle of December, Vatutin, Koniev, and Malinovsky all resumed their pressure, and Manstein, with his reserves dissipated in the local counterattacks of the previous month, was finding it impossible to keep a continuous front in being. The German line was now stretched so taut that Soviet armour which punctured it could roam where it liked in the rear areas. A Russian account of the capture of Pyolichatka tells:

> It was like a garage. Vehicles of all makes and all models were lined up in close ranks on the streets, in the courtyards and the cherry orchards. They had come from all the countries of Europe. From large Demag seven-tonners which carried an entire mechanical workshop to small Renault tricycles, from the luxurious Horch to old Citroëns. All were camouflaged in preparation for the road journey [i.e., the retreat]. In the sidings were long strings of trucks loaded with flour, salt, munitions, tanks, petrol. Before a grain elevator, a train was loaded, ready to depart. The destination was written on the trucks: Köln—Tilsit—Königsberg.

It was now urgent for Manstein to get out of the Dnieper bend for good and retreat at least as far as the line of the Ukrainian Bug. But his mobile reserves had been worn down to extinction, and once the German front began to break up of its own volition, a series of ruinous encirclement battles took place whose outcome Manstein was powerless to influence.

The largest pocket developed in the Kovel-Korsun area of the lower Dnieper, in which SS Viking and the remnants of seven other divisions were trapped. By using the last of his Panzers, Manstein managed to drive a corridor through to the encircled men, but it was possible to keep this open for only a few hours. Leon Degrelle, a Belgian with the SS *Wallonia* Brigade, has described how

> In this frantic race vehicles were overturned, throwing wounded in confusion to the ground. A wave of Soviet tanks overtook the first vehicles and caught more than half of the convoy; the wave advanced through the carts, breaking them under our eyes, one by one like boxes of matches, crushing the wounded and the dying horses . . . We had a moment's respite when the tanks got jammed in the procession, and were trying to get clear of the tangle of hundreds of vehicles broken beneath their tracks.

For six miles the column struggled southeast under continuous fire, dragging its wounded tail, to which the Russian tanks were clinging. Then the Germans were halted by a river

> eight metres broad and two metres deep. The artillery teams which had escaped destruction plunged first into the waves and ice floes. The banks of the river were steep, the horses turned back and were drowned. Men then threw themselves in to cross the river by swimming. But hardly had they got to the other side than they were transformed into blocks of ice, and their clothes frozen to their bodies. Some fell down dead. Most of the soldiers preferred to get rid of their clothes. They tried to throw their equipment over the river. But often their uniforms fell into the current. Soon hundreds of soldiers, completely naked and red as lobsters, were thronging the other bank. Many soldiers did not know how to swim. Maddened by the approach of the Russian armour which was coming down the slope and firing at them, they threw themselves pell-mell into the icy water. Some escaped death by clinging to trees which had been hastily felled . . . But hundreds were drowned. Under the fire of tanks thousands upon thousands of soldiers, half clothed, streaming with icy water or naked as the day they were born, ran through the snow towards the distant cottages of Lysianka.

Still more serious than the defeat at Kovel-Korsun was the threat posed by Vatutin and Koniev to the north. For on 5th February, Vatutin had captured Rovno and begun to wheel his armour south toward the upper Dniester and the foothills of the Carpathians. Once these had been reached, he would have effectively severed Manstein's command in two, with one half bunched in the traditional corridor between the Carpathians and the Pripet Marshes and all the troops on the Bug line dependent for victualling and refit on lines of communication through Rumania.

All the tanks left to the army group were concentrated in the north, where the two Panzer armies of Rauss[4] and Hube were slowly being

[4] Colonel General Hoth, one of the most experienced Panzer leaders in the German Army, and a man who had been continuously in action on the Eastern front since 22nd June, 1941, was dismissed from the command of the 4th Panzer Army after the fall of Kiev.

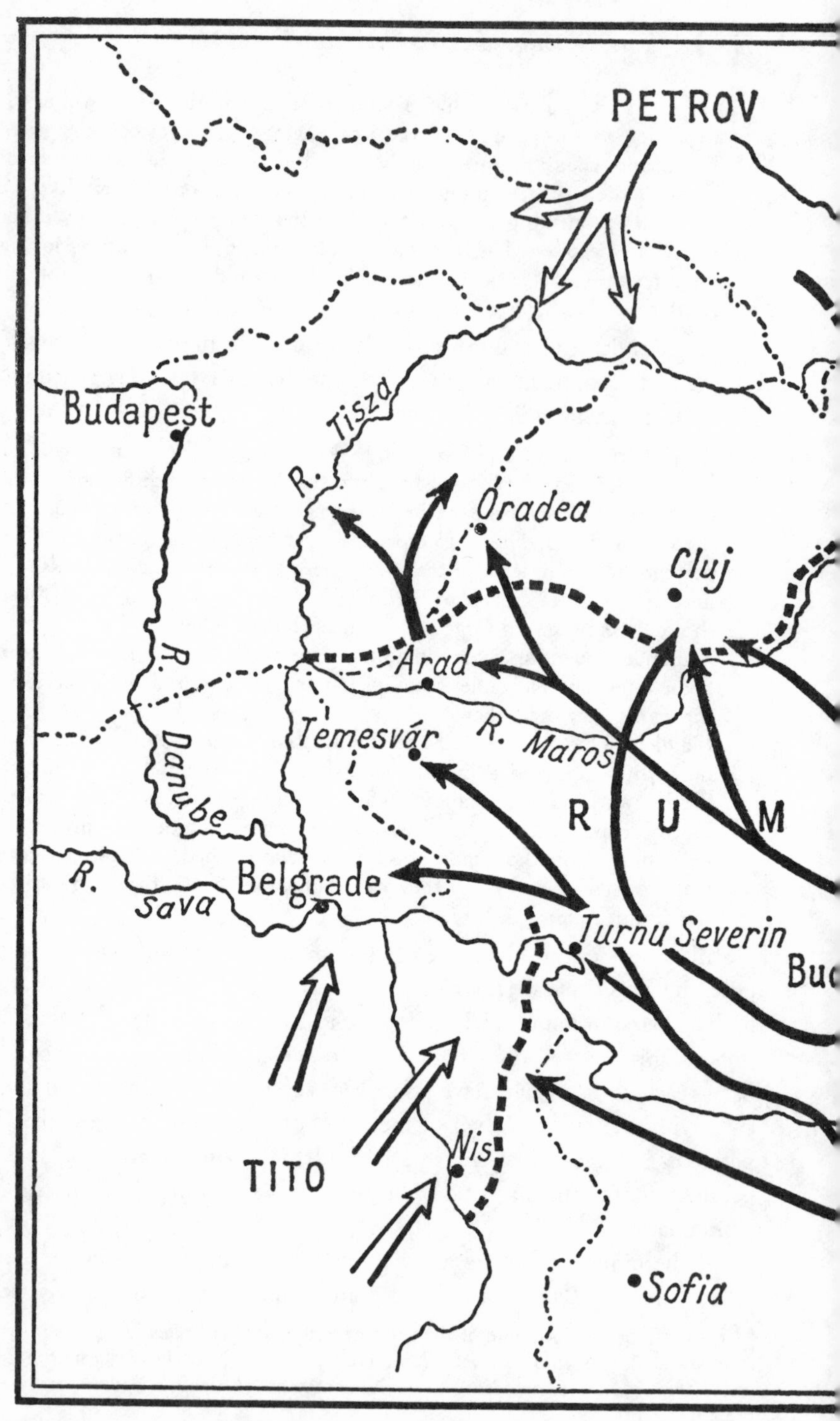

Retreat in South Russia and Collapse in the Balkans

Vinnitsa
MALINOVSKY
Pervomaisk
R. Bug
TOLBUKHIN
Jassy
Kishinev
Nikolayev
R. Prut
Siret
Odessa
Galatz
BLACK SEA
Danube
Varna
Burgas
Front Line, 16th March, 1944
" " 28th Aug. "
" " 4th Oct. "
Scale of Miles
0
50
100

compressed against the Carpathians. But the expected lull which usually allowed a period of recuperation during the thaw never materialised—largely owing to the increased mobility of the Russian infantry in its tracked American carriers—and by the middle of March, Koniev had managed to separate the 4th Panzer Army from the 1st, and plunged Manstein into his last crisis.

It is likely that Hitler had already decided to get rid of Manstein. However, to his credit, Hitler stood by him until the crisis (which was caused by Hube's refusal to obey orders and drive the 1st Panzer Army west) was over. Manstein had been summoned to the Obersalzberg on 25th March. Hitler then accused him of having "frittered away" his forces and of responsibility for "the unfavourable situation in which the Army Group had landed."

Manstein, whose temper cannot have been improved by a telephone call from his Chief of Staff, recounting that Hube was still refusing to break out westward, defended himself "with some asperity" and claims that, after the audience was over, he told Schmundt that Hitler could have his resignation if he wanted it. That evening, though, Hitler went over to Manstein's side against Hube, and even consented to release an SS Panzer corps from the West so as to form a task force for the rescue of the 1st Panzer Army. Emboldened by this, Manstein promptly "followed up with one or two ideas of my own on the future conduct of operations." This was probably the last straw for Hitler. At all events, Manstein had been back at his headquarters barely three days when Hitler's personal Kondor came to take him back to Berchtesgaden. In the aircraft, somewhat apprehensive, sat Kleist, who had already been picked up from his own headquarters. After an uneasy flight the two field marshals were shown into Hitler's presence that same evening (30th March) and the Führer, with notable courtesy, and after presenting them with swords to their Knight's Crosses, dismissed them.

Hitler told him (according to Manstein), "All that counted now was to cling stubbornly to what we held . . . The time for grand-style operations in the East, for which I had been particularly qualified, was not past." Model, whom he had selected to take over the army group, would dash around the divisions and get the very utmost out of the troops.

Manstein's tart reply, "that the Army Group's divisions had long been giving of their best under my command, and that no one else could get them to give anything more"; his expressed conviction that "what we had to pay for, first and foremost, was Germany's failure to stake *absolutely everything* on bringing about a *showdown in the East* in 1943, in order to achieve at least a stalemate"; and the way in which, in his memoirs, he uses Hitler as a convenient sump in which to lay all the blunders of the German Army in the East—these stand jointly in contrast with the Führer's own realistic (and not ungenerous) assessment of Manstein's abilities. In a conversation with Jodl, some time after Manstein's dismissal, the subject came up by chance.

> . . . It's just that there are two different talents. In my eyes Manstein has a tremendous talent for operations. There's no doubt about that. And if I had an army of, say, 20 divisions at full strength and in peace-time conditions, I couldn't think of a better commander for them than Manstein. He knows how to handle them, and will do it. He would move like lightning—but always under the condition that he has first-class material, petrol, plenty of ammunition. If something breaks down . . . he doesn't get things done. If I got hold of another army today I'm not at all sure that I wouldn't employ Manstein because he is certainly one of our most competent officers. But there are just two separate talents . . . Manstein can operate with divisions as long as they are in good shape. (If the divisions are roughly handled I have to take them away from him in a hurry, he can't handle such a situation.) That has to be a person who works completely independent of any routine.

The dismissals of Manstein, Kleist, and Hoth were echoed by certain enforced changes in the ranks of the *Kommissariat.* The loathsome Koch, his Ukrainian kingdom extinct, had returned, fulminating, to East Prussia. Kube, busy with his "blondies" even while the sound of Russian gunfire was audible in the palace at Minsk, had one night returned to his own bed, to find that the welcoming shape of a hot water bottle was in fact an anti-personnel mine. With legs and trunk blown to pulp, he died inside a half hour. Lohse held on in Riga until the spring of 1944, though (as is apparent from his correspondence) in an increasingly nervous condition. Finally the signs of impending defeat and the possibility of assassination proved intolerable, and he addressed a letter to Rosenberg, announcing that he considered it "his duty to act independently, in accordance with the Führer's wishes and his own conscience." Lohse then had a nervous breakdown, and disappeared into Germany. No one could find him (though he surfaced promptly enough after the

Bonn government had been constituted, to lodge a successful claim for a civil service pension), and Koch was despatched briefly, and somewhat reluctantly, to take his place.

In the weeks following Model's appointment the Russian offensive in the western Ukraine gradually died away. The Soviet forces, after nearly eight months of continuous forward movement, had at last exhausted their momentum, and the *Stavka* reserves of men and material were being directed to the Belorussian "fronts" of Chernyakovski, Zakharov, and Rokossovski, in preparation for a massive attack against the German centre, which was to carry the Red Army to the banks of the Vistula.

The Russian offensive began on 22nd June, just over two weeks after D-Day in the West, and was mounted by 118 infantry and 43 tank divisions.[5] By the end of the month Army Group Centre had been forced out of its long-prepared defences and was streaming back across White Russia, abandoning guns, vehicles, dumps of material, wounded, in its haste to reach the old defences of the Polish border. Model was transferred from Army Group South to try to stop the rot, but even he could do little with the battered remnants which Busch had bequeathed him. The Führer moved his headquarters from Obersalzberg to Rastenburg, and in spite of the mounting danger in Normandy, all reinforcements were henceforth directed to the East.

So now we come to the fateful month of July 1944, when the waters were rising along the whole periphery of the Nazi empire, where everywhere, in Speidel's words, ". . . the floodgates are creaking," to the day, the 20th, of the *attentat;* a climacteric in the history of the Third Reich of Hitler's relations with the Army, and of the rational direction of the German war effort.

The details of that dramatic affair are so well known, and have been so often described,[6] that it would be repetitious to present them again here, save where they impinge on the direction of military affairs. So we will do no more than allude to the failure to blow up the telephone ex-

[5] These divisions, though they would have been near full strength at the start of the attack, were still much below the (nominal) strength of German ones. At a late stage 36 more divisions joined them. On a comparative basis, the Russian strength was about four to one in men and six to one in armour.

[6] The best description is given by Sir John Wheeler-Bennett in his masterly work on the German Army in politics—*Nemesis of Power*. Another excellent account can be found in Chester Wilmot's *The Struggle for Europe*.

change at Rastenburg; the hesitation and scruples of the plotters waiting in the Bendlerstrasse; the failure to shoot Fromm and Remer, and to deploy the "loyal" (i.e., disloyal) Berlin garrison immediately, and proceed straight to the appointment of Germany's most brilliant soldier as Chief of the General Staff of the Army.

The appointment was made by Hitler, whose manner less than twenty-four hours after the *attentat* was one of "astonishing calm." After their meeting Guderian walked over to the office block assigned to OKH at Rastenburg and found things in a most un-Prussian state:

> I found the buildings empty. There was no one there to meet me. After looking through various rooms I came across a private soldier by the name of Riehl, sound asleep. I sent this splendid fellow off to find an officer . . . I then attempted to telephone to the Army Groups in order to find out the situation at the front. There were three telephones in the Chief of Staff's office, and no way of telling what purpose each one served. I picked up the nearest one. A female voice answered. When I said my name she screamed and hung up on me . . .

It was an inauspicious opening to the final period of the Wehrmacht's decline.

twenty | EASTERN EUROPE CHANGES HANDS

The collapse of the 20th July plot illustrates a particular facet of the German character—a deep-seated reluctance to act against the established authority, an aversion toward usurping responsibility or appearing to belong to a minority, regardless of right or wrong. Those wasted hours at the Bendlerstrasse while Stauffenberg was flying south! No one had the conviction, or the courage, to really seize power with both hands, and to use it. Even the "winning" side, Fromm and Remer, were fumbling and hesitant until they were sure of the position.

But these defects of character (if such they are) can be seen also as virtues in the days which followed. For as soon as the current of authority and retribution began to flow outward from Rastenburg, everything fell back into place. Resistance there was none; abject confession was profuse. There is no better testimonial to the discipline of the German Army, or to the callous efficiency of the Nazi Party machine, than the way in which the Reich withstood the multiple stresses of the last week in July 1944. At a time when the Russian tide across Poland seemed unstoppable, when the floodgates in Normandy were creaking under the rising weight of Patton's and Bradley's armies, Germany was saturated by a wave of denunciation, imprisonment, and murder. The breach between officers of the Army and the SS (of every rank) came into the open with Bormann's order to the Gauleiters to "arrest Army officers on suspicion [sic] as practically the whole of the General Staff is in league with the Moscow Free Germany Committee." Jodl, ever quick to disclaim association with his own caste, endorsed a memorandum of Burgdorf's from OKH to the effect that ". . . the whole General Staff should be abolished." All the quasi-independent empires of the Third Reich—the SS, the Gauleiters, the Labour Front, the civil police, the Armaments Commission, the Propaganda Ministry, the Hitler Youth—all had wavered momentarily as the tremor ran across their foundations, yet in the shrill chorus of recrimination came to realise that their very existence, even their prerogatives of internecine quarrelling, depended on the life of the Führer.

The Army, its prolonged but timorous flirtation with the conspirators so abruptly terminated, now immersed itself in the demands of professionalism. Guderian moved OKH headquarters back from Zossen to East Prussia, built up a new and personal staff, drew off the reserves Schörner had been collecting on the southern front, and set about repairing the breach in the centre. And then a second miracle, as far-reaching, it seemed, as the salvation of the Führer, occurred. At the very moment when the Western front was on the point of giving way, the Russian advance slowed down. Soon it became apparent from the manner in which quite small German units were succeeding in holding their positions that the impetus of the Red Army had died away, absorbed by one more *tranche* (the last one) of conquered territory.

There now occurs one of the most tragic episodes of the whole Eastern campaign, the rising of the Warsaw Poles and their desperate, hopeless, two-and-a-half-month-long battle in the city streets. The Warsaw uprising has a place in the purely military history of the campaign. But its great importance is political—both as an illustration of the plight of the Polish nation, that strange, gifted, and romantic people, doomed forever to be crushed between the callous monoliths of Germany and Russia—and as an incident of immense significance in the shaping of postwar Europe.

In its essence "the Polish problem" can be simply stated, for time has not altered it. The state of Poland is Western Europe's traditional buffer against Russia, but its security in this role is perpetually threatened by the greed and cruelty of the German landowners in Prussia and Pomerania. It has never been possible for the Poles to make a political deal with either of their neighbours, for each covets their territory and prefers to assimilate rather than to protect it. But in 1939 a disinterested protector appeared. The British Government guaranteed Polish integrity simply because its violation seemed to be the next step in the process of German expansion, and the British were ready to make it a *casus belli.* The Poles thus became the stake in a power game at which both the players were determined to call their opponent's bluff. For Hitler was eager "to blood the German nation," and believed that as the British were strategically incapable of implementing their guarantee they would accept a *fait accompli;* the British, even more fatalistic, thought that their guarantee would stop Hitler by itself—and if it didn't, well, that meant that they would have to fight him sometime, so why not then and "with honour"?

In the result the Poles fought with great gallantry to the end—which was itself accelerated by a Russian invasion of their eastern border under terms agreed between Molotov and Ribbentrop that August. By the end of 1939 the Polish state had once again been extinguished by the predatory giants on its borders, and the men of the Polish Army who had not died in battle languished in prison camps. The Russians made some attempts to "indoctrinate" those they had captured, but the officers proved intractable and were moved to a camp in the Katyn forest, where, after a period, they were all shot. The Germans never even bothered to start POW compounds—the Poles were sent straight to concentration camps and liquidated. The same differences are perceptible in the governments of the two occupied halves of the country. The Russians made some effort to assimilate the inhabitants into a Communist society; the Germans set out, systematically, to exterminate the entire Polish population and to substitute German immigrants.

But the seed of Polish nationhood, bred for centuries under conditions such as these, has a Darwinian tenacity, and now, scattered by default on the chilly soil of wartime London, it began to flourish. London became the seat of the "Polish Government," the goal of *émigrés* and escapees, the focus of all the energy and patriotism of this sad and gifted people. Gradually the tenuous strands of underground communication, which can operate under the most repressive alien regimes, were woven into a chain of command and intelligence which retained its strength right up until the tragic events of the autumn of 1944. The British provided arms and training, a separate Polish Army was created; Polish flyers flew in special squadrons; most important of all, they returned to their own country by parachute, with arms, radios, and instructions from the "Government."

But of course, nobody is so susceptible to the corruption of doubt, the corrosive influence of personal jealousy and intrigue, as a government in exile. And as the war progressed, its difficulties grew no lighter, for with the metamorphosis of the Soviet Union, first as ally then, by 1944 the most powerful army in the coalition, and thereby in the world, an enforced change in the direction of its host's policy threatened. By July 1944 the Red Army occupied all of eastern Poland, the very boundaries, to the metre almost, which they had seized in 1939. But why should they stop there? Indeed, there was not the slightest likelihood that they would. The harsh impulses of strategic necessity and the disintegration of the Wehrmacht would combine, it seemed to the London Poles, to place their whole country under the domination of one of its two traditional enemies. It was a situation in which diplomacy was valueless, for diplomacy means pressure (however gracefully concealed), and there were no longer any pressures to which the Russians were susceptible. Their armies were all-powerful; they had drawn their fill of aid from the West—and in any case its delivery was an irreversible process, subject, like the many other concessions the Soviet Union enjoyed since 1942, to a powerful current of popular emotion in the democracies. For Russian policy was now benefiting from a remarkable change of image, sedulously fostered by the Communist parties of the West and unwittingly promoted by the democracies' own propaganda services. On an international level it was the counterpart of the new emphasis of patriotism over Party loyalties which was inspiring the citizens of the Soviet Union; class warfare and revolution were played down, and in their place were depicted two fresh images: the brave Red Army man, personification of a country steadfast in battle; and that of "Uncle Joe," pipe in mouth, the epitome of trustiness in conduct and negotiation.

In the diplomatic context the position of the London Poles was still further prejudiced by the emergence of the United States as the pre-eminent force in the Western coalition and the gradual shift of the centre of power (for purposes, at any rate, of political intrigue and lobbying) from London to Washington. For if the British leaders, in contrast to the man in the street, had preserved a certain cynicism in their assessment of the new Russian character, the reverse was true in the United States, where politicians (and many soldiers also) had fallen for the new Russian line. At Teheran, when the first discreet British approaches attempted to warn Roosevelt of the dangers of allowing too deep a Russian penetration in the Balkans, the President had confided to his son Elliott:

> I see no reason for putting the lives of American soldiers in jeopardy in order to protect real or fancied British interests on the Continent.

Indeed, American policy[1] was already beginning the reorientation which was to come out in the open at Yalta the following year, whereby Russian "security" was backed against the aspirations of Britain and the lesser nations of Eastern Europe. Roosevelt was determined to get Russian co-operation in the war against Japan; he was determined, too, that Russia be persuaded to join a collective security organisation (the United Nations), which, he believed, could "control" her. The effect was that what the United States wanted from Stalin was of greater value to her than what she was offering him—a state of affairs which the Russian dictator saw and exploited earlier than he might have done on account of Roosevelt's pathetic diplomatic *gaffes*.

In this situation the London Poles had to play their hand alone. A clear indication of the climate in which they were going to operate had come in 1943, when the Germans accidentally uncovered the grave of four thousand Polish officers at Katyn. Stalin had refused an inde-

[1] Roosevelt's betrayal of Eastern Europe, whether out of calculation or gullibility, is so notorious as to need no further recapitulation. But two examples should be cited, in corroboration of the worst fears of London Poles.

When he accepted the reimposition of the 1940 Russo-German frontier (resurrected now as the "Curzon Line"), Roosevelt, perhaps with one ear cocked to the reaction of his own people, suggested that the town of Lvov should be "granted" to the new Poland, "as this would have a salutary effect on American public opinion." However, what slight weight this consideration might have carried was discounted by the President's hurried disclaimer that ". . . he was merely suggesting this for consideration rather than insisting on it."

Two days later, when Churchill was fighting alone to prevent the Russians from foisting the Lublin Committee, a puppet government of Polish Communists which they had set up, on the country, Roosevelt went behind the Prime Minister's back and sent Stalin a private letter, asserting:

"The United States will never lend its support in any way to any provisional government in Poland which would be inimical to your interests."

However, to offset the impression that the Americans were universally affected by this attitude, we should remember the judgment of Major General John R. Deane, head of the U. S. military mission in Moscow, who had written to General Marshall in December 1944, "We never make a request or proposal to the Soviets that is not viewed with suspicion. They simply cannot understand giving without taking, and, as a result, even our giving is viewed with suspicion. Gratitude cannot be banked in the Soviet Union. Each transaction is complete in itself, without regard to past favours."

pendent inquiry by the Red Cross and, after a prolonged and abusive diplomatic bombardment, had taken the opportunity to "sever" relations. Matters had deteriorated steadily in the ensuing twelve months, with attempted Communist subversion in the ranks of the Polish forces in the West, coupled with a steady propaganda campaign (in which certain British publications were not blameless) to the effect that the London Poles were anti-Semitic—in the language of fellow travellers, a recognised halfway stage to being "Fascist"—and "unrepresentative." Then on 24th July, 1944, the Russians, well across the old Curzon Line and the 1939 frontier, had captured Lublin and installed there a "National Committee of Liberation"—an obvious nucleus for a puppet Communist administration. If the London Poles were to assert themselves, time was running out.

Stalin's classic rejoinder to some fulsome Western diplomat who was holding forth about Catholic "good will" had been, "How many divisions has the Pope?" And the same question could be posed with almost as telling effect of the London Poles. Their divisions were as few, as scattered, and as powerless to intervene as had been those of the British five years before, at the time of the German invasion. But they did have a widespread and well-organised underground, responsible to them and controlled by radio from London. This force—the "Home Army," or AK—was centred on the capital, Warsaw; but its authority was already being threatened as the hour, if not of liberation then of change in occupation, approached, by various splinter groups. There was the "People's Army," (AL) of independent left-wing sympathisers; the Communist-dominated PAL; and the "Nationalist Armed Forces," (NSZ) an extreme right-wing force which had broken away from the AK at the first sign of impending compromise with Russian power.

It had become urgently necessary for the AK to show its strength, so that the London Poles could at least assert some sort of armed presence in their own country—they were already getting reports that AK units which had co-operated with the advancing Russians were being disarmed and their officers taken away. This opportunity seemed to present itself in the last week in July, for as Rokossovski approached Warsaw the German administration began to close down, and many of its departments ceased to function. On 27th July the military government issued a proclamation calling up a hundred thousand civilians forthwith for work on fortifications, and still greater dislocation of the Home Army was threatened by a Russian broadcast of 29th July, which spoke of the city's impending liberation and urged the "workers of the Resistance" to rise against the retreating invader. This last development led to great confusion, because although the Home Army, which comprised 80 percent of the armed Resistance, took its orders from London, premature action by the AL and the PAL could well make it impossible for the AK to control its own members. On 1st August, therefore, Bor-Komorowski, the Polish Cavalry General who commanded the AK, issued a proclamation, copies of which were scattered throughout the city.

> Soldiers of the capital!
> Today I have issued the orders so long awaited by all of you, the orders for an open fight against the German invader. After nearly five years of necessary underground struggle, today we are taking up arms openly. . . .

At first the timing looked perfect. It seemed as if the AK would be able to step into the vacuum caused by the German withdrawal and precede Rokossovski in declaring the liberation of the capital. The R.A.F. would then have flown in the London government, which would have been able to install itself in the administrative centre of its country with the prestige of military achievement and backed by a powerful local force. But in fact the Russian offensive had reached the end of its tether. At that very moment when Bor issued his call to arms, the Russians' right wing in the Baltic states was being roughly handled by a counterattack from East Prussia and Courland which recaptured Tukums and Mitau (Jelgava), and which diverted reinforcement from the centre. The customary difficulties of supply and the exhaustion of men and machinery combined to dictate a halt on the Vistula. From the Russians' point of view, the Warsaw uprising could not have come at a better moment (and thus as a political threat it could be discounted). For it did not have the strength to succeed without their help, yet it promised, while burning itself out, to distract German attention and to deny to their enemies the respite which the Russians themselves so badly needed.[2]

All the same, the Poles very nearly brought it off. By 6th August they were in control of almost the whole town, and had greatly enlarged their

[2] It has been widely contended that the Russians deliberately held off their attacks to allow the Germans to do a job (the liquidation of the AK) for them. This seems to be attributing motives (which may well have been held) to circumstances which were largely accidental. Rokossovski's attitude to the struggle of the AK was, from the start, callous and unco-operative. But once he intended to resume a strategic offensive he would not have allowed its existence to deter him, and would doubtless have meted out the same sort of treatment the E.L.A.S. suffered from the British in Athens, in 1945.

armament with captured German material. So confident were they of victory that the rival splinter groups were already exchanging fire with one another, and it was proposed to fly in the first representatives of the London administration on that Sunday. Then, on 8th August, came the first portent of their eventual fate, with the appearance of the sinister Gruppenführer SS von dem Bach-Zelewski.

Bach had been selected for this task because of his special experience in anti-Partisan operations, and because by making the suppression an SS affair it was intended to leave the Regular Army free to face the Russians. It is also clear that the SS wanted to have a completely free hand —free from observation, much less interference, by possibly "squeamish" elements. And for those who may have wondered what, at this late stage in the war and after so much horrific brutality, could possibly make *anyone* squeamish, the answer was not long in coming.

Bach-Zelewski deployed two formations against the AK, the Kaminski Brigade, consisting of turncoat Russian prisoners and general riffraff from Eastern Europe, and the Dirlewanger[3] SS Brigade, made up of German convicts on probation. The impact of units such as these in street battles, always the bitterest kind of infantry fighting, and in an area where the whole civilian population was *in situ,* can be imagined. Prisoners were burned alive with gasoline; babies were impaled on bayonets and stuck out of windows like flags; women were hung upside down from balconies in rows. The object, Himmler had told Goebbels, was that the sheer violence and terror of the repression would extinguish the revolt "in a very few days."

The SS had already mounted one "operation" in Warsaw, in the spring

[3] SS Oberführer Oskar Dirlewanger was an old friend of Gottlob Berger, who had got him a commission in the Kondor Legion as far back as 1935. When Dirlewanger came back from Spain two years later, it was still not easy to find employment for him as he had already served a two-year service for offences against young girls in Germany. However, some further string-pulling got him a transfer to the Waffen SS, and the job of training the first battalion of convicted criminals to be incorporated in the *Totenkopf* Division. As the war progressed, the passage and growth of Dirlewanger's *Kommando* can be traced in SS records, particularly those of the (hardly oversqueamish) Judge Advocate's office. He had to be hastily transferred from Cracow, then from Lublin—where his experiments on Polish girls are hardly printable even today, combining as they did the indulgence of both sadism and necrophilia. He was awarded the German Cross of Gold for his part in suppressing the "Partisan Republic of Lake Pelik" in 1943, in which 15,000 "Partisans" were killed but which yielded only 1,100 rifles and 326 pistols as the "Partisans'" armament. Dirlewanger, incidentally, bribed his way out of the Allied net after the war, and is living in Egypt (1963).

of 1942. Then it had cleared the ghetto with grenades and flame throwers, and succeeded in killing about fifty thousand Polish Jews. The despatches of the commanding officer, together with illustrations, had been sumptuously bound and circulated privately among the higher Nazis, and the action was classified as an SS "battle honour." But in August 1944 the SS found the going very much harder. A considerable quantity of arms had been dropped by the R.A.F. in the spring of 1944, including piat guns, which could knock out tanks at close range and were useful for blowing holes in houses. The Poles were well disciplined, and held their fire until the last moment. They were adept and industrious at making grenades, mines, and detonators. The fighting dragged on; days, weeks, passed; August became September. Four extra "police battalions" were brought from the Reich to stiffen the wavering ranks of the Dirlewanger convicts in a strange alliance of traditional enemies united by their taste for cruelty and violence.

With each day that still brought news from their transmitters in Warsaw, the London Poles became more frantic. For besides their anguish at watching and hearing the slow extinction of their gallant compatriots, there was the fading prospect of their carefully laid plans for asserting their own claims to the country. Yet once again, as in 1939, Britain was powerless to help. A few aircraft from Foggia could get through each night, but their cargo was limited to the barest essentials as the Russians refused them refuelling facilities, and as the AK area gradually contracted it became increasingly difficult to drop with any accuracy. Approaches by the British in London and Moscow, urging that some effort be made by Rokossovski to relieve the pressure on the AK were acknowledged, but nothing was done. One of the Poles has described how from the tallest building, when the smoke cleared, they could see German and Russian soldiers bathing on opposite sides of the Vistula River in apparent amity, or as if in tacit acceptance of a truce which was to last while Polish gallantry was extinguished.

But the tenacity of the Warsaw Poles did have one effect. It caught the imagination of the world, and it began to make a deep and uncomfortable impression on the Germans themselves. The first to act had been Guderian, who questioned Bach-Zelewski about the rumours he had heard when Bach approached him with a request for more heavy equipment with which to renew the assault. Bach admitted that as a result of "desperate street battles where each house had to be captured, and where the defenders were fighting for their lives . . . [the SS brigades] had abandoned all moral standards," and tried to excuse himself by saying that he had "lost control" of them. However, Guderian's sense of chivalry shocked. He has written:

> What I learned from [Bach-Zelewski] was so appalling that I felt myself bound to inform Hitler about it that same evening and to demand the removal of the two brigades from the Eastern Front.

As Hitler had all along been privy to Himmler's intention to terrorise the Poles into submission, this "demand" must have been most unwelcome, and it is hardly surprising that ". . . to begin with he was not inclined to listen." However, SS Gruppenführer Fegelein, who enjoyed a privileged position at Hitler's court because (among other reasons)[4] he had married Eva Braun's sister Gretl, spoke up on Guderian's behalf. Fegelein's intention was primarily that of discrediting Himmler, for he was one of the Bormann-Kaltenbrunner cabal, whose aim was to extend its own empire at the expense of the Reichsführer; but he could also claim a certain *ancien camaraderie* with General Bor, as the two men had competed at horse trials before the war. He may also, like several other senior Nazis at this time, have begun to look over his shoulder at the possibility of arraignment for "war crimes" by the victorious Allies. In the end Bach-Zelewski, never one to let the grass grow under his feet, changed his "approach" to the Warsaw battle, removed the Kaminski Brigade to the rear, and had Kaminski arrested and shot.

By now the Poles in the city were at their last gasp. Ammunition, food, water, medical supplies, all were withering away. The suffering of the civilian population was frightful, and the proximity of the soldiers' families, which had at first been a source of desperate inspiration, now gave rise to a harrowing sense of grief and personal responsibility. On 16th September, Rokossovski had managed to penetrate the German positions at Praga, the suburb of Warsaw on the east bank of the Vistula. Judging that the AK had shot its bolt, Stalin had ordered his own force of indoctrinated Poles, under General Zymierski, to enter the battle and fight their way into Moscow. But the Germans had now enjoyed ample time to prepare their defences, and after a week the Russo-Polish effort died away, with the attackers having learned, as the Anglo-

[4] Some authorities (see Reitlinger, *The SS: Alibi of a Nation,* 376) maintain that Fegelein was the first to draw attention to Bach-Zelewski's atrocities.

Saxons were learning that same time at Arnhem, that the last five or six miles can be critical when a beleaguered garrison has to be relieved.

With the failure of the Zymierski attack, activity by the AL and the PAL stopped and its members tried to go back into hiding. Hopeless shortage of everything required either to fight or to sustain the population impelled Bor to try to negotiate terms with Bach. And it was at this point that the Poles drew the first dividend on the incredible bravery with which they had fought.

Not only did Bach agree to treat the members of the AK as combatants entitled to the honours of war and to status as prisoners under the Geneva convention, but he digressed so readily and so effusively on other subjects that it was plain that further concessions could be extracted from him. He was keenly interested in politics, the Gruppenführer told the AK delegates, of which, after all, war was only the instrument. It was never too late to correct errors. The menace from the East was, or should be, the concern of everyone today, "as it might very well bring about the downfall of Western culture." Further rambling followed about "the necessity of fitting into the framework of German relations after the war." The Poles tended to brush aside these generalities, which were plainly going on to the record with the Nuremberg Tribunal in mind,[5] and insisted that the status of prisoners of war be extended to Resistance fighters throughout Poland instead of being confined to the AK in Warsaw. They also asked for an amnesty on all "crimes" committed by the AK up to that date. (It was an accepted German practice to charge prisoners with war crimes when they wanted to alter the conditions of their detention.) After some days of negotiation, interspersed with speechmaking, the Poles entered the conference to find Bach in a maudlin and reproachful mood. The BBC had just made a broadcast naming a number of SS leaders who would have "to answer for their crimes against the people of Warsaw" at the end of the war, and—monstrous injustice—he, Bach, had been named!

Had he not personally ordered the execution of Kaminski? Bach asked. Was it not by his intervention that the Luftwaffe had been prevented from saturating Warsaw with a fire storm? [6] Had he not gone out of his

[5] In fact, Bach-Zelewski subsequently became a prosecution witness at Nuremberg, and was thus exempted from extradiction to Russia or the Eastern bloc.

[6] It was not. The idea was discussed at OKW and rejected because of the proximity of the front line and the shortage of suitable aircraft.

way to express his admiration for the bravery of the Poles? Bach began a long story of how an AK courier had been captured, a young girl, whose beauty had so impressed him that he had ordered her release. She had reminded him of his daughters, he said; pictures of his daughters were produced and handed around.

We can be sure that Bach's contrition was neither as complete nor as sincere as he claimed. Yet his attitude is interesting as showing the new attitudes that were beginning to develop in the medium echelons of Nazi leadership, and especially in the SS, under the shadow of defeat and retribution. The exultant *furor Teutonicus* of the first years of the war, the *Totenkopf* deification of Germanic violence, were being hastily submerged beneath a new façade—that of the "defenders of Western culture."[7] Hitler, of course, had no time for that sort of nonsense, and had already drafted an order that Warsaw was to be razed to the ground the moment the charade of the surrender negotiations had been enacted. It was the renewal of the Russian offensive and a shortage of local labour which frustrated this intention rather than any compunction of the SS officers entrusted to see it through.

After several days of leisurely, almost conversational parleying, the terms were agreed between Bach and the AK, and the surrender was formalised by speeches from both sides (delivered, it may be thought, with almost equal fervour) on the need for magnanimity toward a defeated foe. Fegelein looked after General Bor, and the brave, starving Poles were herded off into prisoner-of-war camps, postponing, at least for some months, the hour of their execution. But the fact remained that the AK had been dealt a blow from which it never recovered, and thenceforth the Resistance fell increasingly under the control of the Communist-orientated groups. Gone, too, was any hope of reconstituting the country under other than Russian terms.

The story of the Warsaw uprising illustrates many features of the later history of World War II. The alternating perfidy and impotence of the Western Allies; the alternating brutality and sail-trimming of the SS; the constancy of Soviet power and ambition. Above all, perhaps, it shows the quality of the people for whom nominally, and originally, the war had been fought and how the two dictatorships could still find common ground in the need to suppress them. Professor Trevor-Roper had said,

[7] After the war former SS Generaloberst Hausser declared, "The SS was really the NATO Army in prototype, in ideal."

"It is sometimes supposed that Hitler and Stalin are fundamentally opposite portents, the one a dictator of the extreme right, the other of the extreme left. This is not so. Both, in fact, though in different ways, aimed at similar power, based on similar classes and maintained by similar methods. And if they fought and abused each other it was not as incompatible political antipodes but as closely matched competitors. They admired, studied, and envied each other's methods: their common hatred was directed against the liberal 19th century Western civilisation which both openly wished to destroy." No clearer illustration of this truth can be found than in their joint attitude to Poland from 1939 to 1944.

The month of August 1944 saw the munitions output of the Third Reich attain a peak figure, with a production of 869 tanks and 744 assault guns—enough to re-equip ten new Panzer divisions, and by the end of September the factories had produced enough not only to make up for the losses in Normandy but also to overtake the "wastage" rate in the East. In September the British paradrop on the Arnhem bridges failed, and Patton's tanks coughed to a standstill in Alsace, with their fuel tanks dry. Rain fell in the Reichswald and along the length of the Siegfried Line. The boundaries of German territory began once again to solidify; they were still—almost miraculously, it must have seemed—outside the frontiers from which Hitler had started the war in 1939.

The continuous rise in German weapons output, even under the hammer of Allied air attack, was a remarkable feat, achieved at God knows what price in cruelty and privation inflicted on the slave labour force. There were two sectors of the war economy where resources were running out—fuel to run the machines and men, fighting men to operate them, but even here the autumn brought improvement. Stocks of oil, which had stood at a million tons in April and sunk to less than a month's supply during the summer, slowly began to pick up again as the static conditions at the front reduced demand and the fog and thunderstorms of autumn protected the refineries from air attack. At the same time the last of German manhood was conscripted into the Army; by lowering the call-up age from seventeen to sixteen and a half years and a ruthless selection from "essential" operatives on the home front, the intake was raised to over seven hundred thousand in the months of August, September, and October.

Yet, impressive as these figures are, a closer inspection reveals the same malign interplay of personal rivalries and private empires waxing fiercer, it seems, as their period of power visibly diminishes. After the plot Himmler had been appointed to the Reserve or Home Army, in place of the luckless Fromm. He was also commander of Army Group Upper Rhine (an organisation, Guderian contemptuously observed, "for catching fugitives and deserters"), Minister of the Interior, Chief of the Civil Police, and Reichsführer SS. From a purely executive standpoint he was the most powerful man in Germany, but at this stage in the war, and his country's decline, *Treuer Heinrich* was a man of troubled mind. The extension of his personal domain from the SS, where all was tidy and the staff was (or had until lately seemed to be) obedient and loyal, to the rough and tumble of military affairs was proving to be a mixed blessing. For Himmler was in a field where new elements, hitherto outside his experience, were making themselves felt. There was his equivocal position in the command infra structure; as an army commander he owed a nominal obedience to OKW; he was not so much *primus* as (at least in the eyes of his professional colleagues) *minor, inter pares*. There was also the prospect, novel and unwelcome, that he might be engaged with enemies who would actually be *able to shoot back*—a development which opened all sorts of alarming possibilities.

At this stage in the war Himmler seems to assume a double image, one whose contrasts may be most effectively illustrated by the different ways of pronouncing his title. The Reichsführer—the very syllables make the listener tremble—the man who controlled the whole retributive machinery of the Reich, from the humblest policeman on point duty to the 88-mm. Tigers of the Waffen SS; but in translation, the "National Leader," it has a faintly ludicrous air, evoking youth camps and *Lederhosen,* one well suited to the puzzled, troubled face behind the rimless pince-nez spectacles, to the man who threw a screaming fit when the brains of some hapless victim were spattered on his uniform. We will attempt, then, in these last months of Germany's decline, to alternate these titles in keeping with Himmler's role.

Certainly no one, except Hitler himself, could have successfully handled all of Himmler's business. For it was not simply a question of administering his army commands and the affairs of his police empire. Himmler's constant preoccupation was to resist the efforts of the National Director[8] (Bormann) to diminish his powers and erode the frontiers of his influence.

The disputes over the *Volkssturm* are one more example of the mul-

[8] Reichsleiter, Bormann's official title.

tilateral squabbling that sapped the vitality of the German war effort. The idea had originally been Guderian's (or, to be strictly accurate, that of General Heusinger, of the OKH Operations Department, who had first suggested it in 1943, and had it turned down), and in scope had been confined to the Eastern provinces which were immediately threatened by Russian invasion. Guderian had suggested to Hitler that a kind of "Home Guard" be raised locally, drawing its staff from former members of the SA. Hitler had agreed, but evidently discussed the idea during the night with Bormann, for the following day he told Guderian that he had changed his mind, that it would be better if the organisation was made a Party affair, and that he proposed to entrust it to Bormann. Bormann then proceeded to enlarge the scope of the *Volkssturm* to cover the whole country, and made its raising and administration the responsibility of the Gauleiters, who were, of course, directly answerable to him. The Reichsleiter was now, at this late hour, in sight of one of his most cherished ambitions—a private army which would lend the substance of power to his undoubted ascendancy in the privy counsels of the Führer. Meanwhile the National Leader, not to be outdone, was administering the Home Army in a most peculiar way. Instead of directing the new conscripts into the various manpower pools of the fighting fronts for allocation to existing units, Himmler was busy forming yet another kind of division—the *Volksgrenadier*.

These formations were based on the skeletal remnants (often existing as little more than numbers in the OKW *Kriegstagebuch*) of units that had been "burned out" in earlier campaigns. They were now filled out with a motley collection of Hitler Youth, Luftwaffe ground staff, "reserved" middle-aged businessmen, invalids, and naval cadets. If their training period was dangerously short—sometimes as little as six weeks—their *esprit de corps* was very high, and helped by the priority Himmler granted them in the distribution of new weapons from the factories. Indeed, with so many new tanks and assault guns going to the SS, whose divisions were being raised to *Panzergrenadier* establishment as fast as possible, and all the infantry weapons being acrimoniously divided between the rival claims of the Reichsführer and the Reichsleiter, it is a wonder that the regular German Army got any new equipment at all.

Bormann's power over the armed forces was developing in another way also. In the weeks following the *attentat* he had developed and greatly expanded a system first introduced by the odious General Reinecke in 1943, of "National Socialist Leadership Officers," who were to be attached to regular units in the field and "indoctrinate" (i.e., spy on) the officers and N.C.O.'s. Bormann had fixed things so that these characters were responsible to, and reported directly to, himself. Nothing daunted, the National Leader chose to address them himself in his capacity as commander of the Home Army, and cast his speech in a suitably ruthless mould:

> Put the best, the most brutal officers of the division in charge. They will soon round up such a rabble [those who advised retreat]. They will put anyone that answers back against a wall.

Guderian, in the meantime, was dealing with the National Socialist Leadership Officers in his own style. After complaining that "their conduct had led to a number of gross breaches of discipline," he took to transferring the offenders to the OKH staff, where

> . . . I did retain certain limited disciplinary powers. There I allowed the very self-confident young men in question to cool their heels for several weeks, and to consider their manners . . .

When Guderian told Hitler about this, the Führer "looked at me in astonishment but said nothing." However, Bormann had an ally in General Burgdorf,[9] the head of the Army Personnel Office, who was in control of all postings. A running battle developed between the weary OKH staff and Burgdorf's department, with the rival nominees shunted back and forth to the adagio rustle of forms in triplicate.

Meanwhile Himmler, suffering from the overt contempt and indifference of the General Staff, the unrelenting opposition of Bormann, and the pernicious intrigues of Kaltenbrunner and Fegelein, had found a new ally. There had been a day when Dr. Goebbels and the National Leader were barely on speaking terms. "He'll be careful in future not to send me insolent teletyped messages," reads a petulant entry in Goebbels' diary in 1943, but of late each, sensing a growing exclusion from the intimacies of the beloved Führer, had found himself drawn toward a

[9] Burgdorf has a particular niche in the mythology of the German Army, for it was he who had driven to Rommel's house on the 14th October and handed the Field Marshal a box of poison ampoules. Burgdorf had offered Rommel the choice of taking his own life or returning to Berlin and "helping" with an investigation into information which Cesar von Hofacker had revealed under torture, concerning Rommel's part in the 20th July plot.

marriage of convenience. Following the *attentat* Goebbels had been appointed "Plenipotentiary for Total War," with exceptional powers over the home front and, in particular, the redistribution of manpower. It was natural that he and Himmler should work together—indeed, a few days before 20th July, in a conversation with Werner Naumann, Goebbels had spoken yearningly of the possibility:

> The Army for Himmler, and for me the civilian direction of the war! That is a combination which could rekindle the power of our war leadership, but it will probably remain a wonderful dream!

Yet no amount of personal enthusiasm could "rekindle" the German war effort while it was burdened with that rigid departmentalisation, bounded by the frontiers of privilege and personal jealousy which had been multiplying since 1938. Sometimes the different departments pulled in diametrically opposite directions. In September, when fuel stocks were barely adequate for a fortnight's operations, Speer's factories delivered more single-seater fighters to the Luftwaffe than during any previous month of the war. Even if the fuel had been available, to get them airborne the ground crews required to service these aircraft were being pressed into the *Volksgrenadier* divisions at the rate of fifteen thousand a week. Still more ludicrous, although the directives for the comb-out of civilians emanated from Goebbels' office, the apparatus for implementing them was operated by Bormann through the district Gauleiters. When Speer objected to Hitler that he could not part with the three hundred thousand workers Goebbels was proposing to take from industry, and the Führer hesitated, neither Himmler nor Goebbels could acquire his support by arguing on the basis of military requirements. Instead Himmler had to work on Bormann, by suggesting that his dignity as Reichsleiter was being flouted by Speer's resistance. Bormann then approached Hitler, and Goebbels got his conscripts.

While the Diadochi squabbled, Hitler brooded in Olympian isolation on the next stage of the war. Half deaf, trembling down one side of his body, and subject to uncontrollable attacks of paranoiac violence, Hitler was still far from being mad. His spirit had been broken by disappointment and treachery, and his physique was being steadily eroded by the medications of Dr. Morell, but the Führer's intellect still held all its old range and scope. He recognised the change in the balance of power, he saw that Germany's war aims had to be trimmed, that the old cry of

"Security," so shamelessly employed to cover the aggressions of the thirties, had now become a painful reality; that the problems and the strategy of World War I were resurrected. "Wars are finally decided by one side or the other recognising that they cannot be won." Hitler coined this aphorism, and it then became the touchstone of the new "two-front" strategy of the winter of 1944-45.

> Never in history was there a coalition like that of our enemies, composed of such heterogeneous elements with such divergent aims . . . Ultra-capitalist states on the one hand; Ultra-marxist states on the other. On the one hand a dying empire, Britain; on the other, a colony bent upon inheritance, the United States. Each of the partners went into this coalition with the hope of realising his political ambitions . . .

The solution, Hitler believed, was a violent "shock" blow against one of the partners, which would sap their will to continue a struggle to which no end was in sight. The secret weapons were inadequate for such an assignment, but thanks to the efforts of Speer, Goebbels, and Himmler, the "conventional" forces might just do the trick, for by late autumn he had accumulated a reserve of seven Panzer and thirteen *Volksgrenadier* divisions, and by the end of November this figure had risen to twenty-eight. This force, at the earliest possible moment, was to be thrown against the Anglo-Saxons. For they were the weakest, Hitler believed, both morally and physically, and a sharp defeat would "bring them to their senses."

The history of the Ardennes offensive is outside the scope of this work except where it impinges on the Eastern campaign, but where it does so it is extremely revealing. The German plan went wrong, first in its timing by a small margin and then (again by a small margin) in its execution. But in conception it rested on one basic premise—that the Russian front in Poland and East Prussia would stay quiet throughout the autumn. This assumption was shown to be valid. The Eastern front, between the Carpathians and the Baltic, barely shifted at all from August 1944 until the end of the year. Why? Why did the Russians stop when one more concentrated drive could (and later did) carry them to the outskirts of Berlin?

Certainly it was not for want of the necessary strength. At the start of the June offensive the five Soviet "fronts" concerned had deployed forty-one armoured brigades—a force equivalent at the very least to twenty full-size Panzer divisions, or more than three times the German tank strength. In numbers of artillery and infantry their superiority was of the order of six and four times that of the Germans. Furthermore, if the Red Army's communications had been stretched, it is equally true that the Germans' losses in their precipitate retreat had been very much the heavier. If the *Stavka* had allowed its central "fronts" to draw reinforcement from the Baltic and the Ukraine, it should have been able to deploy an even greater superiority on the Vistula by the middle of September, a moment when the whole German position in the West was crumbling and every reserve was being packed into the Siegfried Line.

On a comparative basis of strength, then, the Russians could have ended the war in 1944. No documents will reveal the background of their failure to do so—if only because it is highly unlikely that orders defining the long-term Soviet objectives were ever committed to paper. But it is hard to avoid the impression that purely military considerations, which weighed with the Western Allies until the very end, had already forfeited priority in Stalin's mind. War, the Russian dictator must have reflected, is not only "diplomacy carried on by other means," it offers on occasion a very acceptable substitute for normal diplomatic procedure, being, to an all-powerful nation, both faster in operation and more generous in yield.

The "second front," which had been so essential an adjunct to Soviet policy while it promised to weaken the West and relieve the pressure on the Red Army, was now an impediment to the advance of communism in Europe. Stalin was determined to secure the Balkans for himself, and to push the Russian "frontier" as far to the west as possible. A direct march on Berlin before the Balkans had been overrun might mean that the war would be ended with a large area of Europe still under nominal German occupation. The governments of Hungary, Rumania, and Bulgaria had all been in touch with Western agencies in 1944; in Yugoslavia the spectre of Mikhailovitch still haunted Tito. A sudden collapse of German resistance could be followed by a number of bourgeois "centre" administrations that could appeal to the Allies for diplomatic support at least until "free elections" had been held—and Stalin was realist enough to know that the Communist parties were so closely identified with Russia as to make the outcome of these elections, in countries which had lived for centuries in dread of Russia, a foregone conclusion.

Still more serious from the Soviet point of view was the likelihood of a last-minute switch by the German High Command. Stalin could not believe that once they saw their homeland to be in immediate peril of Russian occupation the German generals would not pack their Eastern defences at the price of "letting in the West." This in turn might mean that the Russians would have to rely after the war on treaties to hold the Germans down, rather than on military force, and that a complete sealing off of the recalcitrant Poles would be impossible.

No Western statesman could match Stalin's icy realism, and Roosevelt's "charm,"[10] which had so astonished and gratified the Russian at Teheran, had been partially offset by the disagreeable sound of the British plan for landing in Istria and striking north to Vienna. In 1944, Stalin could hardly have foreseen Yalta and the American attitude that ". . . if I give him everything I possibly can and ask nothing from him in return, *noblesse oblige,* he won't try to annex anything and will work for a world of democracy and peace."

Thus it was to the Balkans that the centre of gravity of the Eastern front shifted in the autumn of 1944. A fresh "front" was formed by the *Stavka* (the 4th Ukrainian) and allocated eighteen divisions under General Petrov. Petrov's task was to advance into Hungary across the Carpathians and at the same time to maintain contact between Koniev and the two "fronts" charged with the occupation of the Balkans, those of Malinovsky and Tolbukhin. Between them these two commanders had thirty-eight full-strength divisions against a nominal German strength of twenty-five. In fact, the German units were all down to under ten thousand men, and the five strongest divisions (including two Panzer), which Schörner had been holding in reserve, had been taken north on Guderian's orders, following his appointment on 20th July; on the 29th, Schörner himself had been ordered to Courland by Hitler, to take command of the German forces—the rump of the old Army Group North—which were threatened with isolation by Govorov and Chernyakovski.

Colonel General Friessner, who arrived to take Schörner's place, found himself bequeathed with the same sort of classically vulnerable deployment that was a recurring feature of the German dispositions in the

[10] For one who had spent a lifetime in American politics, Roosevelt's attitudes in diplomacy are well-nigh incomprehensible. It is exemplified by his note to Churchill ". . . I think I can handle Stalin personally . . . he hates the guts of all your top people. He thinks he likes me better, and I hope he will continue to do so." (Churchill, *History of the Second World War,* IV, 171.)

southern theatre. Ironically, even the army numbers were the same, for the "resurrected" 6th Army, supported by three infantry divisions of the 8th, was dug into the right bank of the Dniester, covering Jassy and Kishinev, with those two Rumanian armies of ill omen, the 3rd and 4th, on its flanks.

The Germans had had nearly four months in which to prepare their positions, but they did not have the numbers to cover the whole length of the Dniester, and the Red Army, too, had had ample time, not only to accumulate supplies and refurbish its armour, but to seek out the vulnerable sectors which had been entrusted to the Rumanians.

At this stage in the war the satellites had lost interest in everything except working their passage with the Allies and getting the German Army out of their country in the shortest time and with the least damage. King Michael of Rumania had been in touch both with the CIA and with the Russian Legation in Turkey, and made preparations for a *coup d'état* which was to be followed by the "internment" of the troops of his erstwhile ally; he awaited only the signal, which was to be the passage of the Dniester by Tolbukhin.

On 20th August the two "fronts" of Malinovsky and Tolbukhin fell on the submissive Rumanian divisions opposite them and within hours were driving across the open, undamaged fields of Bessarabia. The majority of the Rumanians simply laid down their arms and melted into the countryside. Others hitched on to the advancing columns (which included the heavily indoctrinated Tudor Vladimerescu Division, made up of former prisoners of war, whose approach was a warning to King Michael of the kind of tiger he was riding) and were soon exchanging shots with the Germans. Within forty-eight hours the bulk of the 6th Army had been surrounded, and the few remnants which had been able to escape were heading at breakneck speed from the Iron Gates and the Hungarian frontier. Antonescu had been placed under arrest, and with him General Hansen, chief of the German military mission.

The whole German position in Southern Europe was now on the point of disintegration. And the task of repairing it was made practically impossible by the crippling shortage of troops. There were only four divisions left in the whole of Rumania south of the Transylvania Alps, and one of them, the 5th Anti-Aircraft, had very little transport and was tied down at Ploesti by the zeal of local Rumanian forces that were attempting to "intern" it. Even if greater strength had been available, its deployment would have been impeded by the fact that the OKH command net, which had hitherto exercised absolute sovereignty over the Eastern front, had been pushed back so far that it overlapped the province of OKW and Army Group F, under Weichs, which was responsible for Yugoslavia and Thrace.

To send additional troops into the Balkan whirlpool was clearly hopeless, but OKW made one brief reflex attempt to evoke the formula of 1941, when the *coup* of another young king, the Yugoslav Peter, had upset the German timetable and so aroused the wrath of the Führer. On 24th and 25th August the Luftwaffe attacked Bucharest, and three battalions of the Brandenburg Division were flown in from Vienna to cow the populace. But the Luftwaffe could no longer mount the terror strikes of the old days. A few Heinkels dropped their loads at random, some were caught by prowling Russian fighters, others, landing short at Rumanian airfields, were shot up and their crews made prisoners. The Brandenburgers found themselves facing the whole of Managorov's 53rd Army, advancing at the rate of thirty miles a day. They commandeered what transport they could lay their hands on and set off south to the Bulgarian frontier.

Here the elderly Weichs was attempting to disarm the Bulgarian Army, whose "general behaviour made their reliability suspect."

At the beginning of August, in the teeth of Guderian's bitter protests, OKW had sent a substantial quantity of armour—eighty-eight PzKw IV's and fifty assault guns—to the Bulgarians, in the belief that they were the most reliable of the Balkan allies because of their hatred of the Greeks and their fear of the Turks (whose entry into the war on the Allied side was now regarded as inevitable). But Colonel von Jungenfeldt, in charge of training the Bulgarians, had been reporting directly to Guderian in his capacity of Inspector General, and not to Weichs or OKW. Jungenfeldt's opinion of the situation had been so gloomy that Guderian had ordered the return of the equipment to Belgrade, where it was to be issued to the 4th SS, itself practically the only mobile unit left in the Balkans and one which, having spent the previous six months burning down Yugoslav villages, was in a relatively fresh condition. However, Jodl (who has a record of disastrous uniformity whenever he interferes in the East) had heard of the order, and countermanded it at the last minute. It was not until the 25th that Weichs, on his own initiative, began to take "certain precautionary measures."

By now, though, things were moving too fast for corrective action by individual commanders. On 25th August, Michael's new government had declared war on Germany, and units of the Rumanian Army were attached to Malinovsky and Petrov, to guide them over the Carpathian foothills. Two days later the Bulgarians began to evacuate Thrace and Weichs was asked formally to leave the country. On the strength of the "good faith" thus displayed, Bulgarian plenipotentiaries began frantic negotiations with the British (with whom they had been at war since the Greek campaign of 1941) in Cairo, in the hope of getting some sort of settlement recognised before the Russians (against whom they had taken the precaution of never declaring war) were in possession of their country. But time was running with Stalin. Tolbukhin's armour crossed the Bulgarian frontier on 5th September, to the accompaniment of a formal declaration of war, and the following day the Black Sea fleet disembarked Russian marines at Varna and Burgas. On 9th September a "patriotic front" government, heavily loaded with Communists, was formed and straightway opened negotiations with Moscow, declaring an armistice forty-eight hours later, on the 11th.

There now began for Weichs's army group a long and tortured retreat northward. The railways were barred to them, and Jungenfeldt's precious armour was left behind—many of the tanks still resting on their flatcars at the Sofia goods siding. Along dusty side roads, winding through endless ravines in the arid mountains of Serbia and Montenegro, constantly harassed by guerillas, the retreating Germans struggled homeward.

> That retreat was a terrible affair. The roads would be mined sometimes in the passes, for twenty or thirty kilometres at a stretch, and after the first week we had lost most of our vehicles. Many of the men had worn out their shoes, and discarded everything except their rifles. At night we had to mount half the company on guard for the partisans would allow us no rest. Every village that we passed through bore testimony to the unbelievable ferocity of this partisan warfare . . .

Meanwhile Tolbukhin's forces had diverged, and his right wing was swinging northwest, parallel with the Danube in a race to join forces with Tito at Belgrade and cut Weichs's bedraggled column in half. Capturing Turnu Severin on 9th September, they were delayed by the 4th SS, which took its stand forty miles south of the Yugoslav capital in the isthmus of land between the Danube and the Transylvanian Alps. But three days later the Russians were across the mountains farther north and descended into the valley of the Maros, capturing Temesvár on 19th September and Arad two days later. The whole of Hungary was now open to the Russian advance.

At the end of August, Guderian, cast by Hitler in the role of diplomat, had travelled to Budapest with a letter for Admiral Horthy and instructions to discuss matters with him "as one soldier to another" and to "form an impression of his [Horthy's] attitude." Formal courtesies were maintained throughout the meeting, but Horthy had given the game away the moment they were left alone, for drawing up his chair, the old Admiral had told Guderian, "Look, my friend, in politics you must always have several irons in the fire . . ." The two men continued their discussion for several hours, but from that opening sentence, as Guderian wrote in his diary, "I knew enough."

At the end of September the Germans' position in the Balkans was as close to complete disaster as it had been the previous month in France. The "front" had simply fallen apart; their order of battle was made up of a motley assortment of units, almost decimated, dispersed over several thousand square miles of hostile countryside, united only in the resolve to get within the frontiers of the Reich before the enraged civilian population could exact personal revenge. Indeed, had the two disasters occurred simultaneously, it is probable that the war would have ended in 1944, without the Allies ever having to have fought inside Germany at all, and in spite (perhaps) of the Russian reluctance to force the issue at that time. In any event, by the time that Malinovsky and Tolbukhin reached the Hungarian frontier the crisis in the West had passed, the Allies had been defeated at Arnhem, and the Warsaw uprising had been mastered. Friessner's forces in the Carpathians had managed to hold off Petrov, and although his right "flank" had been bent back in disorder it became increasingly feasible for light scratch forces to hold up the Russian advance, for by the end of September the spearheads of both Malinovsky and Tolbukhin had travelled over two hundred miles from their starting lines.

Incredible though it may seem, the "treachery" of their allies, and the explosions of hatred and violence that took place as German authority in the occupied territories slackened, came as a shock to the Wehrmacht, and even to the SS. Hitherto replete in the enforcement of Machiavelli's maxim, "It is better to be feared than loved," the Germans nonetheless believed that as they were the *Herrenvolk* no one could think of opposing them unless he were a Bolshevik or a Jew. Now they were faced with a frightening prospect. As the burning fringe of battle approached the Fatherland, it compressed not only the Wehrmacht but the whole apparatus of terror within the frontiers of the Reich. Four million foreign workers, a floating concentration-camp population of over one and a half million, a whole collection of "national" legions of one kind and another—even an "Indian Brigade"—many of them carrying arms—what if a spark from the battle front should ignite this mass of tinder? No great powers of derivative argument were required to see that if the Rumanians nurtured so powerful a dislike for the Germans, and had been so quick to take advantage of their discomfiture, little mercy could be expected from the slave immigrants of the Reich itself.

The SS, ever fertile in schemes for administration where these offered scope for an extension of its own powers, now veered around (with a lurch that must have induced, even in the National Leader, a faint twinge of nausea) to a pro-Vlasov standpoint. Vlasov, between bouts of alcoholic intoxication, had been making a thorough nuisance of himself. Bandied from one department to another, humiliated, reimprisoned, released, harangued, he had preserved his dignity and his policies throughout. He refused to identify himself with the German cause, and continued to argue that his purpose was to save Russia from Stalin and to "reconstitute" the Russian state. He lectured German officers on the right way to treat the Russian people, and the entire audience (Himmler had been alarmed to see) "hung their heads in shame." By September 1944 a section of the SS under Gunter d'Alquen, the editor of *Die Schwarze Korps,* and one of the movement's intellectuals, had formulated Operation *Skorpion,* whereby Vlasov was to assume authority over all the captured Russians, including those still in the prison camps and the slave-labour factories, and be allowed to raise fighting divisions. The National Leader. to his marked distaste, was compelled to stage an interview with Vlasov "as between equals," at which the renegade Soviet General had been granted everything he asked for. (It was after agreeing to see Vlasov that Himmler had made his celebrated remark to d'Alquen: "Who compels us to keep the promises we make?")

The theory behind *Skorpion* was that the Russians would be more amenable to discipline from their own compatriots. It was a development of the policy which had led to the inclusion of the scum of Europe in special

SS battalions. A process which had started with the "racially pure" and allowed blond Scandinavians into the Viking Division was now extended to the Flemish, the Dutch, Latvians, Walloons, Uzbeks, Bosnians, Estonians, and even Arabs, whose one qualification was a taste for the dirty work of the SS. It was the old jailor's technique of making the prisoners fight one another, and it ran the same risks. Once the central authority was shaken, all the hate and brutality the jailors had released would recoil upon them and their kin.

The Red Army had reached the Danube on 5th October, and a fortnight later joined forces with Tito in Belgrade. A hundred and fifty miles due north Malinovsky forced the line of the Tisza, and by the end of October the German "front," still little more than a patchwork of *ad hoc* formations of widely varying quality, was back on the upper Danube. Meanwhile Finland had dropped out of the war and the Russians had broken out into the Baltic with the capture of islands of Dago and Ösel. It was only in the centre that the front still held, or, rather, that the *Stavka* maintained its game of cat-and-mouse. And Guderian, for one, knew when its blow would fall. It would come, he reasoned, after the Ardennes offensive opened and when its outcome could be predicted. The last hand the Wehrmacht could deal itself would have been shown, and played; thereafter its extinction would be as simple and calculable as the outcome of the chess problems with which Vlasov and his colleagues would while away their useless hours.

The running down of German strength in Poland and East Prussia had gone far beyond the danger point by Christmas 1944. In November and December, out of a total production of 2,299 tanks and assault guns, only 921 went to the Eastern front. The number of divisions had shrunk to 130; this was 27 fewer than the total with which the Soviet offensive of June had been checked. And of these just under half were deployed where they could play little part in the decisive battle, for there were still nearly 30 infantry divisions in Memel and Courland, guarding the Baltic shore for Doenitz's U-boat exercises; and 28 south of the Carpathians, where the Russians' pressure in Hungary and their encirclement of Budapest was gradually drawing off some of the best divisions in the OKH reserve.

Indeed, it is an extraordinary tribute to Guderian's single-minded dedication and efficiency that he had been able to amass any reserves at all. Yet he had done so. In spite of the demands of the Ardennes offensive, the recurrent crises in the Balkans, the deliberate withholding and diversion of weapons by the Home Army, and persistent obstruction at the highest administrative level by Jodl, at least twelve Panzer divisions had been taken out of the line by Christmas 1944 and were being held in readiness to take the shock of the Russian spearheads. Guderian knew, though, that on a front of nearly six hundred miles and having neither the fuel nor the orders, nor indeed the space in which to fight a mobile battle, his armies were soon to face the risk of annihilation. In December, OKH intelligence reported over sixty rifle divisions and 8 tank corps (or slightly more than the total German strength along the whole front) in the Baranov bridgehead alone. Two other major Russian concentrations, fifty-four divisions and six tank corps north of Warsaw, and approximately the same number on the East Prussian border, gave warning that the blows would be successive. "By the time the last one comes," said Reinhardt, ". . . we shall be debris."

It had originally been Hitler's intention to start the Ardennes offensive in November. If he had done so, and even if the results had been no more substantial than they were in actuality, the return of the Panzer reserve to Guderian's command would have been possible before the Russian winter offensive started. But in fact Manteuffel and Sepp Dietrich did not start until 16th December, and their progress was watched with as great anxiety by OKH at Zossen as it was by the staff of Hitler's temporary headquarters in the West, at Ziegenberg.

By 23rd December, Guderian's nerves could stand the suspense no longer, and he drove across Germany to Ziegenberg, resolved to "request that the battle, which was causing us heavy casualties, be broken off and that all forces that could be spared be immediately transferred to the Eastern Front."

Although the Chief of the General Staff was ultimately proved right, his intervention at this stage was premature, and this misjudgment probably cost him several divisions, for it gave the "Western" school the opportunity to show him up as ill informed and needlessly apprehensive ("far too worried," as the National Leader soothingly expressed it). For on that day, in the words of the U. S. Official Historian, ". . . the attempt to plug the gap had been converted to a struggle for survival, as every division sent to First Army in a counter-attacking role . . . was forced into the defensive fighting to prevent a new German break-out."

It was not until 24th December that Model decided even to substitute the "small solution" for the original plan of breaking through to Antwerp, and it was unthinkable that Hitler should start to disengage at a point which seemed to be the crisis of the battle. "Who's responsible for producing all this rubbish?" Hitler shouted when Guderian showed him a memorandum reciting the accumulation of Soviet divisions in Poland. Guderian was entertained at dinner, after which Jodl gloatingly "confided" that the Ardennes would be followed by another offensive, in Alsace. "We must not lose the initiative that we have just regained," he lectured Guderian, "the enemy's operational timetable has been gravely damaged." Indeed, the sole result of Guderian's visit was that OKW briefly turned its attention to the Eastern front and ordered (while Guderian was returning to Zossen and without informing him) the two SS Panzer divisions of Gille's corps which were in reserve behind Warsaw to proceed to Hungary and "raise the siege of Budapest."

Realising that he would never make any headway with the OKW sycophants at Ziegenberg, who were intoxicated by the unaccustomed approbation of the Führer and by the unfamiliar sensation of handling a major offensive, Guderian now directed his efforts along the traditional channels of Prussian freemasonry—and was better rewarded. On New Year's Eve he returned to Ziegenberg, but this time took the precaution of calling first on Rundstedt, the C. in C. West, and General Siegfried Westphal, his Chief of Staff. These cool professional colleagues, who held nothing but contempt for the "Nazi soldiers" around the Führer, were gravely impressed by the dangers in the East. Westphal gave Guderian the numbers of three divisions behind the Western front and one in Italy which were immediately available and located near railway stations, and even went so far as to send their commanders a warning order to stand by for entrainment. Guderian then went personally to the Field Transport Office and arranged for the necessary rolling stock to be made available, and then, finally, to Hitler's conference room, where he repeated his request of the previous week for reinforcement.

Jodl immediately said there was "nothing available."

"But this time I could contradict him . . . When I gave Hitler the number of the divisions available, Jodl asked me angrily where I had got them from; when I told him—from the Commander-in-Chief of his own front—he relapsed into sulky silence." Guderian was duly allowed to transfer the divisions, but the time and effort wasted to get this slight reinforcement, and the divided state of the German High Command which the incident exemplify, were a poor augury for the critical battles that lay ahead.

Following this minor personal triumph, the Chief of Staff spent the first week in January on a tour of the Eastern army headquarters. What he learned from them and from his own observation was so alarming that he decided to make one last appeal to Hitler, this time both for more troops and for permission to make a pre-emptive withdrawal which would allow a thinly defended buffer of land to take the first shock of the Russian attack. For Guderian rightly saw that in the months which they had spent in preparation the Russians had accumulated such strength that it was no longer within the power of the German Army to stop them in their tracks. The only hope was, by ducking at the last moment, to make Zhukov "hit air," then to fight a flexible battle across western Poland until exhaustion and the spring thaw would take the impetus out of the Russian advance.

But even Manstein in his heyday had been unable to persuade Hitler to a strategic philosophy of this kind. Now, as Guderian wryly observed, ". . . whenever Hitler heard the word 'operational' he lost his temper, knowing it to be a prelude to a withdrawal." When Guderian presented him with the revised estimates of Russian strength, Hitler declared that they were "completely idiotic"[11] and "pure bluff." He who had for so long deployed Panzer "brigades" two battalions strong and "divisions" barely the size of regiments was now attributing the same deceptions to his enemy. Hitler went on to order General Gehlen, who had drawn up the appreciation, to be committed to an insane asylum. This Guderian managed to deflect,[12] but he achieved little else. At the close of their interview Hitler told him, "The Eastern Front has never before possessed such a strong reserve as now. That is your doing, and I thank you for it." Unmollified, Guderian retorted, "The Eastern Front is like a house of cards. If the front is broken through at one point all the rest will collapse . . ."

[11] German estimates were slightly exaggerated, at a total of 225 divisions. In fact, Russian strength was 180 infantry divisions, with 4 tank armies, each of around 1,200 tanks, and an additional 23 independent tank brigades. Increasing mechanisation had reduced the cavalry, and there were only three cavalry corps deployed.

[12] Guderian's high opinion of Gehlen, "one of my best staff officers," was evidently shared by the Americans who put him in charge of a special department in Germany after the war, with almost the same title as his old "Foreign Armies East." This organisation was subsequently taken over by the Bonn government, and at the time of writing (1964) it is an offence to try to take Gehlen's photograph.

On 12th January, Koniev began his attack from the Baranov bridgehead, and within thirty-six hours had broken clean through this, the "Hubertus position," and allegedly the strongest sector of the German line. Within twenty-four hours first Zhukov, then Rokossovski opened their attacks, and by 14th January every one of the precious Panzer divisions in Poland had been committed, grouped in two corps—the 24th (Nehring) was trying to seal off the breaches in the Hubertus, while the 46th struggled to prevent Zhukov wheeling north from the bridgeheads at Magnuszev and Pulavy and encircling Warsaw.

Guderian's strongest reserve was situated in East Prussia, and consisted of the Hermann Goering Division and *Gross Deutschland,* grouped as a corps and under the command of General von Saucken. But on 15th January, OKW (either Hitler or Jodl) had interfered and sent a message from Ziegenberg, ordering that Saucken be sent down to Kielce (a distance of some 150 miles, involving a flank "march" over the Partisan-infested Polish railway system). Guderian at first refused point-blank, arguing that there could be no possibility of holding the present positions on the Vistula, but that the Russian advance could be slowed down if East Prussia were held as a "balcony" overlooking their front from the north. How far this was a real tactical proposition, how far influenced by the fact that Guderian had himself been born in East Prussia and was brought up in the tradition that the cradle of German militarism must be preserved inviolate at any price is a matter of conjecture. In any case, the Führer overrode his Chief of Staff, and Saucken's powerful corps spent the first week of the Russian offensive in railway sidings loading and unloading its brand-new Tiger and *Jagdtiger.* Hitler himself left Ziegenberg on 16th January and set up his new headquarters at the Chancellery in Berlin, where he was to spend the last three months of his life.

Saucken and Nehring together managed to keep up some cohesion at the southern end of the front, but the 46th Panzer Corps in the centre was far too weak to delay Zhukov for longer than a few days. By 18th January it had been forced back to the north bank of the Vistula, where it was caught in the rear by the full strength of Rokossovski's armour. During that week Hossbach's 4th Army was driven back onto the frozen Masurian Lakes, and with Saucken's corps now in south Poland, he had no reserves for their defence. Day by day the hallowed soil of East Prussia quaked under the tread of Russian armour, and the first augurs of a (deliberate) relaxation of Red Army discipline were heard. Thousands of civilians took to the roads, loading their belongings on horse-drawn carts, pushing baby carriages and barrows in a replica of the 1940 scene in Western Europe—but enacted under subzero temperatures.

Over a seventy-mile stretch between the Narew curve and Kielce the German front had been blown to pieces, and with the exception of the debris of the 46th Panzer Corps there were no mobile forces with which to delay their progress. Zhukov's tanks were making thirty and forty miles a day, and on 20th January captured Hohensalza, celebrating their arrival on German soil with a lurid and violent sack of the town which went on for three days.

twenty-one | BLACK JANUARY

With the disintegration of the 46th Panzer Corps the imbalance of the German armies became ever more precarious. On 22nd January, Zhukov's right wing had made contact with Rokossovski a few miles to the east of Graudenz. This meant that besides the encirclement of a number of slower-moving units in the 2nd Army, the flank of the main Russian thrust was now secured for another forward bound. Taking the fullest advantage from the new system of flexible subcommands, Zhukov had already drawn off four of the tank corps with which Koniev had made the first breach at Baranov, on 12th January, and with the identification by OKH of the 1st and 2nd Guards tank armies on 27th-28th January some idea of the size of the armoured wedge that was approaching them began to percolate to the "Wilhelmstrasse circle" in Berlin. Even after allowing for an element of duplication in German intelligence figures, this force cannot have amounted to less than eighteen hundred tanks, of which about six hundred were of the formidable new type T 34/85. The Russians had fully assimilated the techniques for penetrating the *Pak-front,* which they had first encountered, then adapted, at Kursk, and their brigades roamed as independent units of up to three hundred T 34's (T 34/85 in the Guards tank armies) fanned out around a hard core of the slower-moving JS's (usually about two battalions or less in strength, i.e., thirty to forty tanks) with their long-barrelled 122-mm. guns.

Twice daily the situation map at Zossen was marked with the position of the leading Russian formations. Army Group Centre had been literally shattered, with its headquarters staff and the remnants of four divisions forced up into East Prussia. Only the valiant but exhausted 24th Panzer Division and *Gross Deutschland,* struggling westward in an endless series of mobile encirclement battles, formed a link between Reinhardt and the fourteen divisions of Army Group A under Harpe. Then on the 22nd the Russians seemed to be changing direction. They wheeled north, following the left bank of the Vistula as it turns through 100 degrees between Bromberg (Bydgoszcz) and Thorn (Torun). It was flat, open country—the same terrain over which the original Panzers had torn the Polish cavalry to ribbons four and a half years ago. The ground was hard, and lightly covered with powdery snow. Each day the sun shone from a cloudless sky, at night the temperature fell to minus 20 degrees.

The Russian spearhead was heavily weighted—in contrast with the Western armies—along its cutting edge. The tank brigades had a battalion of infantry, riding in American trucks, and a few pieces of artillery, some of it self-propelled, but the mainspring of the thrust, once it hit open country, was the redoubtable T 34. It seems to have been used for every task which in more sophisticated armies was allotted to particular and specialised units and vehicles. Charging ahead in reconnaissance, massed side by side as artillery, dug in as pillboxes; towing, crushing, bulldozing; carrying infantry or, still more perilous, bringing up ammunition chained in boxes to its flat afterdeck, this crude, cramped, poorly ventilated but immensely tough and reliable instrument of war played its many roles. The few early models which had so alarmed the Germans in front of Moscow in 1941 had spawned a limitless progeny. In the one year from January to December 1944 over 22,000 were produced. So extravagant a use of a single type of vehicle to perform a variety of tasks carried general advantages which more than offset the waste of life and effort in particular instances. The Russians limited their supply problem to two items, fuel for the twelve-cylinder diesel engine of the T 34 and ammunition for its 76-mm. gun. Where there was mechanical failure the parts were "cannibalised" from wrecks, and this same policy was adopted toward the trucks. The crews, and to a still greater extent the motorised infantry, were well fed in static positions but obliged to forage for themselves once the advance gathered momentum.

Within a fortnight of the start of the offensive the Russians had left the desolation of Poland and penetrated the frontiers of the Reich itself. Here in Silesia and Pomerania they found a countryside richer and more tranquil than anything they had seen since their enlistment—in many cases they found an abundance such as they had never experienced before. Of all the regions of Greater Germany these were the ones which had felt the impact of the war least. They were the quiet evacuation centres, the prosperous areas of the "new industries," the small factories which Speer had dispersed from the Allied air offensive. Through this peaceful countryside the Soviet columns literally blazed their trail. Shops, houses, farms, were plundered and set alight. Civilians were shot down casually for the possessions they carried with them; it was common for a man to be murdered for his wrist watch. The Russians soon discovered that the inhabitants were hiding their womenfolk in the cellars of their houses and adopted the practice of setting fire to buildings they suspected were being used for this purpose. An incendiary shell from a T 34 proved the quickest way to assemble the occupants for scrutiny.

Yet, barbarous and horrific though it was, the first impact of the Soviet armies on Germany will not stand comparison with the Nazi conduct in Poland in 1939, or in White Russia and the Baltic provinces in 1941. The atrocities of the "Death's Head" (*Totenkopf*) units of the SS which systematically murdered school children and poured gasoline over hospital inmates were the expression of a deliberate policy of terror, "justified" by half-baked racial notions, but implemented with a perverse and sadistic relish. The brutalities of the Russian armies were not so much intended as incidental. Periodically, where overindulgence jeopardised efficiency, they were repressed with ruthless severity by the Soviet military police. The Russian soldiers were illiterate primitives, indoctrinated to hate their enemy, conditioned by years of privation and physical danger, traditionally contemptuous of human life, and many of them with fierce personal incentives for taking vengeance on the Germans. Their impact on the Reich was that of a hard warrior host on a disintegrating civilisation—the very aspect of war in which so many Germans had formerly exulted, now visited on those provinces which had for so long nursed it as a heroic ideal and so seldom experienced it as a reality.

Nonetheless, for Guderian, as for every Prussian in the German Army, the prospect was a terrible one. They could hardly be expected to perceive—much less to appreciate—the retributive irony of the situation. The urgent need for a policy, for some measure to halt the ordeal to which the country of their birth was being subjected, now assumed a desperate personal importance.

The speed of the Russian advance owed much to the extraordinary character of the German dispositions. The full weight of the Soviet blow had fallen, as has been seen, against that sector which was both the most important and the least strongly defended. Along their extended left flank, stretching three hundred miles across East Prussia and into the Baltic states, the Germans had nearly fifty divisions. Yet the majority were infantry, needlessly strong for its protection, yet neither mobile nor concentrated enough to menace the offered flank of the Russians who were pressing across their front to the south, while in the centre the Panzer divisions were grinding to and fro without rest, making up by their mobility for their lack of numbers, wearing out crews and machines when they should have been husbanded for a *riposte*. A few more weeks of this unnatural strain and the front would split from end to end, the whole Reich would be overrun. Now was the last chance to distribute the divisions in an orthodox defensive pattern of concentration and depth. If this could be achieved, there was still a possibility that the Russians' advance might be halted and a sharp defeat administered to them. For the shape of their salient began to look conventionally vulnerable on the map. The maximum force lay at the tip. The base was held down by the German divisions holding out around the Masurian Lakes and in the Carpathians. It was true that the German forces no longer had the weight to converge at the base of the salient and inflict a strategic defeat on the invader, but Guderian believed that the chance might present itself for a pincer counterattack on a shorter axis, with one arm attacking southeast from Pomerania and the other driving to meet it from the Glogau-Kottbus area.

There was some evidence that the Russian offensive was under strain, for units which had fought in Finland and Rumania had already been reidentified in the central and southern sectors.[1] Now, with their communications stretched tight across a frozen and devastated Poland and their tank armies having seen three weeks of continuous action, they

[1] We know today that this last offensive absorbed virtually every weapon and man left in the Red Army. For the first time since 22nd June, 1941, Stalin had released the whole of his strategic reserve. This was partly out of strategic necessity—the Russians wanted to put as much of Western Europe as possible between them and their allies before negotiating the peace—partly from an understandable tactical miscalculation. In conversation after the war Zhukov told Brigadier Spurling, ". . . when we reached Warsaw we could not see how we could get beyond the Vistula unless the German forces on our front were considerably weakened," for the Russians assumed that Hitler would always maintain a maximum concentration against themselves, even if it meant endangering other fronts.

seemed to be vulnerable. Twice in his diary Guderian noted that Zhukov was taking greater and greater risks as his awareness of the German weakness mounted. It would not be the first time that invaders from the East who had the temerity to penetrate into Prussia would be routed by a combination of skill and leadership. Even if Tannenberg or Galicia-Tarnow were beyond their reach, the German armies might still achieve a victory comparable to Manstein's "miracle of the Donetz." [2]

With this in mind OKH conceived the notion that a complete new army group should be created for the north-central region. As the Germans were forced back down the narrowing funnel between the Baltic and the Carpathians, their command structure had fallen into mounting disarray. Guderian's scheme was that the battered Army Group Centre should be broken up, its units divided between Reinhardt (Army Group North) and the new command. Those divisions which had been forced southwest against the upper Warthe could be reinforced from Harpe's relatively intact Army Group South and redesignated as Army Group Centre. The key area, and the most vital command, lay in the new region which was to be known as Army Group Vistula, with responsibility for the front between Poznan and Graudenz. This command would at first consist of a motley of battle-worn units from its two neighbours, but Guderian planned that all the reserves from the West, including Sepp Dietrich's 6th Panzer Army, were to be directed into it and accumulated in close reserve until the moment was ripe for the counterstroke. The 6th Panzer Army, with its high SS content and its profusion of the latest equipment, was in many ways the best instrument the Germans could have selected for this task. It was fresh from a startling "victory" in the Ardennes (it had been withdrawn before the extent of the failure there became apparent), it was well used to operating under conditions of total enemy air superiority, and had trained often over the very ground on which it was now proposed that it should contest Germany's most vital battle. The other prerequisite of success in such an operation—quick and efficient staff work, to ensure rapid concentration of the force and close phasing of the objectives—was a quality in which the German Army had always excelled.

Hence there was more than a likelihood that had he been given a free hand Guderian might yet have achieved his victory and the Soviet advance would have been stopped short. If this had happened, it is highly

[2] See Ch. 16.

unlikely that Stalin would have been able to mount another offensive before the Allies had crossed the Rhine—with consequences to the settlement of Europe and the balance of world power that cannot be calculated. There are many reasons why the fortunes of this particular operation, both in its planning and in its belated realisation, repay study.

It was intended by OKH that Army Group Vistula take over the staff of the army group which had been responsible for the southern Balkans and for liaison with the Rumanian and Hungarian forces which had been forced back over the Prut. This organisation was virtually intact and, its responsibilities having been if not discharged, then evaporated, had lately returned to Germany. Freiherr von Weichs, its painstaking and somewhat elderly head, was selected by Guderian for the new command, but it seems highly probable that Guderian intended to direct this battle personally, with Weichs as a *de facto* Chief of Staff.

On the evening of 23rd January, Guderian had outlined his proposals to Jodl in the course of a telephone conversation. His intention was to enlist Jodl's support at the Führer conference the following day. Jodl apparently "agreed to support the proposal," and at the conference itself everything went smoothly to begin with and Hitler approved the new command zones, which were to become operative from 25th January. But when Guderian suggested Weichs as head of Army Group Vistula, the atmosphere began to deteriorate. Hitler said, "The Field-Marshal seems to me to be a tired man. I doubt if he's still capable of performing such a task." Instead of frankly admitting that Weichs, though nominally in charge, would in fact be acting as a staff officer under his own direction, Guderian concentrated on defending the merits of his choice. Hitler, whose mind was made up, argued with diminishing patience and waited for an opportunity to explode his bombshell. The moment came when Jodl, seeing which way the argument was going, "dropped a sneering remark about the Field-Marshal's deep and genuine religious sense." Thereupon Hitler "refused to sanction the appointment." He was fed up with these professional soldiers, there was no end to their betrayals. Hossbach, Bonin, Rauss, they were fumblers, nincompoops. And there were blackguards, too. Seydlitz's name came up once again. And the loss of Lötzen.

After rambling in this vein for some time Hitler announced his decision. This was to be an SS army. Here in this vital sector the Reich would be entrusted to the Party, and the Party soldiers, whose loyalty was never in question. Furthermore, there could be only one choice for the command of such a force: it must be given to the National Leader, the one officer of the Reich in whom the Führer had implicit trust, *Treuer Heinrich* himself.

Guderian was aghast. Such an arrangement did more than upset his plans for the counterattack. It threatened to undermine the whole new command structure which had already been promulgated, for controlled by a "military ignoramus," (as he would frequently refer to Himmler) this vital new sector would place the whole front in even more serious jeopardy than if the original dispositions had been retained.

It is true that there was no lack of precedent in the history of the German Army for vesting the nominal command in a man of straw and providing a strictly professional staff organisation to guide his hand. But the National Leader was not well cut out for the role of Hohenzollern effigy. He would feel "uncomfortable" if surrounded by regular officers, Hitler said, and it was proposed that he assemble his own staff; as Chief, Hitler favoured SS brigade leader Lammerding. Guderian argued for hours, but the most that he could achieve was that a few General Staff corps officers were assigned to the new headquarters in "purely administrative" roles. The majority of positions were filled by SS officers "who for the most part were uniformly incapable of performing their allotted tasks."

The day after this disastrous conference, 25th January, was that which had been fixed by Dr. Barandon for Guderian's interview with Ribbentrop. It may well be imagined that the Chief of the General Staff, who had enjoyed little more than three hours' sleep and had seen his plans for a military initiative with which to balance the political overtures he favoured dashed, spared no detail in his exposé of the crisis. At the end Ribbentrop, a trifle incredulous, asked if he had been told "the exact truth." He suggested, "The General Staff seems to be losing its nerve."[3] After some further discussion of the position Guderian asked Ribbentrop point-blank if he was ready to accompany him to see Hitler and "propose that we attempt to secure an armistice on at least one front." The exchange is recorded as having continued as follows:

[3] Guderian's comment on this remark has a certain dry humour: "As a matter of fact, a man needed an almost cast-iron nervous system to carry out these exploratory conversations with the requisite calm and clarity of thought."

RIBBENTROP. I can't do it. I am a loyal follower of the Führer. I know for a fact that he does not wish to open any diplomatic negotiations with the enemy, and I therefore cannot address him in the manner which you propose.

GUDERIAN. How would you feel if in three to four weeks' time the Russians were to be at the gates of Berlin?

RIBBENTROP. Good heavens, do you believe that that is even possible?

GUDERIAN. It is not only possible but, due to our actual leadership, certain.

Ribbentrop thereupon "lost his composure." He still refused to approach Hitler, but just as Guderian was leaving said to him, "Listen, we will keep this conversation to ourselves, won't we?"[4]

After Guderian had departed, the Foreign Minister (it appears) brooded on what he had been told. Whether it was that he doubted Guderian's assurance of keeping their meeting secret, or whether out of that loyalty to Hitler which he had mentioned at the time, Ribbentrop decided to break his compact, and sat down to write a report on the conversation in his own hand. He did not refer to the Chief of Staff by name, but as *"an exceptionally high-ranking officer at present in active service in the most responsible position."*

That night chance delayed Guderian on his way to the routine briefing with Hitler, and as he entered the conference room the Führer was already speaking, "in a loud and excited voice." Hitler was in fact holding forth on the subject of Basic Order No. 1 (which laid down that no one was to discuss his work with any other person unless such knowledge was needed for his own official duties). When Hitler saw Guderian at the back of the room he went on "in an even louder voice": "So when the Chief of the General Staff goes to see the Foreign Minister and informs him of the situation in the East with the object of securing an armistice in the West, he is doing nothing more nor less than committing high treason!"

Considering the fate of other members of the General Staff corps who had been thought to have committed this crime in the preceding

[4] Ribbentrop's attitude at this meeting is not without hypocrisy, to put it no lower. For his personal secretary, Fräulein Margaret Blank, giving evidence at Nuremberg (27th March, 1946), told how the Foreign Minister had personally selected a member of his staff, a Herr Birger, to approach the Spanish Ambassador to Switzerland and urge him to "make enquiries as to the possibility of an understanding, if only of a temporary nature," with the West. The date of this project is given as "the winter of 1944," i.e., some weeks, at least, *before* he was approached by Guderian.

months, it says much for Guderian's courage that far from being embarrassed or apologetic, he immediately took up his arguments afresh with Hitler. From the aspect of his own personal safety, this proved to be the best policy, for Hitler was driven over to the defensive, "adamantly refused to discuss the proposal," and discussion passed on to other subjects. But not without the Führer's first declaring the incident to be one more example of the unreliability of professional soldiers, who put their own interpretation of the national interest above the dictates of "loyalty" to Führer and Party.

As the reports of the day's fighting came in, it was plain that sheer force of circumstances was pressing impartially on soldiers and Party members alike. In particular, the news from the National Leader's front was not good. The assimilation of the twenty or so divisions which were to come under the orders of the new army group was proving a heavy burden for the uneven competence of that group's staff organisation, and the many difficulties encountered were leading to local tactical upsets which further eroded the front. Particularly serious was the plight of the 111th Panzer Division, which, together with two weak infantry divisions, was in a position south of Marienwerder on the right bank of the Vistula, and had been caught off balance by a renewal of the Russian attack.

A forty-eight-hour lull in this sector had begun the result of one of those lightning command switches at which the Russians were now so adept: After their junction at Graudenz, Zhukov had passed command of the area east of the upper Vistula to Rokossovski. Rokossovski had taken three armoured corps out of Bogdanov's 2nd Armoured Army, moving them across the German front from the area Orel-Masurian Lakes. On paper this force had a strength of over three thousand tanks. It is likely that wear and casualties had reduced the figure by about half in this, the fourth week of the Russian offensive, and as Rokossovski had opened his attack with the first corps to arrive, four armoured brigades, the total strength was not more than about 450 T 34's. However, the 111th Panzer was also seriously understrength. In origin one of the "weak" two battalion formations, it had only one depleted company of Panthers and five of PzKw Mk IV's (with long 75-mm.). The anti-tank battalion had only nine *Jagdpanzer* as runners, and eleven towed 88-mm.[5] Even after making some allowance for additional artillery support from the two

[5] I am indebted to Major von Wittenberg, who commanded the 2nd Battalion of this division, for these details of its armament.

infantry divisions, it is plain that the Russians enjoyed a superiority of three to one in gun power and an even greater leverage in terms of mobility. This superiority was not such as to give an absolute guarantee of success against a resolute defence, but the power of Bogdanov's follow-through, with two additional armoured brigades coming into action on the two following days, made such a success almost certain unless the defenders reacted very quickly.

Speed of reaction, however, was the last thing that could be expected from Army Group Vistula. The 111th Panzer had reported the start of the Russian attack at four thirty-five on the morning of 26th January, but during the day fought entirely without direction from headquarters. Guderian noted in his diary that Himmler's signals service "failed to function," and that "lack of organisation began to make itself felt," both of which were certainly true, but these failings were aggravated by the fact that on the day of the attack Himmler had decided to move his headquarters back sixty miles to the Ordensburg Croessinsee. By nightfall the 111th Panzer, having been unable to make contact with army group headquarters for fourteen hours, but being in communication with Army Group Centre (which was under a brief interregnum between the dismissal of Reinhardt and the arrival of Rendulic), side-stepped by withdrawing eastward during the night. The move was timely, for at dawn on the 27th the second of Bogdanov's armoured brigades started its attack, on an axis that was in echelon to and northeast of the original thrust. The Russians drove straight through the last of the 111th Panzer's rear guard and forced the two infantry divisions back against the Vistula. By midday the Soviet tanks had passed through Marienburg and their leading elements were nearing Mülhausen—less than twenty miles from the Baltic.

Proper staff work and a cool assessment of the day's intelligence would have allowed the Germans to establish two things. First, that the attack was under the direction of Rokossovski, not Zhukov; second, that it was not a broadening of the westward advance into Pomerania, but a thrust on a northerly axis, designed to isolate the German forces in East Prussia. But during this critical day of 27th January the SS officers who held the majority of key positions on the army group staff were bumping along the road between Dirschau and the Ordensburg Croessinsee in haste and confusion. Their situation was not such as to encourage reflection and analysis. The result was that during the night, when Himmler was told of the depth of the Russian advance, he had ordered that the whole position, the line running north to south from Thorn to Marienwerder, be evacuated. This meant that the strong northern anchor, the position on the lower Vistula, was given up in the face of an imaginary threat. For in fact the Russian advance was not moving against the front of Army Group Vistula, but parallel to it. Had it not been for Himmler's precipitate retreat, the Russian corridor to the Baltic would have been so confined that the extrication of the bulk of the divisions in East Prussia might have been achieved, and with it the possibility of roughly handling the Russian spearhead which had probed so deeply. As it was, the German front had once again been fragmented, and Army Group Vistula was forced back into a position which saddled it with the responsibility for the northern flank as well as the vital centre.

The plight of the National Leader's command was worsened by a simultaneous reverse which had occurred at the southern extremity of the army group boundary. Here the area between the Oder and Warthe rivers was protected by a line of permanent fortifications, many of which dated back to the late twenties, when they had been constructed by the *Arbeitskommando* as a protection against Poland, during the period before the repudiation of the Versailles restrictions. The emplacements were well laid out and stoutly built, but the majority of their guns had been stripped and moved to the Atlantic Wall in 1943. During November and December four "regiments" of *Volkssturm* had been drafted into the position, where they occupied themselves with clearing fields of fire, and some mine-laying. The fighting value of this garrison, even if the position had been properly armed, was highly doubtful. The majority consisted of invalids and old men, with little training or discipline. They were responsible not to the Wehrmacht but to Party officials who themselves were without military experience and not conspicuous for their personal courage.

Nonetheless, on a situation map, or as a ration-strength figure, the position and its garrison looked reassuringly solid, and in those critical last days of January both Army Group Vistula and Army Group Centre came to rely on the "Warthe position," the one hanging its southern, the other its northern flank upon the "line" there. While Poznan, some thirty miles due east, was holding out, it was reasonable to assume that the Russian wave would be broken against the city and would not land with its full force against the unsteady *Volkssturm*. However, by the 27th it was plain that Zhukov's columns were bypassing the city and that their communications, running on the frozen ground, were hardly affected. On 28th January the Russians gained their first bridgehead over the upper Oder at Lubin, and it was plain that a full-strength assault would fall upon the "Warthe position" within forty-eight hours.

Into this position an SS corps, the 5th (in reality a heterogeneous collection of anti-Partisan units lately arrived from Yugoslavia), was drafted. But before it had time to consolidate, its commander, Gruppenführer Walter Krueger, was surprised while on reconnaissance, and in his car a complete map of the position was found. The Russian attack swept through the *Volkssturm* position by daybreak, and during the afternoon of the 30th the Soviet tanks had passed through both Schwiebus and Züllichau.

The effect of this new collapse was that Army Group Vistula again had both flanks "in the air," while Schörner was compelled to draw in his own left. The gap between the two army groups was reopened, and there was little to stand between the advancing columns of Zhukov and Koniev, now approaching in tandem, and the east bank of the Oder along its whole length between Küstrin and Glogau. Some idea of the German disorder can be gauged from the fact that at Oels airfield the Russians captured 150 aircraft in serviceable condition, including 119 four-engine F-W Kondors—the entire strength of the "U-boat support group" which was being husbanded for a renewed offensive in the Atlantic.[6]

But even at this time, as the Russians poured into the neck of land between the Oder and the Warthe, their salient was assuming a classic pattern of vulnerability—comparable only to that of the Germans who had filled the bend of the Don in their lunge toward Stalingrad. The question was, had the defenders the strength, had the attackers the *hubris,* to allow an inverted historical coincidence to set its stamp upon the battlefield?

Guderian saw clearly what must be done. He sensed the Russians' exhaustion, even in their moment of victory. He could see how far they had come, how many islands of resistance remained in their wake. He knew better than anyone the life of a caterpillar track, the stage at which tank crews reach utter exhaustion, the safety point below which the level of supplies cannot drop.

[6] See p. 437.

As we watch this man, a superb technician struggling with worn-out machinery and malicious individuals, it is impossible not to feel sympathy for him. He knew that if the counterstroke was to be effective time was of the essence. Yet he was to suffer one disappointment after another in his efforts to bring this about. First, there had been the appointment of Himmler to command the army group he had selected for the task, thus minimising his own opportunity to exercise personal direction of the battle. Then, at the conference of 27th January, a new blow fell. Hitler told Guderian that he had decided to send the 6th Panzer Army (whose first elements would be arriving from the West the following day) to relieve Budapest instead of committing it on the Oder. There was no discussion of the subject. That evening Hitler was far away on the wings of fantasy and had ear only for the chief of the Army Personnel Office, General Burgdorf.

Burgdorf was one of that disagreeable race of courtiers who thrive by their assiduous cultivation of the mood and taste of the tyrant—to the exclusion of any objective performance of their nominal duties. On that evening he had been regaling the Führer with the measures adopted by Frederick the Great to deal with "insubordination," and had collected details of some of the sentences. These delighted Hitler, who declared, "And people are always imagining that I am brutal! It would be desirable [and there cannot have been many among his audience who heard this with satisfaction] if all the prominent men in Germany were to be informed of these sentences."

While he was in this kind of mood Hitler was unapproachable on matters of concrete fact. And Guderian turned for support, for the second time in as many weeks, to the civil side of the administration. In Speer, the Chief of the General Staff, he found someone both more convinced and more fearless in expression than Ribbentrop. Speer drafted a report, which was based solely on economic premises and thus avoided any taint of criticising by implication the Führer's military genius. The report opened with the flat assertion, "The war is lost." When Guderian handed it to Hitler, "as a document of extraordinary importance, submitted at the urgent request of the Minister of Armaments," Hitler glanced at the opening sentence, then took it over to his safe, where he locked it away without a word. Some days later, having received no reply, Speer asked to see Hitler alone after the evening conference. Hitler refused, complaining, "All he wants is to tell me again that the war is lost and that I should bring it to an end." Speer thereupon handed a copy of his memorandum to one of the SS adjutants, who brought it across to Hitler. Without looking at it Hitler said to the adjutant, "Put it in my safe." He then turned to Guderian and said, with one of those rare glints of humanity that occasionally illumine this titanic and devilish figure, "Now you can understand why it is that I refuse to see anyone alone any more. Any man who asks to talk to me alone always does so because he has something unpleasant to say to me. I can't bear that."

Hitler's inertia at this stage of the battle was particularly unfortunate since the evidence shows that it was during the first ten days of February that the Russians were at their most vulnerable. Zhukov's follow-up of the Warthe break-through was weak numerically and reached the Oder more because of a total collapse of the defence than from its own power. On paper he was disposing of four independent armoured brigades,[7] but the real strength cannot have amounted to more than two—about six hundred tanks, of which the majority were probably in urgent need of service and attention.

On Zhukov's left Koniev, who had travelled less far and had been less severely resisted, was in greater strength. He was rapidly breasting the upper Oder along its whole length with the tanks of Rybalko's 3rd Guards Armoured Army, and his infantry (Koroteyev's 52nd Army) was in close support. Even here, though, the speed of Soviet reaction had suffered. For the Russians had been unable to close the trap on the 1st Panzer and 17th Army. Both of these formations had been cut off from Harpe in the Katowice area, but managed to shoot their way south and slip across the Carpathians in the last days of January. And Rybalko had not been able to eliminate all the German bridgeheads on the east bank of the Oder. Breslau was to remain a thorn in Koniev's side, just as Glogau was in Zhukov's.[8]

[7] Zhukov was leading with the 5th Assault Army (Berzarin) and reinforcing with Bogdanov's 2nd Guards Armoured Army, which had now been disengaged from East Prussia and brought across the lower Vistula, though not, it can be assumed, in any strength at this stage. His infantry component (the 47th Army) was still around Poznan.

[8] Glogau was to hold out until 17th April, Breslau until 6th May. The Soviet Official History claims that the escape of the 1st Panzer was intentional, as "to have liquidated them in this mining area would have risked serious damage to the welfare and economy of our Polish ally." But considerations of this kind had not affected the Soviet commanders when fighting in their own territory—as for example, the Donetz basin—and it is unlikely that they would have paid much heed to them on foreign soil had they disposed of the means and ability to destroy the enemy on the spot.

As early as 23rd January, Koniev had ordered Rybalko to probe only with light forces along the axis Liegnitz-Bunzlau and to swing the mass of his armour back in the southeasterly direction to clear the German forces that were believed to be grouped along the left bank of the Oder. The same preoccupation with the flanks is evident from the orders to Rokossovski and Chernyakovski to clear East Prussia. No fewer than five infantry armies were allotted to this task, which was co-ordinated by Vasilievski in his capacity as the personal representative of the *Stavka,* while Zhukov and Koniev had only four infantry armies between them. Ease of communication was a factor, for it was simpler to direct the slow-moving infantry northward than due west across the devastated territory of Poland and Pomerania. But equally there can be little doubt that these dispositions were the result of a policy conceived by the *Stavka* before the opening of the offensive and to which it continued to hold in spite of the experience and, presumably, the recommendations of the forward commanders.

The same caution which had prompted Stalin to keep a ten- or twenty-division reserve under his own hand even in the darkest days now led him to tread more warily than he need in his last approach on a cornered foe. Three factors probably contributed to this. First, the very tenacity of German resistance at the extremities of the front, in Courland and Hungary, suggested a sinister master plan far more credible than the lunatic and desperate irrationality which was the truth. (There must have been many in the *Stavka* who recalled their own satisfaction in the autumn of 1942, when the Germans began to thin their line before the armies of Golikov and Vatutin at Voronezh.) Second, the Soviet armies had twice been caught and suffered local defeats when overextended. They had been roughly handled by Hossbach in East Prussia in October of the previous year, and still more seriously defeated by Manstein's counteroffensive in February 1943, when the Germans had recaptured Kharkov.[9] Both setbacks could be directly traced to overconfidence and outrunning of supplies, and both played a part out of all proportion to their strategic significance in generating an "inferiority complex" which led the *Stavka* to overestimate German capacity right up until the end of the war. Third, there was the political element in the Russians' calculation. If they were to suffer even a light rebuff—to be compelled, for example, to fall back onto Polish soil again—they believed that their influence at the peace table would be seriously diminished. The Western

[9] See Ch. 16.

Allies had not yet begun their own offensive. If the Russians could consolidate on the Oder, less than forty miles from Berlin, they could afford to wait until their rivals had crossed the Rhine before making a final effort. The strength shown by the Germans in their Ardennes offensive had surprised the *Stavka* less than it had SHAEF, and confirmed them in their view that henceforth the Eastern and Western fronts must apply concerted pressure.

Such reasoning—which had as its outcome a deliberate reduction in the strength at the tip of the Russian salient—played into Guderian's hand. Or, it would be more accurate to say, into the hand which he still hoped that Hitler would deal him. For Guderian's aims were strictly limited. To strike off Zhukov's spearhead; and to teach a cautionary lesson, gaining not territory so much as time; a couple of months, perhaps, in which the Eastern front might be regrouped, the Western Allies might "come to their senses." This was what he hoped to achieve.

During the first week of February the means to achieve this came tantalisingly close to the Chief of Staff's grasp. Two more weak Panzer divisions were drawn off the Western front and concentrated at the training barracks at Krampnitz. Here they were equipped with sufficient Tigers and even a few *Jagdtiger,* to make up an extra company in each battalion. Each division thus had a strength of four Panther and two Tiger companies in addition to a large number of self-propelled artillery attached to the *Panzergrenadier* battalion. Furthermore, the SS divisions of Sepp Dietrich's 6th Army had not yet been definitely committed to the Hungarian sector, but were "resting" at Bonn and in the Wittlich, Traben-Trarbach areas.

Whether Guderian believed that he could wear Hitler down by argument into changing his mind over Sepp Dietrich's assignment or whether he wished to strengthen his reserves from the most obvious source, he began to return to his favourite grievance, the continuing purposeless and costly occupation of Courland. On 8th February there was another disorderly scene, taking its occasion, as was usual, at the Führer's evening conference.

Guderian claims that he spoke to Hitler as follows: "You must believe me when I say it is not just pig-headedness on my part that makes me keep on proposing the evacuation of Courland. I can see no other way left to us of accumulating reserves, and without reserves we cannot hope to defend the capital. I assure you that I am acting solely in Germany's interests."

Thereupon Hitler, "trembling all down the left side of his body," jumped to his feet and shouted, "How dare you speak to me like that? Don't you think that I am fighting for Germany? My whole life has been one long struggle for Germany."

He still continued to scream at his Chief of Staff in such an "unusual frenzy" that Goering was compelled to separate them, which he did by taking Guderian by the arm into another room and giving him a cup of black coffee. While they were sitting there, with the door open, Guderian caught sight of Doenitz passing in the corridor and called to him. The Grand Admiral had his own reasons for discouraging the evacuation of Courland,[10] but Guderian managed to extract from him the admission that the shipping space was available. However, Doenitz insisted that the heavy equipment would have to be abandoned (which he knew Hitler would refuse to sanction). While they were arguing, Hitler, who had been left in the conference chamber with Jodl, Goebbels, and two SS adjutants, but could hear Goering's voice in the adjoining room, began to shout that he wished the company to reassemble. This they did (but with some reluctance, it may be thought), as it was plain that the Führer's ill temper showed no sign of abating. Sure enough, the moment that Guderian returned to the subject of the proposed evacuation there was a new outburst of rage: ". . . he stood in front of me shaking his fists, so that my good Chief of Staff, Thomale, felt constrained to seize me by the skirt of my uniform jacket and pull me backwards lest I be the victim of a physical assault."

Thus was the Nazi hierarchy reduced: shouting, pulling and shoving, comforting one another with hot drinks; they seem to resemble less the myrmidons of a despotic court than inmates of the Servants' Hall on an off day. But the image is deceptive. These were the same men who had held, and still held, positions of absolute power; the gift to steal or "liquidate" at the lightest whim. Now they were frightened. A chill draught could be felt and it blew from the grave. *"Wenn wir diesen Krieg verlieren, dann möge uns der Himmel gnädig sien."*[11]

The worsening of his relations with Hitler had precluded the possibility of Guderian's being allowed to use the 6th SS Army for his counteroffensive, and during the second week of February, Koniev and

[10] See p. 437.
[11] See p. 16 and 16 n.

Zhukov had jointly completed the encirclement and isolation of Glogau and driven Army Group Centre back to the line of the Neisse.

Guderian therefore decided to give up for the time being his scheme for a convergent attack, and to concentrate on a single blow from the Arnswalde forest area against Zhukov's long right flank. The quadrilateral between the Neisse, the Oder, and the Carpathians was only lightly held, and Guderian may well have thought that Koniev's intrusion there could be turned to the German advantage. For if the Arnswalde attack should prosper, he would find it easier to get permission for the employment of Sepp Dietrich's army on the southern flank and the Soviet retreat would be confined at the Oder by the garrisons of Glogau and Breslau. Here, too, the pattern of the Russian dispositions seemed temptingly vulnerable.

For the Arnswalde attack Guderian had managed to assemble, in spite of all his difficulties, a formidable reserve. Rauss, together with three and a half divisions of the 3rd Panzer Army and most of the staff, had been evacuated from Pilau. Into this command were placed the two re-formed Panzer divisions from Krampnitz. At the same time the National Leader and his staff had been scouring the country for Waffen SS in obedience to Hitler's order that the army group should be an exclusively SS formation. Helped by the lull which had descended on the battlefield since the end of January, they had accumulated sufficient to constitute a new SS "army" (the 11th), which was placed under the command of Obergruppenführer Steiner.

But time was still of the essence. If the phasing of the German attack was right, it would catch Zhukov before his reserves were in position. At the same time it would achieve a double delaying effect because when the Russians had recovered their balance the first thaws of spring could be expected to impede their recoil.

The Russian front opposite the Arnswalde was now held entirely by infantry (the 47th Army) for the tank brigades of Bogdanov's 2nd Guards Armoured Army had been pulled back to, and south of, the railway between Landsberg and Schneidemühl for rest and refit. The Soviets were also replacing their tank formations with infantry along the Oder from Küstrin up to the confluence with the Neisse. (OKH estimated the rate of arrival at four divisions per day in this sector, but was probably a substantial overestimate. Certainly their supporting artillery would not have been arriving at that rate.)

When Guderian went to the Führer's conference on 13th February, he had in his mind two objectives, mutually dependent, whose achievement he regarded as vital if the war was to continue at all. The first was that the attack should start no later than the following Friday (15th February). The second was that he exercise some measure of personal control over its course. This he proposed to achieve by the attachment to army group headquarters of his principal assistant, General Wenck, who was "to be in charge of the actual carrying out of the attack."

The conference was held in the principal assembly room of the New Chancellery. As well as Keitel and Jodl, there were present Himmler and Sepp Dietrich and the usual sprinkling of Party hangers-on, SS messengers, and so forth. At one end of the room Lenbach's portrait of Bismarck looked down from the mantel, opposite stood a bronze bust of Hindenburg, larger than life. Instead of Thomale, Guderian had brought with him his protégé, Wenck. They were not long in discovering that, as Guderian shrewdly observed, ". . . both Hitler and Himmler opposed [the plan] . . . since they were both subconsciously frightened of undertaking an operation which must make plain Himmler's incompetence."

Himmler maintained that the army group was not ready to go over to the attack as the necessary fuel and ammunition had not yet arrived. (In his memoirs Guderian claims that this was true of "a small portion" only.) Hitler took the side of *Treuer Heinrich,* and the following exchange took place:

GUDERIAN. We can't wait until the last can of gasoline and the last shell have been issued. By that time the Russians will be too strong.

HITLER. I don't permit you to accuse me of wanting to wait.

GUDERIAN. I'm not accusing you of anything. I'm simply saying that there is no sense in waiting until the last lot of supplies have been issued and thus losing the favourable moment to attack.

HITLER. I just told you that I don't permit you to accuse me of wanting to wait.

GUDERIAN. General Wenck must be attached to the National Leader's staff, since otherwise there can be no question of the attack succeeding.

HITLER. The National Leader is man enough to carry out the attack on his own.

GUDERIAN. The National Leader has neither the requisite experience nor a sufficiently competent staff to control the attack singlehanded. The presence of General Wenck is therefore essential.

HITLER. I don't permit you to tell me that the National Leader is incapable of performing his duties.

GUDERIAN. I must insist on the attachment of General Wenck to the army group staff so that he may ensure that the operations are competently carried out.

Soon Hitler had completely lost his temper. Guderian has described how after each outburst of rage "Hitler would stride up and down the carpet edge, then suddenly stop immediately before me and hurl his next accusation in my face. He was almost screaming, his eyes seemed about to pop out of his head and the veins stood out on his temples." The "conference" continued in this way for two hours. No one else spoke, except Guderian, who had "made up my mind that I would let nothing destroy my equanimity, and that I would simply repeat my essential demands over and over again. This I did with an icy consistency."

Then, quite suddenly, Hitler stopped in front of the National Leader (who had remained silent throughout the scene) and said to him, "Well, Himmler, General Wenck will arrive at your headquarters tonight and will take charge of the attack." He then walked over to Wenck and told him that he was to report to the army group staff forthwith. Returning to his usual seat, Hitler took Guderian by the arm and said, "The General Staff has won a battle this day," and smiled "his most charming smile."

The conference then picked up its normal course, but Guderian had been so exhausted by his victory that he was compelled to retire to a small anteroom. Here, seated at a table with his head in his hands, he was found by Keitel, who at once began to reprove him. How could he contradict the Führer in such a manner? Had he not seen how excited the Führer was becoming? What would happen if as a result of such a scene the Führer were to have a stroke? Soon other members of Hitler's entourage started to come in and add their voice to Keitel's. The Chief of the General Staff found himself in a minority of one, and "had a further hard passage before these anxious and timorous spirits were calmed." He then issued orders by telephone to Army Group Vistula, confirming that the preparations for an attack on the 15th were to go ahead, and without waiting for Wenck, left before Hitler could change his mind.

Now the hour had come for the last offensive the German Army was to undertake in World War II.

The weight of the *Schwerpunkt,* six Panzer divisions in Rauss's recon-

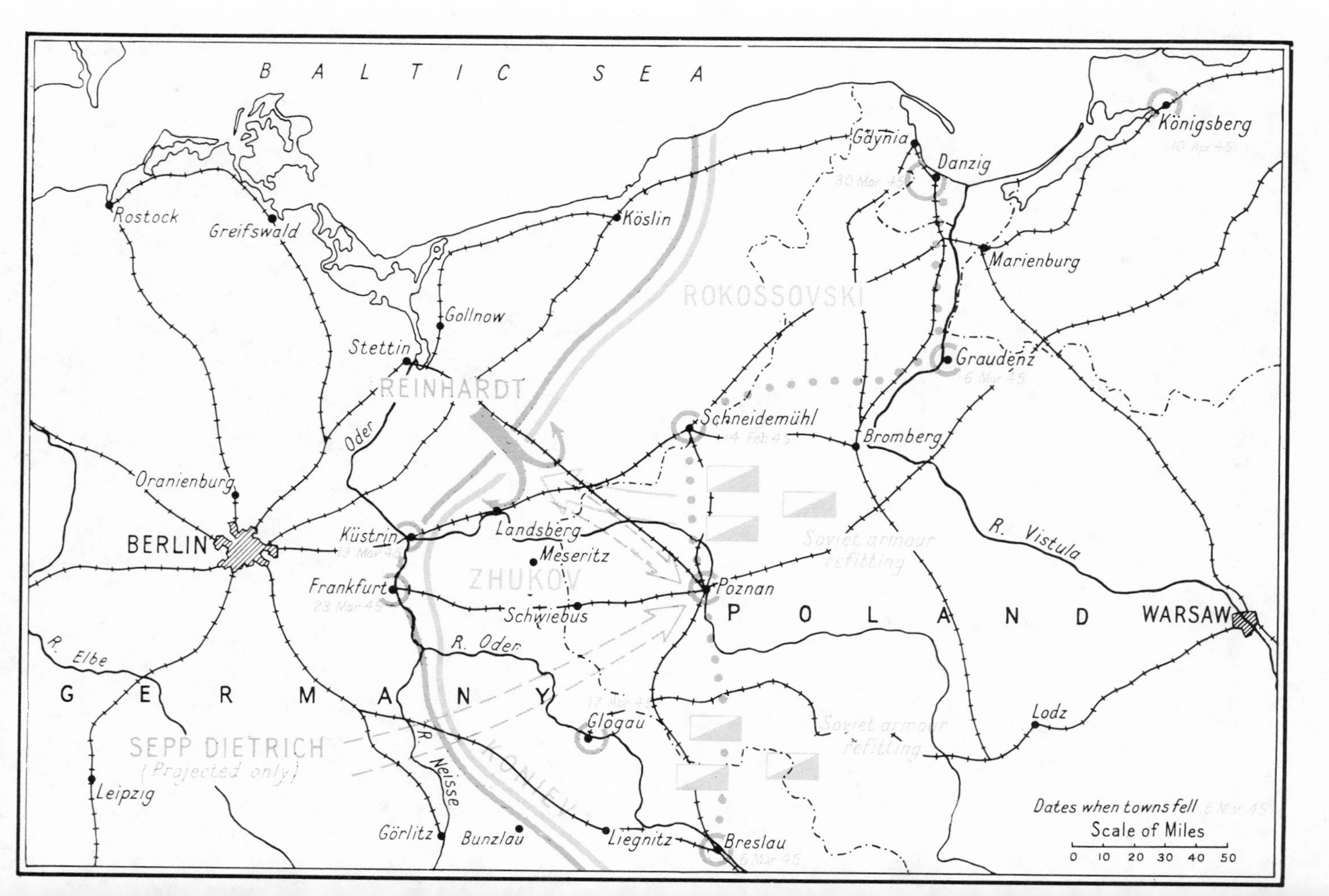

The Last German Offensive

stituted 3rd Panzer Army, was but a shadow of Wehrmacht strength in earlier, less critical battles or, indeed, of the force it represented on paper. For none of the Panzer divisions were above "two-battalion" tank strength, and only three of them were predominantly Panther-equipped. The other three were made up of PzKw IV's with a stiffening of worn-out Tigers.[12] There was no air support whatever—the Panzers needed all the fuel available. Crews had standing orders that when a tank was disabled their first task was to drain off its fuel "as soon as enemy fire slackens." Although some of the equipment was new, many of the crews had been continuously in action for weeks. Others, those who had come from the West, had been so conditioned to perpetual air attack that they showed "a marked aversion to operating in masses during the daylight hours." Still others, replacements hurriedly drafted from Luftwaffe or police duties, had neither combat experience nor the technical ability to keep the machinery in service under battle conditions. Nonetheless, their morale, if not "high" in the conventional sense, had that element of final desperation which can produce a superb élan—and, just as suddenly, a complete fragmentation.

The Russians opposite them had been told to expect an attack, but this warning had already been current for ten days and during that time several units had been changed over. Their line was lightly held by infantry with a thin screen of anti-tank guns. The main mass of the 47th Army was held farther back with its tank element (T 34), while the T 34/85 and JS tanks of the 2nd Guards Armoured Army were concentrated at Meseritz and Schwiebus. Two of Bogdanov's armoured brigades had been pulled back as far as Poznan. Of the majority of Soviet soldiers it was true to say that, after their achievements and losses of the previous month, *"Ils . . . prenaient possession du pays conquis, s'adonnant sans retenue à la récupération individuelle."*[13]

Consequently on the first day the German attack made good progress. The Russians had only scattered a few mines around villages and road junctions, and from these positions they managed to take the edge off the attack with *Cheristikye*[14] and their 76-mm. anti-tank guns (now firing hollow-charge ammunition). But in open country, and where

[12] The SS divisions also had a tank element, usually about two companies of fourteen to seventeen Panthers.

[13] Garder 303, *Une Guerre pas commes les Autres,* Paris 1963. A loose translation might be, "They were letting themselves go."

[14] The Russian equivalent of the Bazooka anti-tank rocket.

the forest ran out into plain, the Panzers drove deep. By nightfall on the 16th three thousand prisoners had been taken. The vital task for the Germans was to confront the Soviet armour and to destroy it in the first days. Their own limitations in fuel, training, and endurance made it essential that the issue be decided before Zhukov could develop his reaction. On 17th December there were some signs that this might be achieved. Tanks of the 4th SS attempting to cross the Warthe southwest of Landsberg were engaged in a gun duel by a group of JS 2's which appeared on the opposite bank and forced them to retire. Probing attacks in company strength across the Oder below Küstrin indicated that Zhukov had no intention of pulling back because of the threat to his right flank.

At the *Stavka* news of the Rauss-Wenck offensive brought an immediate reaction. Rokossovski was ordered to intensify his pressure against Weiss in Pomerania, and Kozlov's 19th Army was taken from Chernyakovski's command [15] to reinforce 2nd ARB. Co-ordination between Zhukov and Rokossovski had been greatly simplified by the surrender of the German garrison at Schneidemühl on 14th February, for this important rail centre provided both north-south and east-west communication. Had the German attack opened but three days earlier, the town would have been relieved and, with Graudenz and Dirschau still holding out in the Russian rear,[16] Rokossovski's power of rapid concentration would have been severely limited.

As it was, fate took a hand in the affair before either the Russians had time to revise their disposition or the Panzers had managed to bring the 2nd Guards Armoured to battle. In spite of Guderian's contention that the evening "briefings" with Hitler were "simply chatter and a waste of time," Wenck was compelled to attend them at the end of each day. This involved a round trip of nearly two hundred miles between Berlin and army group headquarters outside Stettin. On the night of 17th February, Wenck, dead-tired himself, took the wheel of the car so as to relieve his driver, who had collapsed. He fell asleep, and the car went off the road, crashing into the parapet of a bridge on the Berlin-Stettin *Autobahn*.

Wenck was seriously injured and taken to a hospital, and with him went Guderian's last chance of retaining personal control of the Arns-

[15] Chernyakovski was killed by a sniper on 18th February.
[16] Graudenz held until 6th, Dirschau until 13th March.

walde operation. The following day Hitler, at the recommendation of Burgdorf, appointed General Hans Krebs (a close friend of Burgdorf's) to take Wenck's position. *Treuer Heinrich* was now more or less permanently quartered in Berlin, and had given up all but nominal responsibility for his army group, but Krebs, like Model, whose Chief of Staff he had been, was a "Nazi general."

Thus did control of Army Group Vistula return to the hands of the Party, and from that time the attack, lacking continuity of direction, petered out. It had lasted four days—the shortest as well as the least successful offensive undertaken by the German Army. To the many factors which, from its inception, had militated against success—personal jealousies, administrative obstruction, deficiencies of men and equipment—fate had now delivered the coup de grâce. As in the case of General Billotte four and a half years earlier, a motor accident had decapitated an army poised on the brink of defeat.[17]

Hitler, the repository of all blame for Germany's military defeats in World War II, must certainly bear a large proportion for the collapse of Army Group Vistula and the abortion of Guderian's plan for a counterattack. But before condemning this simply as feckless irresponsibility, the product of spite or unsound mind, we should try to estimate what the Führer's strategic intentions were at the time. We have evidence of two, one military and the other political. Doenitz had persuaded Hitler that with the electro U-boat he would once again place in Germany's hands an instrument of war-winning importance, and that retention of the Baltic was vital so as to provide a quiet area where the new boats could be "worked up" by their crews.[18]

The second element in the Führer's calculations was less improbable, although the element of fantasy is still perceptible. Hitler had always conceived of himself as comparable to Frederick the Great, combining military resolution with political guile, and the accumulating successes of the coalition arrayed against him had, if anything, lent force to this illusion. Although he never showed it to his generals (by any manner other

[17] General Billotte was the commander of the Anglo-French armies in Belgium at the time of the German break-through at Sedan in May 1940.

[18] In fact, this was an unsound strategic appreciation, for the war at sea was no longer of fundamental importance; indeed, the Allied High Command considered it "more than likely that the enemy would succeed in recovering the initiative at sea—during the spring."

than his conduct) surviving recorded fragments of discussions with his Party cronies suggest that denuding the front which protected Berlin was an act of deliberate political calculation. This conversation took place on the evening of 27th January:

HITLER. Do you think the English can be really enthusiastic about all the Russian developments?

JODL. No, certainly not. Their plans were quite different. Only later on perhaps will the full realisation of this come.

GOERING. They had not counted on our defending ourselves step by step, and holding them off in the West like madmen, while the Russians drive deeper and deeper into Germany.

HITLER. If the Russians would proclaim a national government for Germany, the English will start to be really scared. I have given orders that a report be played into their hands to the effect that the Russians are organising two hundred thousand of our men led by German officers and completely infected with communism, who will come marching into Germany . . . that will make them feel as if someone has stuck a needle into them.

GOERING. They entered the war to prevent us going into the East; not to have the East come to the Atlantic. . . . If this goes on [the Russian advance into Germany], we shall get a telegram in a few days.

The rest of the evening's discussion on this subject has not been recorded, but it is significant that the following day Hitler announced to Guderian that he had decided to send Sepp Dietrich's 6th Panzer Army to Hungary instead of putting it under Army Group Vistula.

If we realise that the Führer and the Chief of the General Staff had conceived and were trying to implement diametrically opposed policies, each of which they believed to be in the urgent national interest, then the curious and dramatic events of January 1945 are more easily understood.

Vanity and self-delusion are among the lesser vices of despotic courts —indeed, they are to be welcomed, for it is in such ground that the seeds of self-destruction flourish. Thus the Nazis held seriously to this conviction, and even after four years of war they could indulge in extravagant fantasia in which the members of the alliance negotiated either singly or together with them. After all, they were the constituted government of the Reich. There was no alternative form, nor since 20th July were there any "shadow" *personae*. They still held (or so they reasoned) cards of a certain strength: a powerful and disciplined army, the loyalty of some eighty million people, and the lives of fifty million more as hostage. Their power of destruction, if no other, remained.

Above all, they, who had thrived, who had been helped up every rung on the ladder by the bogey of communism, which had so perplexed and divided the counsels of their enemies, could not rid themselves of their pose as the sole alternative regime to the chaos of Bolshevism.

But in the last days of February the conclusions of the Yalta conference began to percolate to Berlin. It was plain that the Allies intended to preserve at least the semblance of unity of purpose—and that purpose, which they had declared at Casablanca, was the "unconditional surrender" of the Reich. The corollary was plain. The Nazi leaders were fighting for their lives. From now on their orders that not another yard of ground was to be yielded have a quality of desperation that is next to pleading, and the fighting in the East entered its final, and most bitter, phase.

twenty-two | THE FALL OF BERLIN

As the weary Panzers drew back into the Arnswalde they gathered up in their tracks a whole mass of civilian flotsam. Infants and aged, wounded, slave labourers, *Hiwi*'s, deserters in various disguises, crowded the roads and byways, huddled in broken carts, spawned over the bleak white-streaked countryside like maggots on an open wound. German soil, which for so long had escaped retribution for the sins of its children, was now visited by scenes whose horror evokes the Thirty Years' War, and which seem to rise straight from the sketches of Goya.

Rape, pillage, and random destruction seethed on the crest of the advancing Russian wave. To Soviet soldiers killing was incidental; the very fecklessness with which they valued human life made its taking, or sparing, a trivial matter. In contrast, the German blood lust was a positive and accelerating cancer which, having devoured so many subject peoples, was now beginning fast to consume the *Herrenvolk* themselves. Men of the *Volkssturm,* hurrying up to the Oder, could see the bodies of "malcontents," their former comrades in arms, swinging from the twisted girders of the blown-up bridges, where they had been hanged by special flying courts-martial which ranged the military zone, pronouncing and executing sentence at will. Every tree in the Hindenburg Allee in Danzig had been used as a gibbet, and the dangling soldiers kicked and threshed, sometimes for hours, with placards pinned to their uniforms: "I hang here because I left my unit without permission."

Many of the "deserters" had been schoolboy flak gunners who had gone to visit their parents for a few hours, proud of the opportunity to display their new uniforms. But their protestations went unheard in an atmosphere where

> It is an act of racial duty according to Teutonic tradition to exterminate even the kinsmen of those who surrender themselves into captivity without being wounded.

At Zossen, Guderian recorded that deserters' families were not the only class of Germans to be "exterminated" in that terrible spring of '45. Black marketeers, spreaders of rumours, hoarders of food, people on the move with unsatisfactory papers, even those who changed their addresses without notifying the Gauleiter, stood in jeopardy of their lives.

Underlying the brutal and chaotic enforcement of the new penal "code" were the first tremors of impending disturbance from the slave population. The camps in the East were opened as the Russians advanced, and in a welter of bureaucratic confusion their occupants suffered a variety of fates, often starting out into the freezing wind with a few guards who soon lost their patience and, as the Red Army approached, their nerve, and simply did their charges to death in some wood or quarry on the route before making their own getaway. Charged with obliterating the camp sites and removing every trace of the satanic practices which had pervaded them, the men of the SS were usually too jumpy to achieve this properly—although the giant crematoria which had been put into Auschwitz in 1943 were totally destroyed by high explosives in an expert demolition job, which extended even to destruction of the architects' drawings in the files of SS *Hauptamt* and of the firm which had built them, a subsidiary of Kammler's WVHA.

Slaves in other categories, less crippled by disease and starvation, often succeeded in overpowering their guards and breaking away into the country, where they roamed for weeks in the shifting boundaries of no man's land between the two armies, ravaging the deserted townships and taking vengeance on any civilians who remained. As the superstructure of civilisation fell away into rubble, war reverted even to the colour it had carried down the centuries. One German has described

> . . . A group of *Hiwi*'s mounted on horseback. They had broken into a deserted Hohenzöllern castle which has been preserved as a museum and looted it indiscriminately. They were all drunk and had draped themselves with golden tapestries, carrying spears and armour, and pulling a covered wagon loaded with priceless pictures and *objets d'art* into which the snow was blowing in gusts . . .

At the end of February the thaw started, and the ice on the Oder began to break up. Within a few days the long and precarious frontage of Army Group Vistula, as it was still optimistically termed, enjoyed the protection of the fast-flowing river, while to the north the battered remnants of the 3rd Panzer Army and its ancillaries could detect the weight of pressure against them easing as Soviet tanks and supply lines began to feel the drag of the mud.

The effect on the National Leader was pronounced. He had been suffering (or so he believed) indifferent health for some months, and had been paying periodic visits to Dr. Karl Gebhardt's clinic at Hohenlychen. Himmler's state of mind, already troubled by "conscience" and desire "to do the right thing" (which euphemisms, among a wider variety, he used to cover his desire to take such measures as were necessary to preserve his own survival in power without risking an open breach with the Führer), was in a condition of perpetual inflammation from the proddings of Schellenberg, who had now come out in the open with his urgings to seize power and open negotiations with the West. Schellenberg also believed—and it is unlikely that he managed to conceal this conviction from his chief even if he had wished to do so—that Himmler was suffering from cancer of the bowel, and this added to the gloom. Certainly Dr. Gebhardt, whose only recorded medical achievement is the highly an-Hippocratic feat of deliberately infecting Polish girls with gas gangrene at Ravensbrück, was no man for the valetudinarian. He had the National Leader on a course of strychnine and hormone "tonic," supplemented by that durable standby of the medical profession for hysterical stomach—Beladonal. For spiritual fare Himmler alternated between the mystic (and mystifying) prophesies of Dr. Wulf—"a student of poisons, Sanskrit and other interesting subjects" who had been discovered by Schellenberg and had evolved some precooked tale of the National Leader's impending succession to supreme office—and the more arduous material of the army reports which piled up around his bedside, dutifully brought to him by SS despatch riders every twelve hours.

From time to time, powerfully impregnated with Dr. Gebhardt's medicaments, Himmler made the pretence of returning to his headquarters and directing the battle. His personal regime was leisurely by any standards. He would rise at eight-thirty in the morning, sleep for three hours after lunch, and retire to his room at ten o'clock at night—very different hours from those prevailing at the *Reichskanzlei*. But when the ice melted on the Oder, Himmler dropped even this pose and retired to his favourite room at Hohenlychen. The thaw had affected him profoundly, he told Schellenberg. It had been a miracle. The second miracle (the other, of course, need he say it, had been the Führer's deliverance at Rastenburg) to affect him personally within a year. And it had left him convinced of the existence of the Almighty. There was a God, Himmler told Schellenberg, and they were His instruments.

Like so many in the condemned cell, the Reichsführer, who had forbidden church parades at Bad Tölz, hounded the Wehrmacht chaplains, and permitted his soldiers to desecrate shrines and use church walls for execution the length and breadth of Europe, had seen the Light.

Aside from the spiritual conversion of its chief, the thaw on the Oder had come just in time for Army Group Vistula. For the 9th Army, which was holding the centre and most of the river stretch, was an untidy hodgepodge of *Volkssturm* and SS, many of the latter being strictly "Byzantine" Waffen units, like Krueger's 5th SS Mountain Corps of Albanians and Slovenes, or security forces with little combat experience against regular enemy troops. Among them were those durable villains of the Eastern front, Bach-Zelewski, Dirlewanger, and the "Police General" Reinefarth, all of whom will be remembered for their part in suppressing the Warsaw uprising.[1]

Guderian had noted in his diary that cracks were appearing in SS morale ". . . although the *panzer* troops continued to fight bravely, whole SS units, taking advantage of the cover thus offered, proceeded to retreat against orders," and he now became increasingly concerned with the prospects of the Reichsführer's army group once the Russians had accumulated the strength to mount a set-piece attack on the Oder position. On 18th March he drove to Himmler's headquarters at Prenzlau, where he found conditions "completely chaotic," and was told that Himmler had gone to Hohenlychen "with an attack of influenza." Guderian got back into his car and went to Gebhardt's clinic, where the National Leader was found, sitting up in bed but "apparently in robust health."

Among the variety of Mad-Hatter's-Tea-Party scenes which stimulate historical curiosity during the last weeks of the Third Reich, the interview which followed must have a high place. The bright, impersonal

[1] See Ch. 20.

hospital room; the Chief of the General Staff, jack-booted in his ankle-length *feldgrau* greatcoat, Knight's Grand Cross at his throat; and in pajamas, his countenance puffy, the Reichsführer SS, Chief of the Police, Commander in Chief of the Home Army, and G.O.C. Army Group Vistula. No matter what respect he might have for Himmler's ability, Guderian told him, surely ". . . such a plethora of offices was bound to be beyond the strength of any one individual?" Himmler said nothing. Perhaps by now the Reichsführer had come to realise that the command of troops at the front is no easy matter? Might it not be appropriate for the Reichsführer to give up the command of his army group and concentrate on his other offices?

Himmler plucked at the sheets. Guderian formed the opinion that he "was no longer so confident as in the old days." Then the National Leader offered an excuse. He could not go and say such a thing to the Führer, he protested. "He wouldn't approve of my making such a suggestion." Guderian, however, was equal to the situation, and immediately suggested that he make the proposal to Hitler on Himmler's behalf. The National Leader gave his assent and sank back into the pillows, to ruminate undisturbed on the sibylline utterances of Dr. Wulf, and the Chief of Staff returned to his motorcar and hurried over to Zossen. That very evening Guderian told Hitler that the "overburdened" Reichsführer should be replaced by General Heinrici, a meticulous and donnish man, at that time commander of the 1st Panzer Army. Hitler, after "a certain amount of grumbling," agreed, and the relief took place on 22nd March.

On the day after Guderian's visit the National Leader roused himself and travelled, not to his headquarters at Prenzlau, but to Berlin, "to catch up on the political situation." Then, on the evening of the 21st, he returned, thoroughly exhausted, to Prenzlau for the handing over to Heinrici the following day.

But at this point the military situation began to make its influence felt once again, and soon its intolerable pressure, intensifying the claustrophobia which pervaded the Nazi leadership, was to trigger a new round of dismissals and realignments. For Zhukov had massed twenty-seven armoured brigades above the confluence of the Warthe and the Oder, and Rokossovski, deploying nearly as great a strength, had reached the Baltic on either side of Kolberg and was bombarding Stettin. It was plain that a full-dress assault on the Oder position was not many days off.

The meeting between Himmler and Heinrici (who recorded that the Reichsführer was looking "unusually white and puffy") was interrupted by a telephone call from General Busse, commanding the 2nd Army, with the news that the two small Russian bridgeheads on either side of Küstrin had sprung to life and linked up to the west of the town, isolating the garrison. "You command the Army Group now," Himmler told Heinrici, handing him the receiver, "please give the appropriate order." [2]

But if Himmler was now too preoccupied—with his health, with his conscience, with "global matters" of politics and reconciliation—to concern himself with the fate of the Küstrin garrison, the Führer was not. For the defenders of Küstrin were an exceptionally "reliable" force. Of such a character, that is to say, that their reliability was guaranteed by the fate which awaited them at the hands of the Red Army. Cooped up inside the town were none other than Reinefarth and four battalions of his "police" army corps, wearing uniforms which ensured their immediate execution should they surrender. Hitler addressed himself to the problem of their relief. He told Guderian to see to it that Busse achieved this immediately, and also to commit five divisions in a counterattack from the Frankfurt bridgehead, as a "diversion."

After a certain amount of shuffling Busse launched his relief attack on 26th March. But both Guderian and he refused to waste their strength in attempting a pointless sortie from the Frankfurt perimeter. By now the Russian footholds on the west bank of the Oder were building up at such a rate that Busse's relief force was heavily outnumbered, and within twenty-four hours he had been beaten back with heavy casualties. At the Führer conference on 27th March, Guderian defended Busse with difficulty, and by frequently referring to the figures of killed and wounded he had sustained. The Chief of Staff undertook to visit the Frankfurt position the following day and see whether the projected attack was "a practical proposition." However, that morning Krebs brought Guderian a message from the Führer, forbidding him to visit the front and summon-

[2] In his account (Thorwald, *Das Ende an der Elbe* II, 25) of this meeting, Heinrici states that Himmler ended by volunteering the opinion, "The time has come to enter into negotiations with our western neighbours. I have initiated steps. My agents have established contact." Yet this is very hard to believe, for only the previous afternoon Guderian had walked with Himmler in the Chancellery garden in Berlin and suggested that the two of them go to Hitler and urge him to conclude an armistice. To which Guderian alleges (p. 426) Himmler replied, "My dear Colonel-General, it is still too early for that."

ing him and Busse to attend the noon conference (instead of the usual night conference).

To avoid interruption from air attack, it had been customary for some time for these afternoon "briefings," as they were called, to be held in the corridor of Hitler's personal underground bunker, and into this confined space there crowded, at 2 P.M. on 28th March, Guderian and Busse, Keitel, Jodl, Burgdorf, Hitler, Bormann, and sundry adjutants, staff officers, stenographers, and men of the SS bodyguard. Soon the conference took on the character, which was to be a recurrent feature of the "bunker period," of a hysterical multipartite shouting match. Busse had barely started on his report when Hitler began to interrupt him with the same accusations of negligence, if not cowardice, which Guderian had protested against the previous day. Guderian then began to interrupt, using unusually strong and dissenting language, drawing in turn murmurs of reproof from Keitel and Burgdorf. Finally Hitler brought the company to order by dismissing everyone except Guderian and Keitel, and turning to Guderian, he said, "Colonel-General, your physical health requires that you immediately take six weeks' convalescent leave."

With the dismissal of Guderian the last rational and independent influence was removed from the direction of military affairs in Germany. Only the "Nazi soldiers" remained, all of them now in timid conformity with Brauchitsch's "office boy" image and tied to the execution of the Führer's wayward policies. It is one more paradox of the Russian campaign that at the end, when Hitler had mastered the General Staff and finally extinguished the evasions and insubordinations which had persisted among them (albeit in diminishing strength) since 1941, he began to take on all the characteristics which the generals had so long ascribed to him, and which they had used to excuse their own intermittent disobedience.

For in April 1945, Hitler was living the reality of a dream he had foreshadowed to Hermann Rauschning back in 1934:

> Even if we could not conquer, we should drag half the world into destruction with us, and leave no-one to triumph over Germany. . . . we shall never capitulate, no, never! We may be destroyed, but if we are we shall drag a world with us—a world in flames.

Yet to describe the atmosphere in Berlin as pure *Götterdämmerung,* a sort of nihilist fantasy where rational considerations no longer applied, is a gross oversimplification. Hitler, certainly, was obsessed. His ". . . terrible appetite for blood, like his appetite for material destruction, seemed rather to grow when the price was to be paid not in inferior currency but in good Aryan coin." This attitude was shared by Goebbels and professed (though not, it was later to emerge, held) by many of the lesser personalities at Hitler's court.

There were others who gave thought to their predicament, and they reasoned in different ways. First, the Diadochi closest to and most heavily dependent on the Führer: Bormann, the Fegelein-Kaltenbrunner cabal, Ribbentrop, Koch, and some of the senior Gauleiters. These people had to attach their faith to the Führer's wisdom, the new weapons, the prospect of the "diplomatic coup," which Ribbentrop still maintained was possible. Their personal ambitions could best, could only, be served by sticking as close to Hitler as possible, and by watching for opportunities to discredit their more independent-minded rivals, should the fortunes of the Reich change and a redistribution of power become feasible.

Second, there were a few leading executives, notably Himmler and some of his intimates in the SS, like Schellenberg. They saw themselves (ludicrously enough) as acceptable in principle to the Western Allies because of their proven anti-Bolshevism, and necessary in practice because of the military and administrative power which they controlled. There can be little doubt that it was this conviction, as much as considerations of personal vanity, which prompted Himmler to acquire so many extra titles and provinces, just as it was Bormann's idea to edge *Treuer Heinrich* away from the Führer's personal entourage by loading him with an accumulating burden of responsibilities. For the successful exercise of all Himmler's various powers was, as has already been shown, fast receding from the scope of his capability—a state of affairs to which malign and pointed reference could be made in the privacy of the Führer's intimate circle.

The third group in the Nazi hierarchy was the technocrats, military and civil. These were men who saw themselves as trustees of Germany's greatness. They chose to look no further than appearances (itself an alarming enough prospect), to ignore the lunatic and bestial cruelty that underlay the regime, and to regard their duty in simple terms—those of protecting the German people from their enemies without and, of late, the unpredictable savagery of edicts from within. In the first months of the Third Reich we can see this group—generals like Guderian, Model,

Heinrici, civilians like Speer, struggling doggedly in their private efforts to preserve the still towering edifice of the Thousand-Year Reich while, around them, the state façade was crumbling away to reveal a rotten and verminous substructure.

These three groups were all, while united by the focus of the Führer's headquarters, pulling in approximately the same direction. And although members of each coveted Hitler's succession there was little that they could, or dared, do. With the deteriorating military situation and the isolation of the Führer in Berlin, many people came to equate physical separation with administrative impotence. They soon learned that "The power of the Führer was a magic power and no profane hand might reach out to touch it until the reigning priest was really dead," a profound truth, which each of the Diadochi (except Goebbels) sought to evade at some point in the spring of 1945, and all to their peril. It is this which causes the history of the German collapse to stratify at three levels. First, the sequence of defeat in the purely military sphere, the accumulating strain and final rupture of the bedraggled armies along the Oder. Second, the train of events at the Führer's headquarters in Berlin; and third, the crude and hesitant efforts of a few leading Nazis to assume and exercise power in their own narrow interest.

All through March the Russian guns had been audible in Berlin. Across the city, dark grey, windowless, ravaged by four years of bombing, glistening wet in the rain and melting snow showers that masked the ebbing winter, sirens wailed day and night. Order remained, but law had ceased to function and the "flying courts-martial" of the SS were both the source and the embodiment of authority. Eastward toward the "front" streamed a continuous procession of reinforcements, *Hitlerjugend,* apprentices, foreign "brigades," the sweepings of prisons and hospitals, to take their place in the flimsy patchwork that still offered a screen against the gathering armies of Zhukov and Koniev. In the West, both Patton and Montgomery had crossed the Rhine during the last week in March, and it was plain that the Wehrmacht no longer had the capacity to defend a true north-south line from the North Sea and that in military terms the war was lost.

But to the people of Berlin this meant little—except to those, and they were not few, who had come to look on the Anglo-Americans almost as a relieving army. They knew that one battle remained to be fought, and with their beloved Führer in their midst (and there is no doubt that even at this late stage of the war nine out of every ten Germans *did* still love and venerate Hitler), they were preparing for the final ordeal.

During March the Russians had confined their activities to enlarging and reinforcing their bridgeheads across the Oder, and to clearing their right flank and the Baltic coast between Stettin and Danzig. The three "fronts" of Rokossovski, Zhukov, and Koniev together disposed of more than seventy armoured brigades, but of this strength only some twenty-five were in action. The remainder were being massed in two battering-rams which were to strike—one above, one below the Berlin latitude—and meet west of the city.

This, the *Stavka* was determined, was to be the last battle. Like the Western Allies on the Rhine, it seems to have overestimated German strength and the degree of force which it would need to break it. The stop lines and the zones of occupation had been agreed upon in principle at Yalta, so the "race to Berlin" no longer had any reality. But once the Allies were across the Rhine in strength the start of the Russian offensive could be delayed for only a few days.

If the Russians overestimated the difficulty of their task and the Germans themselves still hoped for a miracle, the neutrals had no illusions, and viewed the prospect of Communist irruption into Western Europe with barely concealed alarm. In February 1945, Ribbentrop, as a preliminary to his "diplomatic coup," had prepared a memorandum which had been submitted to both the Vatican and the Swiss Government. This document seems to have been rather muddled, simultaneously "threatening" to "hand Germany over" to the Russians and offering to surrender to the West and switch the whole weight of the Wehrmacht to "stemming the Bolshevist flood." Anxious though they must have been for some arrangement of this kind to be successful, the neutral embassies harboured few illusions concerning its acceptability in the West, and Ribbentrop found himself compelled to attach a series of codicils promising (although on what authority is not clear) that ". . . the National Socialist Government would resign" and that ". . . the persecution of Jews and political opponents would cease."

This last did arouse a flicker of interest from the Swiss, who asked for "stronger assurances on the matter of Jews and concentration camps." As the Swiss had particularly asked that these assurances come from the SS, Ribbentrop had been compelled, to his chagrin, to approach the

Reichsführer, and had taken the path, trodden by so many other senior Nazis that March, to Dr. Gebhardt's nursing home at Hohenlychen.

Himmler's reaction to the Foreign Minister's approach is not recorded, but it seems safe to assume that it amounted to nothing more than the usual placatory and evasive pattern which the Diadochi adopted to one another (and with which both Ribbentrop and Himmler had themselves replied to Guderian when he approached them that winter)[3] when the subject of "independent" peace negotiations was raised. For the last thing Himmler wanted was Ribbentrop interfering in negotiations which he, the Reichsführer, had already well under way. All that Himmler would consent to was that a certain Fritz Hesse (styled as "Referent on British Affairs" at the Wilhelmstrasse) should proceed to Stockholm and maintain contact.

Whether or not Himmler had, as he asserted to Schellenberg, become convinced of the existence of God, it is plain that he had not yet absorbed the Christian ethic to the full, for throughout the winter months of 1944-45 he had been bargaining with the Swiss and Swedish Red Cross concerning the lives of the million-odd Jews who were incarcerated in concentration camps, midway between the ghettos and final "resettlement." The first approaches had been made in the summer of 1944, when the mass deportation of Hungarian Jewry had taxed the capacity of the Auschwitz crematoria to such an extent that the idea of selling their lives for cash and goods, took hold as an acceptable proposition in the higher echelons of the SS. The negotiations were handled by Eichmann through a colonel in the SS named Becher, who had earlier proved his competence in acquiring the Baron Oppenheim racing stud for the SS Cavalry School. A price of about seven hundred Swiss francs per head was demanded. (Curious how down the ages the price of a human being as merchandise remains constant. This figure is almost exactly the same as two hundred Confederate dollars, the price of a slave before the Civil War in the United States—or the forty talents used to buy one in the Roman Empire.) The first deal was for thirty thousand Hungarians against a payment of twenty million Swiss francs, to go into the SS numbered accounts in Zurich. However, things moved very slowly because, although international Jewry had been quick enough about collecting the ransom, they were reluctant to pay it over without proof that the Germans would keep their promise. Only two trains of Jews were "de-

[3] See p. 421.

livered" to Switzerland, one in August and one in December 1944. The deliveries then stopped altogether, as the SS bankers reported that none of the ransom had yet been received. Nonetheless, Himmler's main purpose, that of direct contact with foreign heads of state, was achieved, as in February 1945 the first five million francs was paid over by the President of the Swiss Republic, "on the understanding that it would be used to finance further emigration through the Red Cross."

The concept of the Red Cross taking a hand in affairs was one about which Himmler could not but have held mixed feelings. As early as January, Professor Karl Burckhardt, head of the Swiss Red Cross, had suggested that the concentration camps be opened to Red Cross inspection. This, clearly, was a very delicate matter indeed, as the death rate in the camps was running at about four thousand a day. But Himmler had partly brought it upon himself by his practice of putting individual Jews who had had ransom promised, or deposits paid, "on ice" at camps at Strasshof, in Austria, and the notorious "Belsen" (Bergen-Belsen), soon to be ravaged by an outbreak of typhus. Eichmann's bureaucratic machine, functioning with its habitual *punctilio,* had recorded all this on punch cards, and the names of Jews in this category had in some cases been acquired by the Red Cross, which had entered *in loco custodiis* to them and could therefore claim the rights of contact and, in due course, inspection. Very reluctantly a few Swiss officials had been admitted to the inspectorate at Oranienburg—itself more of a "labour" camp, though perilously close to the gas ovens and torture chambers of Sachsenhausen —and as further evidence of "good faith" a third train of immigrants was despatched to Switzerland.

By now, though, rumours of the whole deplorable affair were starting to get back to Hitler, and matters came to a head when someone[4] gave him a report from the Swiss newspapers of the train's arrival. How Himmler survived this confrontation, after being once again caught in an act of flagrant and heretical disobedience, will forever remain a mystery. Yet he seems to have done so, just as he had survived the stormy interview after the Langbehn affair, with no more serious a setback than an order—which he plainly had no intention of keeping—that "no

[4] Colonel Becher maintained in an affidavit after the war (Nuremberg Case XI NG 2675) that this "someone" was Schellenberg. A contention not without relevance to the enigma of Schellenberg's "loyalty" to Himmler and his part in the earlier Langbehn affair (see p. 349 and fn. 4).

concentration camp inmate must be allowed to fall into Allied hands alive."

In the meantime Ribbentrop, who had been distinctly put out to find that the Reichsführer was conducting parallel negotiations, had broadened the base of his own standpoint by authorising Hesse to make an offer to the representative of the World Jewish Congress in Stockholm, to hand over "all Jews in German-held territory or to put them under neutral protection." This put the Reichsführer in an exceptionally difficult position. For Himmler's purpose, of course, had nothing to do with the welfare of the Jews—any more than Ribbentrop's had—but was to establish "diplomatic" contact at a high and fruitful level. Ribbentrop's offer had been "most irresponsible." The Foreign Minister's intention had been, by the extravagance of his promises, to return the focus of attention to himself; it was to be *his* coup—whatever the price. But this made things very awkward for Himmler, who had now met Bernadotte (the head of the Swedish Red Cross) at the instigation of Schellenberg, and had indulged with him in a certain amount of cautious shadow-boxing on "real" diplomatic issues.

The danger now loomed that the "subtlety" (i.e., unreality) of these approaches would be jeopardised (i.e., demonstrated) by the highly embarrassing revelations of the German "resettlement policy" which now appeared to be imminent. Frenzied activity began, to put a gloss over the horrors of the camp system. Himmler circularised Kaltenbrunner, Pohl,[5] and Gluecks,[6] and sent a memorandum to Grawitz (who combined the outwardly incompatible offices of head of the SS Medical Service and chief of the German Red Cross), insisting on measures against the typhus epidemic at Belsen—measures which, it seems safe to assume, differed from the usual treatment by *Flammenwerfer* and "isolation." Still more macabre, there remained many in the SS who either from conviction or personal jealousy of Himmler were set on disrupting this plan. Between them, Kaltenbrunner and Eichmann (the latter, as always, obedient to every order which emanated from above, even when one contradicted another) began to turn Belsen into a Dantesque clearinghouse for the whole camp system. Across the tortured railway system of the Reich train after train of battened-down Jewish families were shunted from one siding to another, clogging the through traffic, draw-

[5] SS General Oswald Pohl, head of Wirtschafts und Verwaltungshauptamt, the economic administration of the SS.

[6] Gruppenführer Richard Gluecks, head of the concentration camp inspectorate.

ing the fire of marauding Allied aircraft, using valuable fuel and rolling stock before discharging their three-quarters-dead cargo at Belsen, the "transit camp," where, as Hoess the commandant stolidly recounted to the Nuremberg tribunal, ". . . tens of thousands of corpses lay about everywhere." From the time of the first payment of five million Swiss francs in February nearly fifty thousand inmates and "arrivals" died in Belsen alone. Yet the National Leader felt no compunction about writing to Hillel Storch that his release of the three Swiss trains was "but a continuation of his work to assist Jewish emigration which he had begun in 1936."

In fact, all this energy of Himmler's was misplaced. The real concern of the neutrals was to achieve contact and, preferably, a settlement between the West and some form of German "presence." And the only person who really had the authority to negotiate such a settlement was the Reichsführer. Yet try as Schellenberg might, and discreetly encouraging as Bernadotte was, Himmler would not commit himself. He and Bernadotte met four times, and on each occasion Himmler ducked the issue at the last moment. His indecision was partly an endemic fault; partly, too, it had a rational base. He was the natural heir; why do anything which by anticipating events might go awry and spoil his chances? Like every one of the Diadochi (except Speer), Himmler could not perceive just how close the final collapse was.

Yet even he had his moments of apprehension. After one of their many agonising and circumlocutory "walks" Himmler turned to his aide and said:

"Schellenberg, I dread the future."

The vernal equinox came and passed. The days lengthened, and the last hours of the German Army began to slip away as the Russians perfected their plan of attack. Fresh material and drafts of infantry continued to flow up to the front, embracing the whole spectrum of military quality, from the brand-new *Jagdtiger* with 122-mm. guns and infrared sights (a tank markedly superior to anything which NATO disposed of for the first eighteen years of its existence) to stocks of French Army rifles of World War I vintage, captured and stored in 1940. The replacements included *Volkssturm, Volksgrenadier,* the remains of Luftwaffe field forces, special police companies, out-of-work camp guards, foreign "legions" of every kind, Hitler Youth, Gauleiters and their staffs. At night the Russian patrols gave the Oder position no rest. Periodically during the day a hurricane of fire would sweep across the front as newly arrived Russian batteries registered and then fell silent.

Then, on 16th April, the German front took the full shock of a double assault by Zhukov and Koniev, deploying over three thousand tanks on a forty-mile front between Schwedt and Frankfurt and between Forst and Görlitz. The Soviet armour was fewer in numbers than in some of the monstrous battles of 1943, and of the Vistula fighting the previous January, but it was of the highest quality. Stalins, T 34/85's, and the deadly SU self-propelled 122-mm. guns, which could outrange every German tank except the *Jagdtiger,* overwhelmingly outgunned as well as outnumbered the Panzers, which were scattered defensively along the length of the front in an array of unit numbers that looked impressive on the OKW wall map but in reality represented no more than the feeble husks of once famous and formidable divisions. Within two days the Russian tanks were through to traverse the open countryside of Pomerania, leaving the infantry to fight a rear-guard action whose end was out of sight. *"Ein Volk! Ein Reich! Ein Führer!"*—all three were now going under.

There is no doubt that Hitler's presence in Berlin was an inspiration to the German Army, even if it was a source of anxiety and annoyance to his court and advisers. Originally the Führer had intended to leave for Obersalzberg on his birthday, 20th April, and ten days earlier many of his personal staff had been sent ahead to prepare the house for his arrival. But in the previous forty-eight hours Russian tanks had taken Eberswalde to the north and Kottbus to the south of Berlin. They were roaming almost at will behind the German lines, and with Eisenhower's forces already on the Elbe at five points, it was plain that Germany was on the point of being divided into first two, then several parts.

Many of the birthday visitors to the Führer's bunker were, therefore, understandably apprehensive. They were anxious for Hitler to leave immediately for the mountains, where his personal safety (and, by the transfer of the seat of government, their own) would be assured. They must have been put out, these suppliants,[7] to find that the Führer was "still confident" and that no decision had yet been made about the evacu-

[7] The final authority for the birthday party and the events which followed in the bunker remains Professor Trevor-Roper (*The Last Days of Hitler,* 119 *et seq.,* 1962 ed.). He gives as attending on 20th April, Himmler, Goering, Goebbels, Bormann, Ribbentrop, Speer, Keitel, Jodl, Doenitz, Krebs, Burgdorf, and Axmann (Schirach's successor in the *Hitlerjugend*), besides the usual secretarial and SS staff, doctors, etc.

ation. Hitler was affable, but imprecise. The furthest he would go in recognising the situation was to appoint two commanders, Doenitz for the north, Kesselring for the south, who would hold the supreme authority, under him, for military operations in their areas.[8] The Russians would "meet their bloodiest defeat before Berlin," he told his audience, and he had personally planned a counterattack which would throw them back to and then across the Oder. In the evening the party dispersed with feelings which though they may have been mixed cannot have been agreeable. Trucks loaded with documents, staff cars with personal luggage, a succession of convoys, picked their way through the rubble with dimmed lights and started on the journey south, down the last remaining land route which remained open.

Some had business to attend to. Himmler had first to get through a meeting with Norbert Masur of the World Jewish Congress, then to a soothing night at Dr. Gebhardt's, followed by a breakfast appointment (arranged by the desperate Schellenberg) with Count Bernadotte. Goering, who was going to travel to Obersalzberg by air, had some last-minute supervision of his private baggage, which included a number of startling *objets d'art.* Bormann, Goebbels, and Ribbentrop remained in Berlin. Only one man can have left the bunker with his mind at rest. This was the remarkable and dispassionate Albert Speer, who was devoting himself in the most single-minded fashion to the sabotage of all Hitler's scorch-earth policies. He had taken a day off to visit the forward commanders and enlist their support in preventing demolitions, and also succeeded in persuading Goebbels to preserve the Berlin bridges intact, and to site the garrison with the minimum of risk of street fighting in the centre of the capital.

With the party over, Hitler's future, or rather his decision regarding his future, hung on the fate of the "Steiner attack," ordered for 22nd April. And here the Führer fell victim to his own delusions. For Steiner's forces amounted to very little. Five SS *Panzergrenadier* divisions on the OKW situation map were down, in most cases to little more than regimental strength—and of their number only two were German, the others were hodgepodge units of foreign SS, some not even Nordic in their characteristics, who were only too eager to divest themselves of the once dreaded black uniforms, which now, like the mark of Cain,

[8] In fact, Kesselring was already conducting surrender negotiations through SS Gruppenführer Karl Wolf, another of the *dramatis personae* from the Langbehn affair (see p. 347).

raised every man's hand against them and made the Russians shoot on sight. Of his few reliable Panzer units Steiner had already committed all those that still had fuel in trying to contain the southern rim of the Soviet break-through and channel the torrent of Red armour away from the outskirts of Berlin. His own headquarters had lost touch with most of its subordinate formations, he had no artillery, no contact with the Luftwaffe, and was now receiving alternate (and conflicting) orders from Heinrici and Doenitz in addition to those which emanated from the bunker.

It is no surprise, then, to learn that the "Steiner attack" never materialised. Instead the day passed with a number of disjointed local reports from units down to battalion level showing a gradual, and irreversible, erosion of the front around Berlin. Russian tanks were firing on Oranienburg, whence the concentration camp inspectorate had been hurriedly evacuated and, to the south of the city, had reached the Elbe at Torgau.

With the realisation that the day had passed and his orders—the most crucial orders, it seemed, that he would ever issue—had been disobeyed, Hitler fell into a paroxysm of rage. By all accounts, the scene which was staged at the evening conference at the end of that day must have made all of Hitler's previous outbursts seem negligible by comparison. For three hours his audience trembled, and when, at last, they were dismissed it was forever.

Keitel, Jodl, Berger, Hitler's two adjutants, even Dr. Morell ("I don't need drugs to see me through," Hitler told him), all these left the bunker in the next twenty-four hours, never to return. To the world Goebbels broadcast that the "Führer was in Berlin, that he would never leave Berlin, and that he would defend Berlin to the very last." That evening the members of the staff who remained in the bunker began to burn the papers there.

Worse was to come. For the Reichsmarschall, hearing of the hysterical scene of 22nd April and fastening on some purported phrase about "negotiation" which Hitler had uttered to Keitel and Jodl, bestirred himself to send a telegram from Obersalzberg to the bunker. In spite of the disclaimer, added apparently as an afterthought, that "Words fail me to express myself," the message was painfully clear. Goering was entering into his inheritance, "the total leadership of the Reich, with full freedom of action at home and abroad. . . . If no reply is received by ten o'clock tonight, I shall take it for granted that you have lost your freedom of action, and shall consider the conditions of your decree[9] as fulfilled, and shall act for the best interests of our country and our people."

Speer was to say of Hitler after the war, "He could hate fiercely in some fields, while forgiving almost anything to those he loved." And the bonds of ancient *Kameradschaft* were too strong for him to turn completely on his old crony of twenty years. Bormann handled the affair, ordering the SS at Obersalzberg to arrest Goering immediately, but the Führer insisted that Goering be excused the extreme penalty in view of his earlier services in the Party, and this dispensation was duly communicated to the Reichsmarschall in the reply which Bormann telegraphed to him that night.

Now indeed the Reichsleiter's wishes were being fulfilled with the same chilling exactitude and loaded horror of *The Monkey's Paw.* And within days the second part, the elimination of *Treuer Heinrich,* was to follow. For the National Leader had at last taken upon himself the task of forming a government, a "Party of National Union" (the title had been suggested by Schellenberg), composed largely of senior SS officials who could administrate while Himmler, Schellenberg, and Schwerin von Krosigk conducted the diplomacy.

On 28th April news of Himmler's activities reached the bunker. For Hitler there was nothing left to say. He had already expended the whole of his self-pity to Hanna Reitsch on 26th April.

> Nothing now remains! Nothing is spared me! No loyalty is kept, no honour observed; there is no bitterness, no betrayal that has not been heaped upon me; and now this! It is the end. No injury has been left undone!

Berlin itself had been doomed from the moment of Doenitz's and Kesselring's appointments. For this had been a tacit admission that Germany would henceforth fight, if she fought at all, in two halves, and as these areas contracted the city poised between them would fall into the pit. Heinrici and Army Group Vistula were pulling their right flank back fast, to form the south side of a "box" resting on the Baltic and the Elbe. Schörner, with the strongest forces left in the Reich, was sitting tight in the Carpathians. Wenck, the man whose appointment Hitler had so energetically resisted in January, and who was now commanding an army against the Americans, was the recipient of a personal

[9] The decree of 29th June, 1941, which named Goering as Hitler's deputy.

message sent by hand of the ever obedient Field Marshal Keitel, appealing to him to turn his back on Eisenhower and march east, to the relief of Berlin. He moved extremely slowly. Aside from half-armed irregulars, the Berlin "garrison" amounted to fewer than 25,000 men: Mummert's 57th Corps, with two regular but depleted infantry divisions; SS *Nordland;* a battalion of the French SS (*Charlemagne*); and Mohnke's SS Guard Battalion.

The thud of Russian shellfire was now audible inside the tunnels of the bunker itself, and Soviet tanks had, by 27th April, broken into the Potsdamer Platz. From the Chancellery garden you could hear, through the clatter of small-arms fire, the sound of their tracks. Below ground the man who had made the world tremble himself shook with suppressed hysteria and the pain of emergence from habitual sedative addiction. Only a few units could still be reached with his commands, among them the fanatic Hitler Youth, who continued to defend the Spree bridges against the expected arrival of Wenck's "relief army," and the Frenchmen, three of whom, such is the irony of history, were the last people to be decorated by Hitler with the Knight's Cross. The Führer, who still had nearly six million men under arms, could control scarcely one division. No ascent in history had been so meteoric, no power so absolute, no decline so complete.

But at least one of Hitler's qualities remained—his personal courage. He had said that he would remain in Berlin and die there, and so he did. Hitler may have despised the Prussian aristocracy, but few exits from the stage of history have been so scrupulous in their honouring of the seignorial code.

Hitler wrote his will, he addressed a few words to everyone of his retainers, he poisoned his faithful dog. Then at a formal ceremony he made an honest woman of his mistress, retired to an antechamber, and shot himself.

His epitaph he had already written, twenty years before:

> At long intervals in human history it may occasionally happen that the practical politician and the political philosopher are one . . . Such a man . . . reaches out towards ends that are comprehensible only to a few. Therefore his life is torn between hatred and love. The protest of the present generation, which does not understand him, wrestles with the recognition of posterity, for whom he also works.

For a few hours the spell lingered on. Bormann, rightly apprehensive that the news of the Führer's suicide would finally sterilise the bunker as a source of authority, continued to send telegrams—to Doenitz, to Kesselring, to Schörner—but omitted mention of his death. Hitler's body and that of Eva Braun were taken up the stairs of the emergency exit to the garden, where Biondi had sniffed and padded on her daily exercise jaunt with a sergeant of the SS, and there laid in a bomb crater. Several jerricans of gasoline were emptied on them, and a match struck.

As the flames leaped to the sky, the tension which held the bunker, and the Reich itself, taut with obedience relaxed. There were smoking and laughter in the hallowed conference chamber. And as the news spread outward, the vast geodetic structure of the German military machine collapsed. In a frantic *sauve qui peut* men threw away uniforms, shaved moustaches, buried gold and documents, burned everything that might incriminate them, from flags to home movies. Alone the German Army preserved to the end a kind of discipline and even exchanged shots with the SS on occasion.[10] But the dreaded *Schwarze Landsknechte* went to pieces in the race to save their own skins.

Some succeeded. Others met uncomfortable ends, like SS Gruppenführer Paul Greiser, who was paraded around Poznan in a cage before being hanged, and Hans Pruetzmann, the chief of the SS police in the Ukraine, who died in agonies from cyanide poisoning in a dry stomach. Pruetzmann's superior, Erich Koch, the Gauleiter of the Ukraine and East Prussia, and one of the most detestable figures to rise in the whole Nazi era, survived. He escaped deportation to Poland until 1950, and seems never to have been brought to trial there. Another figure from the period of *Ostpolitik*, Heinrich Lohse, who had tried to found the hereditary margravate of Belorussia[11] and had written, asking if he was to select Jews for execution "regardless of the requirements of industry," received a short term of imprisonment and is currently (1964) drawing a pension from the Bonn government.

Certainly more of the real villains escaped than were caught. They

[10] When Himmler's bodyguard, the Mohnke battalion, arrived in Berlin, it had shown a stronger taste for rounding up deserters than taking on the Russian armour. Mummert, the General commanding the 57th Corps, took the view that "A division that has the greatest number of bearers of the Knight's Cross with oak leaves does not have to be persecuted by these young louts."

[11] See Ch. 3.

had had too much time to prepare their getaways, open their numbered accounts in Zurich, purchase their nominee properties in Spain and South America, Ireland and Egypt.

Among these were Gruppenführer Gluecks, chief of the concentration camp inspectorate; Wirth, who had commanded the extermination network at Plaszow, in Poland; and Heinz Lammerding, Himmler's Chief of Staff of Army Group Vistula and one of the senior SS officers who had made the mistake of not confining his atrocities to the East. (He had been responsible for the "liquidation" of the French village of Oradour-sur-Glane while his division, *Das Reich,* had been moving across France in August 1944.)

Generally speaking, those who had been involved in the East *only* fared surprisingly well if they could side-step the first wave of extraditions, for the immediate outbreak of the Cold War made their calling to account a fairly leisurely affair. (General Reinecke, personally responsible for the death of three million Russian prisoners, got a life sentence.) And if trial could be postponed long enough it usually resulted in their going before a German court, in pursuance of General Clay's theories that German self-respect would be enhanced thereby. Thus Bach-Zelewski drew a ten-year (suspended) sentence from a Munich court. Max Simon, commander of *Totenkopf* until 1943, was acquitted altogether, and Gottlob Berger, sentenced to twenty-five years, was released after serving two.

Dirlewanger, probably the worst of the lot, was surrounded with the remnants of his brigade at Halbe in April 1945. In one of the most gruesome massacres of the Eastern campaign the whole unit and a large number of German civilians were put to the sword by the Russians. Dirlewanger is rumoured to have escaped by hiding under a pile of bodies. He surfaced in Egypt in 1955, and currently lives in fine style in a villa in Cairo, although it is rumoured that the Israelis periodically send him explosive packages.

The sinister Dr. Gebhardt, who had anticipated "trouble" by persuading Hitler to make him head of the German Red Cross and had made a special journey to the bunker in the last days of April, so that this immunity might be officially conferred upon him, found (it is satisfactory to record) that the British brushed it aside and, after a disagreeable two-year period for reflection, hanged him. Sturmbannführer Rascher, who had been in charge of the freezing and altitude "experiments" on

Russian prisoners (including the notorious "human warmth" experiments with naked gipsy girls) had already been liquidated by Himmler in the course of the brief attempt to clean up the camps during the Reichsführer's consultations with Bernadotte. General Blaskowitz, the German Army commander in Poland, and an upright man who had the courage to protest several times to OKH about SS behaviour, had been arrested by the Americans and was garrotted by SS guards employed as "trusties" by the prison camp commandant.

The most ignominious and probably one of the most painful demises, was that of the National Leader. After a few miserable days at Flensburg, trying to make himself agreeable to Doenitz and to assert his importance; suffering humiliations that were a constant source of embarrassment to his staff; and deserted by many of his closest companions who had already set off on their private journeys to ranch cattle in the Argentine or collect butterflies in Switzerland, Himmler slunk away. Clean-shaven, in private's uniform, with a black patch over his eye, he was caught by the British, announced his identity, then lost his nerve and crunched the *Zynkali* capsule of potassium cyanide that was wedged between his gums.

Of the Russians poor Vlasov, who ended his military career as he had begun it, fighting Germans with the Slovakian Partisans, surrendered to Patton. He was returned to Moscow and hanged a year later. Most of his colleagues, the young men who had emerged with distinction from the crucible of 1941, held senior commands in the most powerful army the world had seen. Zhukov, Koniev, Rokossovski, Tolbukhin, Malinovsky, all had "fronts." Others, like Ryabyshev and Katukov had tank armies. Some, like Chernyakovski, had been killed in action. Most were destined for obscurity with the reassertion of Stalin and Party hegemony that was to follow the peace.

The elder marshals, Budënny and Voroshilov, long since shunted into the background, were to be paraded from time to time, bemedalled effigies of some slight ceremonial value. Soldiers entered politics at their peril. Zhukov and Voroshilov both tried to cross Khrushchev's path —albeit from opposite directions—and both suffered ignominy as a result. Malinovsky, venturing later, was more successful and is currently (1964) the Minister of Defence.

For Chuikov, the hero of Stalingrad, fate had a special reward. Not so much the appointment as Commander in Chief, though that fell to him in due course, but the performance of a special role on 1st May, 1945. For it was Chuikov, the expert in house-to-house fighting, and his 8th Guards Army (the old 62nd Army of the Volga battles) who were spearheading the drive into Berlin. And on that day he was visited by General Krebs, who with three other officers had emerged from the bunker under a white flag to parley a surrender.

Krebs, who knew some Russian and at one stage in his career had been embraced by Stalin, was "a smooth, surviving type." And so, with almost incredible effrontery, he tried to talk to Chuikov as an equal, opening the conversation with the general comment:

"Today is the first of May, a great holiday for our two nations . . ."

With seven million Russian dead, half his country devastated, and fresh evidence mounting daily of the unspeakable barbarity with which the Germans had treated Soviet captives and civilians, Chuikov's answer was a model of restraint, a standing testimony to the cool head and dry wit of that remarkable man. He said:

"*We* have a great holiday today. How things are with you over there it is less easy to say."

EPILOGUE

For over a year the embers of the *Ostfront* smouldered in Berlin and in Germany. A hot summer followed the surrender, and plagues of flies multiplied in the stricken cities of the Oder, breeding in the corpses of man and beast. Rats and lice spread disease; food no longer came in from the countryside; the casual brutality of the occupying army showed no abatement. Then autumn came, and those same portents of lengthening darkness and falling temperature which had, for four years, chilled the hearts of the Wehrmacht closed in upon the whole civilian population. How many died in the first winter of the peace will never be known. Children ate cats, raw, for sustenance, and burrowing in mountains of rubble at whose centre there glowed, like that of the earth itself, the warmth of interminable fires, were suffocated there.

Chiefly it was the innocent who suffered. Most of the villains had slunk off the stage or else, with a change of costume, had taken employment with their conquerors—readily accepting posts with the Communist administration of the Eastern or the intelligence organisations of the Western zones. While the Nuremberg trials were in process, a genuine effort was made by both East and West to arraign the guilty. But with the widening of political differences between the victors, enthusiasm for the pursuit of their former enemy declined. So many escaped or disappeared: Borman, "Gestapo" Müller, "Dr." Mengele. When the twenty-year limitation period expires in 1965, they may or may not reappear.

This paradox epitomises the real tragedy of the war in the East. It achieved nothing constructive—not even the orderly application of retributive justice. The very totality of the German defeat made it inevitable that each side would try to resuscitate in its own image the territory it was occupying, and so even the war's most notable result, the elimination of German military power, was of short duration.

For the West the German revival has seemed a matter for self-congratulation, but for the Russians it has provoked an agony of conscience and politics which is at the root of the world's insecurity today. The Communist "government" of East Germany is a cruel and dishonest fabrication. But what are the Russians to do? The further they build up the economy of their zone the more ridiculous and inappropriate will its Communist administration become; but if they allow no amelioration of its repressive character they risk a revolt whose consequences are dangerous and unpredictable. Yet if they withdrew altogether, immediate reunification with West Germany would follow, and that country, already the strongest in Western Europe, would become the third power in the world.

To prevent this most Russians would willingly risk their lives. Their pathological fear and distrust of Germany is all the more dangerous because they will not admit—are indeed prohibited from admitting—what a close-run thing their victory was. At this time (1964) virtually every important military office post is occupied by soldiers who served with or under Khrushchev during the critical three months at Stalingrad: the Minister of Defence (Marshal Malinovsky, who commanded the 2nd Guards Army), the Chief of the General Staff (Marshal Biryuzov, who was Chief of Staff at the 2nd Guards Army), the Commander in Chief of ground forces (Marshal Chuikov), the Commander in Chief of the Air Defence Forces (Marshal of Aviation Sudets, who was commander of the 17th Air Army at Stalingrad), Commander in Chief of the Navy (Admiral of the Fleet Gorshkov, commander of the Azov flotilla, but operating ships on the Volga during the Stalingrad battle), Commander in Chief of Strategic Missile Forces (Marshal Krylov, formerly Chief of Staff to Chuikov in the 62nd Army). The Commander in Chief of the Air Forces (Marshal of Aviation Vershinin) commanded the air forces of the north Caucasus front in the spring of 1943; Krylov's predecessor with Strategic Missiles, Marshal Moskalenko, commanded the 38th Army and is now chief of the general inspectorate; while the head of rear services (Marshal Bagramyan) knew Khrushchev in the bad days of 1941, when he was chief of the operations directorate of Kirponos' staff. Of the Russian nuclear battery over four fifths is deployed against Western Europe, some are trained on China, and only the remaining fraction have United

States targets according to present estimates of the Institute for Strategic Studies.

As we formulate our policies to ensure that these missiles are never fired we should always bear in mind the Russian experience in World War II, the extent to which it colours the hidden mainstream of Soviet strategy, and the memories of it which are borne by every senior Soviet administrator, both military and civilian, at the present time.

Appendices

FACTS ABOUT THE RUSSIAN AND GERMAN LEADERS

Bach-Zelewski, Erich von dem, Gruppenführer SS	German chief of Partisan warfare; commanded SS forces engaged in suppressing the Warsaw uprising August 1944; sentenced in Munich, March 1951, to ten years' "special labour" (suspended sentence).
Becher, Kurt, Standartenführer SS	Head of the SS remount purchasing commission; Himmler's negotiator for the lives of Jews 1944-45.
Beck, Ludwig, Colonel General	Chief of the German General Staff; resigned 1938; nominated by the Resistance circle as future head of the German state; committed suicide 20th July, 1944.
Berger, Gottlob, Obergruppenführer SS	Head of the SS main leadership office and for some months virtual administrator of the Rosenberg ministry in Russia; sentenced to twenty-five years' imprisonment April 1949; released at the end of 1951.
Beria, Lavrenti P.	Chief of the NKVD 1938-53. Circumstances of death mysterious, but believed to have been shot on Khrush-

chev's orders in 1953, the only member of the "anti-Party group" to be "liquidated."

Blomberg, Werner von, Field Marshal	Minister of War, January 1933-February 1938; Field Marshal, April 1936; dismissed because of his scandalous marriage; died in the witness wing of Nuremberg prison 1946.
Bock, Fedor von, Field Marshal	Commanding Army Group North (Poland) 1st September, 1939, to 3rd October, 1939; commanding Army Group B (West) 5th October, 1939, to 12th September, 1940; commanding Army Group Centre 1st April, 1941, to 18th December, 1941; commanding Army Group South 18th January, 1941, to 15th July, 1942. Killed in an air raid near Hamburg 4th May, 1945.
Bormann, Martin	Head of Nazi Party Chancellery (succeeding Rudolf Hess); one of Hitler's closest advisers. Disappeared during the battle of Berlin, at the end of the war; condemned to death in absentia Nuremberg 1946.
Brauchitsch, Walther von, Field Marshal	Commander in Chief of the German Army (Oberbefehlshaber des Heeres) 4th February, 1938, to 19th December, 1941; died in British captivity 18th October, 1948.
Budënny, Semën M., Marshal	Early career with the cavalry in the Russian Revolution; blundered in the 1920 campaign against Poland; promoted during the purges 1937; commanded the southwestern front 1941, with disastrous results.

Canaris, Wilhelm, Admiral	Chief of Amtsgruppe Ausland/Abwehr (intelligence in OKW) 1938-44, when dismissed owing to frequent quarrels with the SS; involved indirectly in the plot of 20th July, 1944; hanged in Flossenburg concentration camp April 1945.
Chuikov, Vasili I., Marshal	Junior officer in the Russian Revolution; military adviser to Chiang Kai-shek 1941; appointed to the 62nd Army and command of Stalingrad defence by Khrushchev, September 1942; accepted the surrender of Berlin, May 1945.
Dietrich, Josef (Sepp), Oberstgruppenführer SS	Commander of Hitler's SS bodyguard 1928; army commander 1944-45; sentenced to twenty-five years' imprisonment 1946; released 25th October, 1955.
Dirlewanger, Oskar, Brigadeführer SS	Commanded a brigade of ex-convicts against Russian Partisans, and in suppression of the Warsaw uprising; protégé of Gottlob Berger; disappeared May 1945.
Fegelein, Hermann, Obergruppenführer SS	Himmler's liaison officer with Hitler from the beginning of 1943; executed by Hitler a few days before the latter's death in 1945.
Fritsch, Baron Werner von, Colonel General	Commander in Chief of the German Army 1934-38; framed by Himmler 1938, forced to resign; sought death in the Polish campaign 18th September, 1939.
Fromm, Fritz, Colonel General	Commander in Chief of the German Replacement Army; tried for his ambiguous role in the July 1944 plot and executed March 1945.

Gebhardt, Professor Karl, Gruppenführer SS	Boyhood friend of Himmler's; head of Hohenlychen Hospital and chief consultant to the SS and police; hanged for his medical experiments 2nd June, 1948.
Goebbels, Joseph	Reich Propaganda Minister and Gauleiter of Berlin; from July 1944 "Plenipotentiary for Total War"; suicide in Hitler's bunker 1st May, 1945.
Goering, Hermann, Reich Minister for Air	Reich Minister for the Four-Year Plan. Commander in Chief of the Air Force throughout World War II; relieved of all posts and commands 23rd April, 1945; condemned to death Nuremberg, and committed suicide 15th October, 1946.
Guderian, Heinz, Colonel General	Commander of the 2nd *Panzergruppe* (later *"Panzergruppe* Guderian") 1941; Inspector General Panzer Forces 1943; Chief of the Army General Staff, July 1944; dismissed by Hitler, March 1945. Died 1954.
Halder, Franz, Colonel General	Chief of the German Army General Staff (OKH) 1939-42; in correspondence with the Resistance circle; arrested July 1944, but not brought to trial.
Hassell, Ulrich von	German Ambassador to Italy till 1938; member of Goerdeler-Beck Resistance circle; executed 8th September, 1944.
Heydrich, Reinhard, Obergruppenführer SS	Chief of the SD 1931-34; chief of the security police and SD 1934-39; chief of RSHA, which also included the criminal police and Gestapo, 1939-42; Protector of Bohemia-Moravia, Sep-

tember 1941; died of wounds from a bomb 6th June, 1942.

Himmler, Heinrich, Reichsführer SS — Reichsführer SS 1929; Police President, Bavaria, 1933; chief of the Reich political police 1935; chief of the German police 1936; Minister of the Interior 1943; Commander in Chief of the Replacement Army, July 1944; Commander in Chief of the Rhine and Vistula armies December 1944-March 1945; committed suicide at the British Interrogation Centre, Lüneberg, 23rd May, 1945.

Hindenburg, Paul von, Field Marshal — Last President of the German Republic; confirmed Hitler's succession to him as head of the German state in his will August 1934.

Hitler, Adolf — Born 1889. Founded the National Socialist Workers' Party (the Nazi Party) 1919-20. Organized an unsuccessful revolt in Munich (Beer Hall *Putsch*), November 1923; sentenced to five years' imprisonment 1924; paroled after nine months. In prison wrote *Mein Kampf*. Named Chancellor by Hindenburg 1933; became President and Chancellor (der Führer) on the death of Hindenburg, August 1934. Occupied the Rhineland, March 1936; annexed Austria, March 1938; the Sudetenland, October 1938; Czechoslovakia, March 1939; concluded a non-aggression pact with the Soviet Union, August 1936. Invaded Poland, September 1939, which started World War II. Committed suicide April 1945.

Jodl, Alfred, Colonel General	Chief of Staff of the High Command of the German Armed Forces (OKW) 1938-45; condemned to death Nuremberg 1946; hanged 16th October, 1946.
Keitel, Wilhelm, Field Marshal	Chief of the High Command of the German Armed Forces (OKW), February 1938-May 1945; hanged Nuremberg 16th October, 1946.
Khrushchev, Nikita S.	Born 1894. Commissar during the Russian Revolution. Elected to the Politburo 1934. Political member of Budënny's Military Soviet from June 1941, also primarily responsible for organising industrial evacuations in south Russia. Remained in the southern theatre for the duration of the war as Commissar to successive military commanders; allegedly responsible for Chuikov's appointment to the command of the Stalingrad garrison.
Kleist, Ewald von, Field Marshal	Commander of the 1st Panzer Army 1941; Commander in Chief of Army Group A, August 1942-April 1944; died in a Russian prison camp October 1954, having been extradited from Yugoslavia 1949.
Kluge, Guenther von, Field Marshal	Commander of the 4th Army, June 1941; Army Group Centre, December 1941-September 1943; in desultory correspondence with Resistance circle from November 1942; committed suicide to avoid arrest at Dombasle 19th August, 1944.
Koch, Erich, Honorary Gruppenführer SS	Gauleiter of East Prussia 1930-45; Reichskommissar of the Ukraine 1941-

44; extradited to Poland 1950, but not heard of since.

Koniev, I. S., Marshal — Early career in the Russian Revolution as Commissar, but graduated from the Frunze Academy in the thirties. Skilful conduct of a brigade group on the central front in 1941 led to a succession of front commands. Personal rivalry with Zhukov, with whom he shared the spearhead of the advance into Germany 1945; Denounced Zhukov after the latter's dismissal 1957.

Lammerding, Heinz, Gruppenführer SS — Commanded SS division *Das Reich* 1944; Himmler's Chief of Staff, Army Group Vistula, January-March 1945; condemned to death in absentia Bordeaux 1951; still wanted for the Oradour massacre.

Lammers, Hans, Honorary Gruppenführer SS — Chief of the Reich Chancellery 1933-37; Minister from November 1937; sentenced to twenty years' imprisonment at Nuremberg, but released November 1951.

Langbehn, Carl — Lawyer and friend of Himmler, with whom he maintained contacts on behalf of the Resistance circle; arrested September 1943; executed 12th October, 1944.

Malenkov, Georgi M. — Successful career in security and political sides of the Soviet military organization, 1920-41; Promoted to GOKO over Khrushchev's head June 1941. An intimate of Stalin's; succeeded him briefly as Premier after Stalin's death. Ousted by Khrushchev and Bulganin,

	and branded as "anti-Party" at the 20th Congress.
Manstein, Erich von, Field Marshal	Principle commands: 11th Army 18th September, 1941, to 21st November, 1942; Army Group Don 28th November, 1942, to 14th February, 1943; Army Group South 14th February, 1943, to 30th March, 1944. Sentenced by British court 24th February, 1950, to eighteen years' imprisonment; commuted to twelve years; freed 6th May, 1953.
Mikoyan, Anastas I.	Origins obscure. In charge of the Light Industries Division of the Soviet of Labour and Defence 1930; appointed to the Politburo 1938; in charge of Evacuation Soviet, September 1941; firm friend and supporter of Khrushchev in postwar political disputes.
Molotov, Vyacheslav M.	Appointed to the Central Committee of the Communist Party 1921; backed Stalin in his dispute with Trotsky 1928; appointed to the vital Soviet of Labour and Defence 1930; People's Commissar for Foreign Affairs 1939-46; Soviet Foreign Minister 1946-49 and 1953-56. A convinced Stalinist, disgraced by Khrushchev after the 20th Party Congress.
Mueller, Heinrich, Obergruppenführer SS	Official of the Munich political police under the Weimar Republic; head of the Gestapo 1935-45; last heard of in Hitler's bunker 29th April, 1945; still alive.
Mussolini, Benito	Born 1883. Duce, or head, of the Italian Government; deposed 25th July, 1943;

	rescued from the Badoglio party 13th September, 1943; head of a puppet Italian government till the end of World War II; murdered April 1945.
Paulus, Friedrich, Field Marshal	Surrendered with the German 6th Army to the Russians at Stalingrad; broadcast for the Russian Free Germany Committee; gave evidence at Nuremberg for the Russian prosecution; permitted by the Russians to live in East Germany 1953.
Philip, Prince of Hesse	Obergruppenführer SA; married to Princess Mafalda of Italy and employed by Hitler as a reliable Party man on missions to Italy; sent to Sachsenhausen concentration camp September 1943; liberated at the end of World War II.
Pohl, Oswald, Obergruppenführer SS	Head of the economic administration of the SS (WVHA), including the concentration-camp inspectorate, 1942-45; hanged at Landsberg 8th June, 1951.
Popitz, Professor Johannes	Acting Finance Minister for Prussia 1933-44; member of the Resistance circle who tried to gain Himmler's adherence; executed February 1945.
Reichenau, Walther von, Field Marshal	Played an uncertain role in the blood purge of 1934 as head of the German Army supply office; a Party general; commander of the 6th Army, June-December 1941; Army Group South, December 1941; died of a mysterious infection 17th January, 1942, soon after being made Field Marshal.

Reinecke, Hermann, Colonel General	Chief of the general Wehrmacht office throughout World War II; responsible for Russian prisoners of war; chief of the National Socialist Guidance Staff (NSFO) from 1943 (the *Oberpolitik*); sentenced to life imprisonment 28th October, 1948; still in Landsberg.
Ribbentrop, Joachim von	German Ambassador to Great Britain 1936-38; Foreign Minister 1938-45; hanged Nuremberg 16th October, 1946.
Rosenberg, Alfred	Chief of the foreign political section in the Nazi Party office 1933-41; Minister for Eastern Territories from April 1941; hanged Nuremberg 16th October, 1946.
Rundstedt, Gerd von, Field Marshal	Commander in Chief of the centre group of German armies against France 1940; commander of Army Group South (Russia), June-December 1941; Commander in Chief of the West, with two short intermissions, from March 1942 to March 1945; died 24th February, 1953.
Sauckel, Fritz von	Gauleiter of Thuringia; made plenipotentiary for labour recruitment March 1942; conducted slave raids in Russia and other occupied countries; hanged Nuremberg 16th October, 1946.
Schellenberg, Walter, Gruppenführer SS	Deputy chief of Amt VI, foreign intelligence section of the SD, 1939-42; then chief till 1944, when he became head of the united SS and Wehrmacht

	military intelligence and Himmler's personal adviser; released from prison December 1950; died 1952.
Schlabrendorff, Fabian von, Major	German staff officer in correspondence with the Resistance circle; arrested after July 1944 plot; acquitted the following March, but kept in a concentration camp till the liberation.
Schleicher, Kurt von, Colonel General	Reich Chancellor for a few weeks, at the end of 1932; planned to get back by intriguing with Roehm and Gregor Strasser; murdered by Hitler 30th June, 1934.
Schörner, Ferdinand, Field Marshal	Commander of German Army Group South in the Ukraine, April-July 1944: Army Group North, July 1944—January 1945; commanded an army group in Czechoslovakia and Silesia from January 1945, which was expected to relieve Berlin; returned from captivity in Russia 24th January, 1955.
Shaposhnikov, Boris M., Marshal	Career officer in the Tsarist army; joined the Red Army 1918; worked chiefly on the staff and planning side; Chief of the General Staff 1928-31 and 1937-43; advised Stalin throughout the crisis of 1941 (though not always heeded); retired November 1942 owing to ill health; died 1945.
Speer, Albert, Professor of Architecture	Hitler's architect; Minister of Armament and War Production 1942-45; convicted of procuring slave labour 1st October, 1946; at present serving a life sentence at Spandau.

Stalin, Joseph	Born 1879. Joined the Russian Social Democratic Party 1896. Several times exiled to Siberia for Bolshevik political activity 1904-13; imprisoned 1913-17; freed after the February Revolution. Took part in the October Revolution. General Secretary of the Central Committee of the Communist Party from 1922; after Lenin's death in 1924 overcame the opposition of Trotsky and others, to be dictator. Made a nonaggression pact with Germany, annexed eastern Poland after the German invasion of Poland 1939. Started war with Finland 1939. Annexed Latvia, Estonia, Lithuania, Bessarabia 1940. Met with Roosevelt and Churchill at Teheran to discuss international affairs December 1943; at Yalta, February 1945. With Truman, Churchill, and Attlee at Potsdam, July 1945. Died 1953.
Stauffenberg, Klaus Schenk von, Colonel	Chief of Staff to Fritz Fromm, commanding the German Replacement Army; tried unsuccessfully to blow up Hitler in his staff conference room 20th July, 1944; murdered by Fromm the same night in the Bendlerstrasse.
Thomas, Georg, Colonel General	Chief of the armaments section at OKW (Wi Rü Amt); member of the Resistance circle; arrested July 1944, but not brought to trial; liberated by the Allies.
Timoshenko, Semën K., Marshal	Distinguished combat record in the Russian Revolution and Soviet-Finnish War; commanded the central front June-October 1941; took over Bu-

dënny's command October 1941 and remained in southern Russia throughout 1942, though fell somewhat into obscurity after the failure of his attack on Kharkov in May 1942; held training commands for the remainder of World War II.

Tresckow, Henning von, Major General	Successive Chief of Staff to the commanders of German Army Group Centre, Kluge and Bock; creator of the Valkyrie plan for a *Putsch* against Hitler; walked into the Russian field of fire after the failure of the plot in 1944.
Vlasov, Andrei A., Colonel General	Russian army commander; captured 1942; allowed to set up a Free Russia Committee and eventually to command two divisions of Russian deserters; executed Moscow 1st August, 1946.
Voroshilov, Kliment E., Marshal	Early member of the Russian Social Democratic Party (1903); fought with distinction in the Revolution; Commissar for Defence 1925-40; commanded the Leningrad armies 1941 (see Ch. 6); removed from operational command 1942 and held various ceremonial posts until disgraced by Khrushchev 1959.
Warlimont, Walter, Lieutenant General	Led the German volunteers for General Franco; chief of the national defence section in OKW 1938-44; sentenced to life imprisonment 28th October, 1948; sentence commuted to eighteen years; released from Landsberg 1954.

Zhukov, Georgi K., Marshal	Served under Timoshenko during the Russian Revolution; commanded Russian forces against the Japanese, Khalkin Ghol, August 1939; Chief of Staff of the Red Army, January 1940; organized defence operation at Leningrad, October 1941, winter counter-offensive at Moscow, December 1941, Stalingrad in November 1942. Thereafter commanded the strongest Soviet front in the advance into Germany. Demoted by Stalin after World War II; brought back by Khrushchev 1955, but again disgraced 1957.

CHRONOLOGY OF DEVELOPMENTS IN THE EASTERN CAMPAIGN 1941-45

1941 June 22nd Invasion starts.
End July Panzers diverted to attack Kiev.
October 2nd Germans resume advance on Moscow.
December 2nd Failure of the "last heave" at Moscow.
December 5th Zhukov's counteroffensive opens.
December 7th (Pearl Harbor)

1942 April Russian winter operations peter out.
May 9th-26th Russian offensive against Kharkov defeated.
July 28th German summer offensive opens.
August 30th Germans reach outskirts of Stalingrad.
September-November Street fighting in Stalingrad.
November 18th Russian winter offensive opens.

1943 February 2nd German 6th Army surrenders at Stalingrad.
February 22nd Manstein's counteroffensive starts; Russian winter operations halted.
July 4th *Zitadelle,* attack on the Kursk salient, starts.
July 10th (Allies land in Sicily.)
July 25th (Mussolini overthrown.)

1944 June 6th (D-Day; Allies land in Normandy.)
June 21st Soviet summer offensive opens.
August 1st-October 2nd Warsaw uprising.
August 20th-October 31st Collapse of German position in the Balkans.
December 16th (Start of German offensive in the Ardennes.)

1945 January 12th Start of Soviet winter offensive.
February Russians reach the Oder.
March 28th Guderian dismissed.
April 25th U.S. and Russian forces join at Torgau on the Elbe, splitting Germany.
April 30th Hitler commits suicide.
May 2nd Surrender of Berlin to General Chuikov.

THE COMMISSIONED RANKS OF THE WAFFEN SS AND THEIR ARMY EQUIVALENTS

Untersturmführer—Second Lieutenant
Obersturmführer—Lieutenant
Hauptsturmführer—Captain
Sturmbannführer—Major
Obersturmbannführer—Lieutenant Colonel
Standartenführer—Colonel (Oberst)
Oberführer—Brigadier General
Brigadeführer—Major General
Gruppenführer—Lieutenant General
Obergruppenführer—General
Oberstgruppenführer—Colonel General (Generaloberst)

There were no field marshals in the SS, but Himmler held the special supreme rank of Reichsführer, equivalent to Goering's rank of Reichsmarschall as senior officer of the armed forces. Many of the higher SS officers had double ranks, as members both of the SS and the police, e.g., Obergruppenführer and General of the Police.

There were four regular SS divisions which passed through the SS training school at Bad Tölz, and these are referred to in the text by their titles: *Das Reich, Leibstandarte, Totenkopf,* and *Viking*. The "Police Division" was also granted the substantive title of Waffen SS, but had not passed through Bad Tölz. After 1943 the rapid expansion of the SS (to twenty-nine divisions in 1945) was achieved by the incorporation of ethnic Germans and other racially "pure" nationalities. Very few of their members passed through Bad Tölz, and in view of this and the constant overlapping of commands and staffs in the last years of the war these units are referred to by their numbers and not by their titles.

GLOSSARY OF MOST FREQUENTLY USED ABBREVIATIONS

AK—Polish "Home Army" (organised by the London government)

AL—Polish "People's Army" (independent left-wing sympathisers)

AWA—General Armed Forces Dept. of the OKW

E.L.A.S.—Communist-dominated Greek guerilla force

GBA—Special German office of labour "allocation"

GESTAPO—German Secret State Police

G.O.C.—General Officer Commanding

GOKO—Soviet Committee of the Defence of the State

GPU—Soviet Secret Police (title held until 1938)

G.S.O.—General Staff Officer

KV—Kirov plant (tanks)

MVD—Soviet Secret Police (1948-55)

NKVD—Soviet Secret Police (1938-46)

NSZ—Nationalist Polish Armed Forces (extreme right-wing force which had broken away from AK at first sign of impending compromise with Russian power)

OKH—High Command of the German Army

OKW—Organisation of the Supreme Command of the German Armed Forces (including Luftwaffe and Kriegsmarine)

PAL—Communist-dominated Polish guerilla force

RSHA—SS Main Office for State Security

SA—"Brown Shirts," Nazi private army under Roehm, purged in 1934

SD—Security and Intelligence Service of the SS

SHAEF—Supreme Headquarters Allied Expeditionary Forces

SS—Elite Nazi armed force concerned with security and intelligence (SD, etc.) and actual combat (Waffen SS). Originally Hitler's personal bodyguard.

Stavka—Soviet equivalent of Combined Chiefs of Staff

WVHA—Economic Administrative Head Office of the SS

TEXT OF FÜHRER DIRECTIVE NO. 34.

I am not in agreement with the proposals submitted by the Army for the prosecution of the war in the East and dated August the 18th.

I therefore order as follows:

1. Of primary importance before the outbreak of winter is not the capture of Moscow but rather the occupation of the Crimea, of the industrial and coal-mining area of the Donetz basin, the cutting of the Russian supply routes from the Caucasian oil fields, and, in the north, the investment of Leningrad and the establishment of contact with the Finns.

2. An unusually favourable operational situation has arisen as a result of our troops reaching the line Gomel-Pochep; this must be immediately exploited through a concentric operation by the inner wings of Army Groups South and Centre. The objective must be not merely to drive the Russian 5th Army back across the Dnieper by means of a solitary attack by 6th Army, but to destroy the enemy before he manages to withdraw behind the line the Desna-Konotop-the Sula. This will give Army Group South the necessary security for crossing the central Dnieper and for continuing to advance with its centre and left wing in the direction Rostov-Kharkov.

3. Without consideration of future operations Army Group Centre is therefore to employ as much force as is necessary for the achievement of its objective, the destruction of Russian 5th Army, while occupying such positions on the centre of its own front as will enable minimum forces to ensure the defeat of enemy attacks in that area.

4. The capture of the Crimean peninsula is of extreme importance for safeguarding our oil supplies from Rumania.

BIBLIOGRAPHY

This bibliography has no pretensions to being a complete catalogue of the primary works on the Eastern campaign. It is a list of the volumes and material upon which I have relied most heavily and to which reference is made in the text and the back-matter notes. Very much more extensive lists will be found in Erickson's *The Soviet High Command* and Dallin's *German Rule in Russia* (the Bibliography being published as a companion volume to the leader text).

In cases where a book or other source is of only tangential importance, that is, where it may contain verification of a single incident or assertion, it is simply identified at the appropriate page in the back-matter notes and is not included in the general list.

The Führer conference transcripts are identified by the date of the conference (and the hour, where two occur on the same day) regardless of whether or not they have been reproduced in Dr. Gilbert's edition.

Documents from the proceedings at Nuremberg are identified by the prefixes NO (Nazi Organisation), NG (Nazi Government), NI (Nazi Industry), and NOKW.

Where a source is identified simply by a name and rank (if any), it refers to correspondence from, and notes taken of conversations with, individuals, which are in my possession. Such items usually relate to eyewitness descriptions, and are not used to substantiate assertions of historic significance.

A.C.

Assmann, Vice-Admiral Kurt, "The Battle for Moscow—Turning Point of the War." *Foreign Affairs,* Vol. 28 (Jan. 1950).

Bayerlein, Lieutenant General Fritz, "With the Panzers in Russia 1941-43." *The Marine Corps Gazette,* Vol. 38, No. 12.

Beliaev, S. and Kuznetsov, P., *Narodnoe Opolchenye Leningrada.* Leningrad, 1959.

Blond, Georges, *L'agonie de l'Allemagne, 1944-45.* Paris, 1952.

Blumentritt, G., essay in *The Fatal Decisions* (ed. Freidin, *q.v.*). London, 1956.

Boltin, E., "The Victory of the Soviet Army before Moscow, 1941." *Voprosy Istorii* 1957, No. 1.

Bor, P., *Gesprache mit Halder.* Wiesbaden, 1950.

Bormann, Martin, *The Bormann Letters* (ed. H. R. Trevor-Roper). London, 1954.

Budënny, Marshal S. M., *Proidennyi put.* Moscow, 1958.

Bullock, Alan, *Hitler, a Study in Tyranny.* London, 1952.

Choltitz, General von, *Soldaten unter Soldaten.* Zurich, 1951.

Chuikov, Vasili, *The Beginning of the Road* (Eng. tr. of his memoirs). London, 1963.

Churchill, W. S., *The Second World War.* London, 1948-54.

Ciano's Diaries, 1939 to 1943 (ed. Muggeridge). London, 1957.

Crankshaw, Edward, *The Gestapo, Instrument of Tyranny.* London, 1956.

Dallin, Alexander, *German Rule in Russia, 1941-45.* London, 1957.

Degrelle, Leon, *Die verlorene Legion.* Stuttgart, 1955.

Documents on German Foreign Policy. Official series, Vols. D1 and D5, 1952 and 1954. Washington, U. S. Dept. of State.

Dulles, Allen, *Germany's Underground.* New York, 1947.

Dwinger, Edwin Erich, *Wiedersehen mit Sowjetrussland.* Jena, 1943.

Erickson, John, *The Soviet High Command.* London, 1962.

Fadeev, A., *Leningrad v dni blokady.* Moscow, 1944.

Flicke, W. F., *Spionagegruppe "Rote Kapelle."* Kreuzlingen, 1954.

Footman, D., "The Red Army on the Eastern Front." *St. Antony's Papers on Soviet Affairs.* St. Antony's College, Oxford (no date).

Freidin, S. and Richardson, W. (eds.), *The Fatal Decisions,* by K. von Zeitzler, et al. London, 1951.

Fremde Heere Ost, Documents of the OKH intelligence section, "Foreign Armies East" (tr. U. S. Army Foreign Military Studies).

Galay, Nikolai, "Tank Forces in the Soviet Army." *Bulletin,* Vol. 1, No. 7 (Oct. 1954), Munich.

Garder, Michel, *Une Guerre pas commes les Autres.* Paris, 1963.

Garthoff, Raymond L., *How Russia Makes War.* London, 1954.

Gilbert, Felix (ed.), *Hitler Directs His War.* New York, 1950. (*See also* acknowledgment to the University of Pennsylvania.)

Gisevius, Hans, *To the Bitter End*. London, 1948.

Goebbels, Joseph, *Diaries* (ed. L. P. Lochner). London, 1948.

Goerlitz, Walter, *History of the German General Staff 1657-1945*. New York, 1953. *Paulus and Stalingrad*. London, 1963.

Golikov, S., *Vydayuschiesya pobedy Sovetskoi armii v Velikoi Otechestvennoi Voine*, 2nd ed. Moscow, 1954.

Goure, Leon, *The Siege of Leningrad*. London, 1962.

Greiner, Helmuth, *Die Oberste Wehrmachtführung, 1939-1943*. Wiesbaden, 1951. Records of Situation Evaluations of the National Defence Branch (tr. U. S. Army Foreign Military Studies): 8th August, 1940-25th June, 1941, MS C-065 xi; Operation Barbarossa, MS C-065 i; War Diary, 12th August, 1942-17th May, 1943; MS C-065 a.

Guderian, Colonel General Heinz, *Panzer Leader* (with Foreword by Captain B. H. Liddell Hart). London, 1952. *Erinnerungen eines Soldaten*. Heidelberg, 1951.

Guillaume, General A., *La Guerre Germano-Sovietique, 1941-45*. Paris, 1949.

Halder, Colonel General Franz, *Diaries* (cyclostyled). Imperial War Museum, London. *Hitler als Feldherr*. Munich, 1949. (*See also* Bor, P.)

Hassell, Ulrich von, *The von Hassell Diaries, 1938-44*. London, 1948.

Hausser, Oberstgruppenführer Paul, *Waffen SS im Einsatz*. Göttingen, 1953.

Hitler, Adolf, *Table Talk, 1941-44*. London, 1953.

Hossbach, Major General Friedrich, *Infanterie im Ostfeldzug, 1941-42*. Osterode, 1951. *Schlacht um Ostpreussen*. Überlingen, 1951.

Hubatsch, Dr. Walther, *Hitler's Weisungen für der Kriegführung, 1939-45*. Frankfurt, 1962.

Istoriya Kommunisticheskoi Partii Sovetskovo Soyuza. Moscow.

Istoriya Velikoi Otechestvennoi Voiny Sovetskovo Soyuza, 1941-1945. Moscow, 1960.

Jacobs, Walter Darnell, *"Limits of Soviet Military Originality." Revue Militaire Générale* (10th December, 1957), 680-90.

Karasev, A. V., *Leningradtsy v gody blokady, 1941-1943*. Moscow, 1959.

Kraetschmer, Ernst, *Die Ritterkreuztraeger der Waffen SS*. Göttingen, 1955.

Krylov, *Journal de Krylov* (*Le Journal de Genève*).

Leyderrey, Colonel E., *The German Defeat in the East*. (Restricted circulation, published by Her Majesty's War Office, 1952).

Liddell Hart, Captain B. H., *The Other Side of the Hill*, 3rd ed. London, 1956.

Malaparte, C., *The Volga Rises in Europe*. London, 1957.

Mannerheim, Marshal Carl Gustaf von, *Mémoires, 1882-1946*. London, 1952.

Manstein, Field Marshal E. von, *Verlorene Siege*. Bonn, 1955.

Mellenthin, F. W. von, *Panzer Battles 1939-1945*. London, 1955.

Messe, General, *La Guerra al Fronte Russo*. Milan, 1947.
Mitteldorf, E., *"Zitadelle," Schaftiche Rundschau, 1953. Taktik im Russlandfeldzug*. Darmstadt, 1956.
Nazi Conspiracy and Aggression, 10 vols. Washington, U. S. Govt. Printing Office, 1946.
Platonov, Lieutenant General S. P., Pavlenko, Major General N. G., Parotkin, Colonel I. V., *Vtoraya Mirovaya Voina 1939-1945*. Moscow, 1958.
Pavlov, D. G., *Leningrad v blokady*. Moscow, 1958.
Reitlinger, Gerald, *The SS—Alibi of a Nation*. London, 1956.
Reinhardt, H. von, *"Der Vorsfoss des XLV Panzer Korper im Sommer 1941." Wehrkunde* III (March 1956).
Rosenberg, Alfred, *Letzte Aufzeichnungen*. Göttingen, 1955. (Part translation, *Rosenberg's Memoirs*. New York, 1949.)
Samsanov, A. M., *Velikaya bitva pod Moskvoi 1941-1942*. Moscow, 1958.
Schlabrendorff, Fabian von, *Offiziere gegen Hitler*. Zurich, 1945.
Schapiro, Leonard, "The Army and Party in the Soviet Union." *St. Antony's Papers on Soviet Affairs*. St. Antony's College, Oxford, 1954.
Schramm, P. E. (ed.) *Kriegstagebuch des OKW 1940-45,* 4 vols. Frankfurt, 1961.
Schultz, J., *Die letzten 30 Tage*. Stuttgart, 1951.
Senger und Etterlin, General Fridolen von, *Neither Fear Nor Hope*. London, 1964.
Shirer, W. L., *The Rise and Fall of the Third Reich*. London, 1960.
Siegler, Fritz Freiherr von, *Die hoheren Dienststellen der deutschen Wehrmacht, 1933-45*. Munich, 1953.
Sirota, F. I., *"Borba Leningradskoi partiinoi organizatsii za sokhranenie zhizhni naseleniia goroda v period blokady."* Moscow, 1960.
——— *"Voenno-organizatorkaia rabota Leningradskoi organizatskii VKP (b) v pervyi period Velikoi Otechestvennoi Voiny."* Moscow, 1956.
"Strategy and Tactics of the Soviet-German War, by Officers of the Red Army and Soviet War Correspondents." *Soviet War News*. London (no date).
Thorwald, Juergen, *Es begann an der Weichsel* (Thorwald I), *Das Ende an der Elbe* (Thorwald II), Stuttgart, 1950.
Tolly, G., *Con l'Armata Italiana in Russia*. Turin, 1947.
Trevor-Roper, Professor H. R., *The Last Days of Hitler* (3rd ed.). London, 1962.
Trial of the Major War Criminals before the International Military Tribunal, 1947-49, 42 vols. Published at Nuremberg.
Trials of War Criminals before the Nuremberg Military Tribunals, 15 vols. Washington, U. S. Govt. Printing Office, 1951-52.
Valori, A., *La Campagna di Russia, 1941-1943*. Rome, 1951.

Vlasov, General A. A., Official Biography, as Appendix I in George Fischer, *Soviet Opposition to Stalin.* Harvard Univ. Press, 1952.
Voznesenski, N., *The Economy of the U.S.S.R. during World War II* (tr. Wheeler-Bennett). Washington, Public Affairs Press, 1948.
Wheeler-Bennett, Sir J. W., *Nemesis of Power: The German Army in Politics, 1918-1945.* London, 1953.
Wilmot, Chester, *The Struggle for Europe.* London, 1952.

NOTE ON SOURCES

(*Asterisk indicates unpublished material*)

3 For Brauchitsch's interview with Hitler see also his evidence at Nuremberg (NR XX 575), Halder's diary for 5th November, 1939, and Halder's interrogation 26th February, 1946.

5 Speer's interrogation by Trevor Roper: *The Last Days of Hitler 13-14*

9 Wheeler-Bennett 320. Also a fuller account in Jean Francois, *L'Affaire Roehm-Hitler,* Paris 1939

10 Wheeler-Bennett 338

14 On the historical origins of the Oath, and its mystical powers, *ibid.* 394-95

17 For the various planning errors in the equipment of the Panzer Force see, generally, the chapter in Ogorkiewicz on "Organisation of the German Armoured Forces," also Guderian 143-44.

22 Hassell 309

23 For further details of the evolution of the Eastern campaign planning, the most up-to-date account can be found in Warlimont, "Inside Hitler's Headquarters," [London 1964] 111-16. However, his evidence conflicts with that of Guderian—in describing Jodl's briefing of Section L, Warlimont claims that he and his colleagues set up "a chorus of objections." ". . . Jodl countered every question and had an answer to everything, although he convinced none of us." *Op. cit.,* 111.

24 Führer Directive of 12th February, 1940 (No. 18), par. 5

25 The Berghof conference: Schmundt's minutes (IMT Doc. PS-1014); Halder's diary 22nd August, 1939; Boehm's notes (Raeder Defence Doc. Book No. 2, Doc. 27, 144)

30 For Tukhachevski's plan see Erickson 308, 321, 351, 406, *et al.*

31 Quoted in Leyderrey 28-29 (source unidentified)

33 On Tukhachevski abroad: Mme. Tabouis, *Ils m'ont appelé Cassandre,* 248-49; G. Castellan, *Les Relations Germano-Sovietiques,* 217-18 and 224, using evidence of Gen. Köstring, German Military Attaché in Moscow before IMT; General Gamelin, *Servir,* Vol. 2, 196

35 For fuller details of Soviet participation in the Spanish Civil War see Erickson 428-31

36-37 The Khalkin Ghol battles: Erickson 451-52, 455-56

43 On Osoaviakhim: Leyderrey 12 n; *see also* Erickson 307-8, 319, 409, *et al.* Rundstedt, interrogation January 1946; Krylov: writing in *Le Journal de Genève,* May 1949

44 Army Group Centre signals intercept 22nd June, 1941

46 Red Army order of 07.15, 22nd June, 1941, Karasev 32-33

49 Kesselring, *Memoirs* (London 1953), 90

54 Halder's diary 27th June, 1941

55 *Ibid.* 29th June, 1941; Bernd von Kleist, quoted Wheeler-Bennett 515

56 Manstein 180-81; Narrative, Capt. Pietl;* Halder's diary 29th June, 1941

57 Halder's diary 30th June, 1941

59 On the *Führerkorps Ost:* Dallin 102 and Rosenberg conference 21st June, 1941 (Doc. 1034-PS NCA iii 693-95); On Rosenberg's views: *Rede des Reichsleiters A. Rosenberg vor den engsten Beteiligen am Ostproblem am 20 Juni, 1941,* Doc. 1058-PS TMWC xxvi 610-27

60 Hitler on small sovereign states: *Führer Naval Conferences* (U.S. edition, Washington 1947) 120; On German goals: *Aktenvermerk* 16th July, 1941, Doc. 221-L TMWC xxxviii 86-94; Wilfred von Oven, *Mit Goebbels bis zum Ende* (Buenos Aires: Dürer-Verlag 1950), i, 217.

61 Letter from Himmler to Bormann, 25th May, 1941*

62 Backe, *12 Gebote* 1st June, 1941, Doc. 089-U.S.S.R. TMWC xxxix 366-71

63 Dallin 204-5; Carl to Kube, 30th October, 1941; Kube to Lohse, 1st November, 1941, Doc. 1104-PS TMWC xxvii 1-8

63-64 Hansel to Rosenberg, *Anlage zum Reisebericht,* 3rd March, 1942; Rosenberg to Lammers 10th March, 1942;* Heydrich to Meyer, 26th March, 1942*

64 Doc. 1104-PS TMWC xxvii 1-8

65 (Goering) Herwarth, *Germany and the Occupation of Russia,** quoted Dallin 123; (Himmler) Eberhard Taubert, *Die Deutsche Ostpolitik**

65-66 Doc. 1517-PS TMWC xxvii 270-72

66 Ribbentrop to Lammers, 13th June, 1941, Doc. NG-1691 NMT xiii 1277-79

67 On the "Adloniada" generally, see Dallin 133-37. Dallin also quotes two unpublished documents, NG-479 and Himmler File 140, which show Rosenberg's consolidation of his victory over Ribbentrop.

68 *Hitler's Table Talk,* London 1953, 617-18; Rosenberg's warning: *Verhalten der deutschen Behörden und die Stimmung der ukrainischen Bevölkerung*

69 Dallin 186-87; Lohse in an address to his staff, 23rd February, 1942, Doc. Occ. E 3-54/56;* On the hereditary principle: Peter Kleist, *Zwischen Hitler und Stalin,* Bonn 1950; 159-64

70 Guderian 147

71 Halder's diary 10th July, 1941

72 *Ibid.* 29th June, 1941

73 Guderian's account of his disputes with Kluge 161-62

74 Guderian 169; Manstein 185

75 Manstein 186

76 Dallin's interrogation of Köstring 62, No. G-1

78 Halder's diary 13th July, 1941

80 Leyderrey 29 (source unidentified)

83 Stalin's personal message; Zhilin 102 (Col. L. L. Dessyatov)

84 Dwinger, *Wiedersehen mit Sowjet-Russland*

87 Halder's diary 23rd July, 1941; Liddell Hart 198

89 Guderian 179

90 *Ibid.*

91 Warlimont (184-85) asserts that Jodl was present at this conference and that the army commanders persuaded him of the importance of Moscow, with the result

that he, in turn, managed to get Hitler to draft the subsequent directive (34) in the form which allowed the army commanders considerable latitude in interpreting it. However, Warlimont's account of the sequel is not very clear, and throughout his memoirs he shows considerable anxiety to protect the reputation of Jodl, his immediate superior. Guderian's description of his relations with Kluge, 189

99 Von Hassell 222

101-2 See Warlimont's version (cited above). My account is based on Guderian 189-90.

103 Soviet Official History is vague about this gap in their line, but see Erickson 601-2; Halder's diary 8th August, 1941

104 Guderian 190

109 *Ibid.* 198

111 Halder's diary for the 23rd August, 1941, and Guderian's account in his memoirs (198-200) of this meeting differ slightly according to the separate individual standpoint of each man, which led to their quarrel immediately afterward.

112 Halder's diary 25th August, 1941

116 Karasev 98-99 and 101

118 The Proclamations, quoted in Goure, 176-78; Leningrad Military Soviet: Karasev 105-6

120 Warlimont's plan (not mentioned in his memoirs, incidentally), *Vortragsnotiz Leningrad,* 21st September, 1941, Doc. NOKW 21

125 Reinhardt, *Vorsfoss des XLV Panzer Korps in Sommer 1941,* Wehrkunde 111, March 1956

126 Halder's diary 17th September, 1941

127 Sirota 139

131 My assessment of Marshal Budënny's character is a personal one, and several authorities on Soviet affairs—in particular Mr. Geoffrey Jukes, who has helped me with much Red Army material—have expressed their dissent. Some sources assert that Budënny's behaviour during the purges was creditable and that he went to considerable lengths to protect subordinates.

134 Malaparte 102

138 Narrative of Corp. Täsch

139 For further discussion of the responsibility for the Kiev disaster see Erickson 610-12

140 For disputation between army group headquarters and OKH see Guderian 206-9

143 Malaparte 100

145 *See also* "Documents on German Foreign Policy," Series D xiii 423-26

146 Malaparte 119; Dwinger, *op. cit.*

147 Liddell Hart 203

150 On the *Rote Kapelle:* Wheeler-Bennett 538-65; On "Sorge": Erickson 631; On "Lucy": Erickson 637

152 Anti-Partisan operations: Hitler: *Aktenvermerk* 16th July, 1941, Doc. 221-L TMWC xxxviii 88; OKW order: *Erganzung zur Weisung,* No. 22 H; Terror: OKH *Behandlung feindlicher zivil Personen und russischer Kriegsgefangener,* 25th July, 1941, Doc. NOKW-182

153 The means of execution: *Erlass des chefs de OKW,* 16th September, 1941, Doc. 389-PS TMWC xxv 531

154 Red Army Archives, letter from Untersturmführer Huber, *Waffen SS Leibstandarte* div. (d. 1942)

154-55 Narrative, anon
159 Leyderrey 42 (310th Motorised)
160 Bayerlein, *An Consontoir*
161 Moscow panic: see Erickson 622-23; Hopkins: *Ibid.* 627; Cripps: I. Deutscher, *Stalin* 464
162 From a letter in the author's possession
163 See Erickson 624 for details of Soviet command changes
164 Narrative, Sgt. Imboden
165 Liddell Hart 206
166 Bayerlein, *An Consontoir*
169-70 Liddell Hart 207
171 Guderian 247
173 Leyderrey 47
174 Guderian 251
176 Liddell Hart 208
178 *Ibid.* 205
181 Augier, *Les Partisans,* Paris 1949
182 Assmann, *op. cit.*
187 Liddell Hart 215 (Blumentritt was now Deputy Chief of Staff)
189-90 Liddell Hart's interrogation of Halder on this topic, 226-28
192 Correspondence with a German soldier, the SA Army Journal
193 Hassell 207
194 The episode of the troop train is described in Konstantin Simonov's novel *The Living and the Dead*
195 German tank strength: Ogorkiewicz, *op. cit.*
199 For further details of the Crimean fighting from the German side see Manstein 225-60
202 A good account of the Kharkov offensive, based on the Soviet Official History, can be found in Garder 121-26; *see also* Paulus' papers 180-83, where the Field Marshal characteristically (and wrongly) complains that ". . . the action cannot be brought to a successful conclusion with the force available."
205 Letter from Huber (Red Army Archives) see 154n above; Mot Pulk: Leyderrey 73
206 OKH on prisoners: Doc. NOKW-182;* *see also* OKW/AWA Kgf-Abt *"Anordnung für die Behandlung sowjetischer Kr-Gef,"* Doc. 1519-PS TMWC xxvii 273-83
207 OKW/AWA Kgf-Abt *Kriegsgefangenenwesen im Fall "Barbarossa"* 16th June, 1941, Doc. 988-PS;* Ciano, Diplomatic Papers 464-65
208 Quoted Leyderrey 73
210 Liddell Hart 220; Paulus, 151-54, describes the "war game" to which the roots of this strategic error may be traced
211 Liddell Hart 223; Diary, S. F. Klein*
212 Chuikov 17
213 *Ibid.* 18-19
214 *Ibid.* 19
215 *Ibid.* 33
217 Diary of Wilhelm Hoffmann, 267 Regt., 194th Inf. Regt. (d. 1942)
221 Doerr, *Feldzug nach Stalingrad*
222 Hoffmann's diary
223 Chuikov 74

224 *Ibid.* 85

225 *Ibid.* 90; Narrative, Michael Koumar (Hungarian attached to 79th Inf. Div.)

227 Narrative of Anton Kuznick, quoted Chuikov 123-28

229 Hoffmann's diary; Narrative of Andrey Khozynanov, Marine Inf. Bde., (quoted Chuikov 99-101)

233 Kleist: Liddell Hart 221

234 Jodl in disgrace: Liddell Hart 228-30; *see also* Warlimont 256-57

236 Schmundt's visit; Paulus 64-66; Chuikov on street fighting 287-88

238 Diary, Lieut. Weiner

241 Rumanian equipment: Dumitrescu's report, Paulus 196-97

243 The sniper: Chuikov 142-43

249-73 I have based my account of the failure to relieve Stalingrad on Manstein's memoirs 289-366. The figures for Russian strengths are taken from the Soviet Official History. The description of the engagements fought by the 11th Panzer Division is taken from the account given by General Balck, its commander, to Mellenthin (*op. cit.* 179-80)

280 For the disagreements between Manstein and Kleist see Manstein 381, *et seq.*

284 Mellenthin 195; Leyderrey 101

285 Narrative, Lieut. Dieter*

286 Narrative, Corp. Täsch*

292 Manstein 406

293 *Ibid.* 407-8

294 See Ogorkiewicz, chapter on "Organisation of German Armoured Forces"; Guderian 299, *et seq.*, on Dr. Porsche and other schemes

297 Guderian 288, *et seq.*, describes his meeting with Hitler

299 On failure of SS *Viking:* Manstein 418

303 Polevoi, *De Bjelgorod aux Carpathes*

307 The primary source for the "brandy bottles affair" is Schlabrendorff, *Offiziere gegen Hitler; see also* Wheeler-Bennett 560-64

308 Schlabrendorff 71

309 *Ibid.* 74

310 The verse is by Emily Dickinson

311 Schlabrendorff 80

313 The Assignment of Duties is reproduced in full in Guderian 289-91

314-15 Guderian, 294-98, for his account of the conference, and his lecture notes

315 Sauckel, *"Ausführung des Generalbevollmächtigten für den Arbeitseinsatz,"* 5th-6th February, 1943, Doc. 1739-PS TMWC xxvii 586-87

317 OKH/Gen Qu *"Arbeitererfassung im Osten,"* 13th July, 1943, Doc. Occ. E4-I YIVO*

317-18 *Einsatzgruppe A* (Report on Exterminations), ND Doc. 2273-PS TMWC xxx 76-79

318 Kube to Lohse, 5th June, 1943; Lohse to Rosenberg, 18th June, 1943: Doc. 135-R TMWC xxxviii 371-75

319 Chef Sipo und SD, decree 17th December, 1942: Doc. 041-L TMWC xxxvii 437-39; OKH on labour recruitment: Doc. 3012-PS TMWC xxxi 488; Sauckel's telegram: Doc. 407 (11)-PS TMWC xxvi 3-4

320 Goebbels' diaries 242; Prisoners: Doc. 1199-PS TMWC xxvii 63

323 Führer conference fragment (undated) March 1943. Warlimont to Liddell Hart, 233. In his memoirs, written eight years after his first interrogation, Warlimont moderates this opinion somewhat.

324 Guderian refers to this incident, 308

325 Guderian 307; Warlimont (333-34) produces some evidence to show that in the last fortnight many of the General Staff, too, lost their nerve. "On the 18th June the OKW Operations Staff submitted an appreciation to Hitler *leading up to the proposal* [*sic*] that, until the situation [in Africa] had been clarified, *Zitadelle* should be cancelled . . ."

326 Leyderrey 117

327 Jukes memorandum*

329 Mellenthin 214

330 Narrative, Sgt. Imboden*

330 Ferdinands, Guderian 311; Panthers, Mellenthin 224

336 Reitlinger 78

337 Narrative, Sgt. Imboden;* Manstein 448

338 Guderian 312

339 The details of the Langbehn affair are obscure. Reliable accounts can be found, *inter alia,* in Wheeler-Bennett, Schlabrendorff and Allen Dulles' *Germany's Underground*

344 Manstein 452

345 Jukes memorandum*

347 The description quoted is from Wheeler-Bennett, 578

349 It is still not easy to find out who was double-crossing whom at the close of the Langbehn affair. Himmler gave his version of things: *Vierteljahreschäfte fuer Zeitgeschichte* 573-76 (No. 4); *see also* Reitlinger 298-301.

351 Jukes memorandum*

362 Manstein 455

363 Ibid. 456-57

364 Schirach 130-34

369 Diary of Major Kreutz*

370 Manstein 459

375 Guderian 325

376 Polevoi, *op. cit.*

377 Leon Degrelle, *La Campagne de Russie,* Paris 1949

380 Manstein's own account 544-45

381 Lohse to Rosenberg, 25th July, 1944: Himmler File 57*

383 Guderian's own account 341-43

385 Guderian 351

386 Katyn. The subject remains controversial and obscure; see Werth, *Russia at War* (London ed.), 416

388 Elliott Roosevelt, *As He Saw It,* 186; Wilmot 647-56

393 Guderian 356; *see also* IMT xv 285; iv 19, affidavit 13

396 Details of German weapons production, Wilmot 355

399 Himmler: Karl Paetel, *Beitrag zum Soziologie des Nationalsozialismus in Vieteljahreschäfte fuer Zeitgeschichte* Vol. 2, p. 50; Guderian 351; Goebbels' diaries 354

400 Oven, Vol. II 94

401 The text of Hitler's address is reproduced in Warlimont 487-88, also a further interesting fragment 495-96

403 Recorded by William Bullitt (U.S. Ambassador in Moscow) reproduced in *Life,* 30th August, 1948

405 Guderian's account of his efforts to cope with the situation 366-67

406 Narrative, Peter Vidler*
407 Guderian 378-79
408 Himmler and Vlasov: *"Rede des Reichsführers-SS bei der SS-Gruppenführertagung in Posen am 4 Oktober 1943,"* Doc. 1919-PS TMWC xxix 110-73; *see also Sicherheitsfragen* (Berlin) OKW NS-Führungstab 1944, Doc. 070-L TMWC xxxvii 498-523
411 Guderian 382-87
419 Guderian's version of his plan for a counteroffensive 406, *et seq.*
420 Guderian 403-4; Warlimont 498
422 Guderian 408
426 Leyderrey 202
428 Speer/Hitler conferences*
431 Guderian 412-13
438 Führer conference 27th January, 1945
440 Thorwald, *Das Ende an der Elbe,* II
441 Racial duty: Pechel, *Deutsche Wiederstand* (Zurich 1947), 170, Hiwis: Narrative of Frl. Diem*
442 On Dr. Wulf: Trevor-Roper 95-96
443 Himmler's routine: Reitlinger 411; conversion to Christianity, Trevor-Roper 94; Guderian's visit to Prenzlau 421-22
444-45 Further testimony concerning the encounter between Himmler and Guderian at Hohenlychen, Reitlinger 412-13 (on evidence of General Heinrici, Thorwald 1125)
446 Guderian 428
447 Trevor-Roper 82
448 *Ibid.* 186
449 Reitlinger 449; *Documents sur l'activité du CRIC en faveur des detenus Civils,* Geneva 1946
450 Reitlinger 414-16; *see also* Fritz Hesse, *Hitler and the English,* London 1954
451 For details of the Jewish bargain see Reitlinger 356-58 (drawing on unpublished mss. of Reszoe Kastner)
452 Case XI NG-2675, and see Reitlinger 357
453 Felix Kersten, *Memoirs,* 228; for the Bernadotte-Himmler conversations, and relations between Schellenberg and Himmler, see Trevor-Roper 117-18, 128-29, 182-83, 246-49
457 Trevor-Roper 167
458 Hitler's epitaph, *Mein Kampf* 231
459 Doc. 3663-PS TMWC xxxii 435-36

INDEX